MATHEMATICS A CREATIVE ART

MATHEMATICS A CREATIVE ART

Julia Wells Bower
Connecticut College

HOLDEN-DAY, INC.
San Francisco
Cambridge
London
Amsterdam

MATHEMATICS
A
CREATIVE
ART

500 Sansome Street
San Francisco
California
94111

Library of Congress Catalog Card Number: 72-83241
ISBN: 0-8162-1074-8

Printed in the United States of America

1234567890 MP 79876543

PREFACE

It is generally conceded that a liberal education should include an introduction to the mode of thought and the precision of language of mathematics. For students not majoring in the physical sciences, this introduction can be made by working creatively with simple mathematical concepts. As their mathematical insights develop, students establish informed attitudes toward the subject and gain confidence in their ability to make it a comprehensible part of their intellectual tradition.

The course for which this text was written is designed to show mathematics not as a tool or technique but as a creative form of abstract thinking. It was introduced into the Connecticut College curriculum in 1954. Topics from abstract algebra, geometry, and set theory were used to demonstrate the nature and the power of the postulational method.

A leave of absence permitted me to hold extensive correspondence with other mathematicians who were also experimenting with subject matter for general cultural courses and gave time for observation of the successful teaching of mathematics in the College of the University of Chicago. The many helpful suggestions obtained were incorporated into a syllabus prepared for use in 1959.

Three years later a grant from the Carnegie Corporation of New York provided a year's leave during which I began the task of transforming the syllabus into a text. The University of Miami extended visiting-scholar privileges, and the first draft was completed during that year. The text appeared in preliminary edition in 1965. To teachers who adopted it, to students who used it, and to critics who evaluated it, I am indebted for valuable criticisms and suggestions, many of which have been incorporated into this present edition. The work of revision has been facilitated by a semester leave granted by Connecticut College.

The original syllabus was constructed for a one-semester course. In writing the text I have expanded the material in such a way that it permits a variety of one-semester choices without subjecting the student to a "take-and-drop" sampling. The instructor is thereby given considerable freedom of choice in the topics that he wishes to discuss. The organization and level of presentation of the material make the text suitable for students who are concerned with the actual mathematical content as well as with the cultural connotations of the study. As a result, the text can also serve as an introduction to modern mathematics for future teachers.

The first part of the work, Chapters 1 through 7, is devoted to a study of progressively more complicated algebraic systems. In the first chapter, the field of residues modulo 5 is introduced. This simple, although new and unfamiliar, material is used to scrutinize the structure of algebra. Most of the results are derived from the study of addition and multiplication tables. The ring of residues modulo 6 is then used to acquaint the students with divisors of zero. The reason for their occurrence in this system, but not in the five-element system, is investigated. In keeping with the intuitive interpretation of the arithmetic in these chapters as the arithmetic on a clock, the symbols 5 and 6 are used instead of 0 to denote the additive identity. After the properties of this identity under multiplication and division have been studied and understood, application is made to the properties of zero in the rational number system. In keeping with the goal of understanding concepts before recalling familiar algorithms for their use, the symbols * and ′ are employed to denote additive and multiplicative inverses, respectively.

The work continues with a discussion of two different algebraic systems of pairs of elements. The first uses component-wise equality, addition, and multiplication. Properties of the pairs system are derived from properties of the element system, and students are taught to make formal proofs of their results. A different system is then developed by methods analogous to those employed in the construction of the integers from the natural numbers. The concepts of equivalence classes and of isomorphism are introduced. The last chapter in this section deals with pairs of pairs in the form of two-by-two matrices. They are a reasonable and natural extension of the systems already studied.

Chapter 8 is concerned with the algebra of logic. Statements, connectives, and statement patterns are introduced; truth tables are constructed; and the relations of logical implication, equivalence, and negation are investigated. Application is made to the reasoning of the earlier chapters. Discussion is extended to open and quantified sentences, which provide a natural basis for the study of sets in Chapter 9. Here the material is handled chiefly by intuitive and graphical methods, although a simple analytical treatment is also presented.

The final part of the work, in Chapters 10 through 12, diverges sharply from the earlier parts. It is intended to use and test the mathematical sophistication of the students, and for this purpose it develops completely new material. After a survey of the first part of Euclid's *Elements*, the effect of changes in the fifth postulate is discussed, and introductory theorems in hyperbolic and elliptic geometries are presented. This study stretches the mathematical imagination of the students and gives them an acquaintance with nonalgebraic systems.

To the professional mathematician the pace of the early chapters will seem slow and the methods cumbersome. Experience has shown, however, that this pace and these methods give students the time and the means whereby they can gain comprehension of the material, get used to the notation, and become adept at this type of reasoning. If, at the beginning, the pace is too fast or the methods too elegant, then the students will fall into bad habits of reading superficially and trying to memorize without understanding. They need time to become acquainted with the concepts, and they need practice in writing in their own words the explanations called for in the problems. Then, when they are shown the concise or ingenious mathematical method, the more gifted among them will comprehend and welcome it, thereby encouraging the others to seek also to comprehend and use it.

During the early part of the course, when so little mathematical content is contained in the assignments, the students will profit from class periods devoted to discussion of answers to assigned problems. The mathematical ideas are so simple that their elucidation and expression are well within student competence. Nevertheless, because many students receive little training in expository writing, it is not easy for them to find the words to express these simple ideas. Lively and interesting class sessions are generated when, in the spirit of good mathematical sportsmanship, students criticize each other's answers and thus begin to free themselves from the helpless illiteracy expressed in the statement, "I understand it, but I can't explain it." The construction of the text offers a second justification for careful class discussion. Results found in the exercise at the end of a section are often the basis for the development in the next section. For this reason, hints as well as answers are given in the back of the book.

As noted above, the text permits a variety of one-semester courses. In all of these, the first two chapters must be covered. I list below a suggested minimum number of assignments for each chapter and some combinations of chapters suitable for one semester of work. There are other good combinations of chapters as well as profitable selections of sections within chapters.

Chapter	*Number of Assignments*
1	5
2	7
3	4
4	3
5	3
6	4
7	6
8	7
9	7
10	3
11	4
12	4

One-semester Courses

Algebra and logic: Chapters 1–8.
Algebra, logic, and sets: Chapters 1–3, 5, 7–9.
Algebra and geometry: Chapters 1–3, 5–7, 10–12.
Algebra, logic, and geometry: Chapters 1–3, 5, 8, 10–12.
Algebra, logic, sets, and geometry can be covered by suitable pruning in Chapters 1, 2, 8–12.

Suggestions for teaching and answers and hints for even-numbered problems are contained in the Instructor's Manual, which is published separately.

It is impossible for me to acknowledge all the generous assistance that I have received from many sources. I am, however, especially beholden to Dr. Rosemary Park (former president of Connecticut College), Professor Alice Schafer, and Mrs. Marguerite Quimby, without whose encouragement, suggestion, and criticism neither the course nor the text would have come into being. I wish also to make special mention of President Charles Shain of Connecticut College whose generosity in granting leave provided the time for the completion of this new edition. I am grateful to the Carnegie Corporation of New York for their grant and to the University of Chicago and the University of Miami for their hospitality. Finally, I am indebted to my students for their enthusiasm for the course as well as for their insistence on their right to understand and enjoy the mathematics they are being taught.

Julia Wells Bower

CONTENTS

SYMBOLS

Page	*Symbol*	*Definition*
16	a^*	additive inverse of element a
31	a'	multiplicative inverse of element a
35	$\circ$	operation circle
35	i	identity for circle
36	$a\dagger$	inverse of a under circle
85, 114	F	five pair or five-matrix
94	$A \sim B$	pair A is equivalent to pair B
96	$[A]$	equivalence set of pairs
96	$A \in [A]$	pair A is a member of $[A]$
128	$\lvert A \rvert$	determinant of matrix A
130	$\tilde{A}$	adjoint of matrix A
145	$p \wedge q$	p and q
145	$p \vee q$	p or q
145	$\sim p$	not p
145	$p \rightarrow q$	if p, then q
145	$p \leftrightarrow q$	p if and only if q
155	F	false
155	T	true
172	$\forall x(Px)$	for every x, Px
172	$\exists x(Px)$	there exists an x, Px

179	$\{x \mid Px\}$	the extension of Px
179	$a \in A$	a is a member of set A
182	$A \subset B$	set A is a subset of set B
191	$A + B$	symmetric difference of sets A and B
192	$A \cap B$	intersection of sets A and B
196	$A \cup B$	union of sets A and B
202	A'	complement of set A

MATHEMATICS A CREATIVE ART

I

ALGEBRAIC SYSTEMS

Mathematics is a creative art. Words and symbols are used in its compositions and, like all art, it is limited only by the potentialities of its practitioners. Few people possess the gifts required to produce great mathematics, but almost every student can share the creative spark.

We will develop some simple mathematics and thereby glimpse the characteristics and the content of certain topics of current interest. We will find that some of our results contrast sharply with others, because the mathematician can construct any world of mathematical imagination he chooses. We will discuss the laws that help to ensure that each system is consistent within itself, and we will consider some of the criteria by which the worth of our conclusions may be judged. Most important, we will have an opportunity to experience the excitement and the satisfaction that come from using our minds creatively.

1

THE FIVE-ELEMENT SYSTEM

ADDITION

Our study begins with a number system so simple that all counting can be done on the fingers of one hand. Yet it has many interesting properties, some of which are neither simple nor obvious. We will learn how to express these properties in symbols and how to make convincing proofs of their validity. Our investigation will help us to understand the reasons for many of the rules taught in arithmetic and algebra. This simple system will be a delight to work with as well as a very profitable subject for discussion.

1.1 RECALLING THE NATURAL NUMBERS

Because words and symbols are used in mathematical compositions, we must be careful to use them with their proper meanings. Consider the word "number." Our first thought may be of a whole number such as 2, 6, or 10. To a mathematician, however, the word would have a much wider range of meaning, and so he needs a special name for the whole numbers. One of their most common names is *natural numbers.* They are the numbers 1, 2, 3, 4, 5, 6, 7, . . ., continuing without end.

We know that we can add these numbers by counting. To add 2 to 6 we can make six marks, then make two more, and then count the total of marks. It is 8. We write, in symbols, $6 + 2 = 8$. If we add 6 to 2 by making two marks and then six more, the total of marks is again 8. Furthermore, we are sure, from our previous experience, that for any two natural numbers (not just 2 and 6) the sum will be the same regardless of the order in which the numbers are added.

We could easily recall many other properties of these numbers. Let us, instead, invent a new number system, discover its properties, and find out whether these properties are the same as, or different from, those of the familiar natural numbers.

1.2 THE FIVE-ELEMENT SYSTEM

Sometimes, instead of counting with the natural numbers, which go on and on, we count only so far and then begin again. We do this with time, counting up to 12 o'clock and beginning again with one o'clock. It is quite natural to set an alarm for seven every morning; yet if we take the midnight between Sunday and Monday to be the zero hour, then by ordinary counting the hours at which we plan to be wakened during a week are 7, 31, 55, 79, 103, and 127. But the 12-hour system is convenient and serves us well in our daily life.

Now suppose that we can count only to five, the number of fingers on one hand, and then must begin again with one. We create a system containing five distinct items, called *elements,* to distinguish them from the natural numbers we are accustomed to using. For these elements, however, let us continue to use the familiar names and symbols 1, 2, 3, 4, and 5. Let us call our new system the number system of five elements or the *five-element system*.

1.3 ADDITION IN THE FIVE-ELEMENT SYSTEM

Let us define addition in our new system by counting, with the condition that whenever we reach 5 in the counting we must begin again with 1. A clockface arrangement like Figure 1.1 helps us to keep this addition rule in

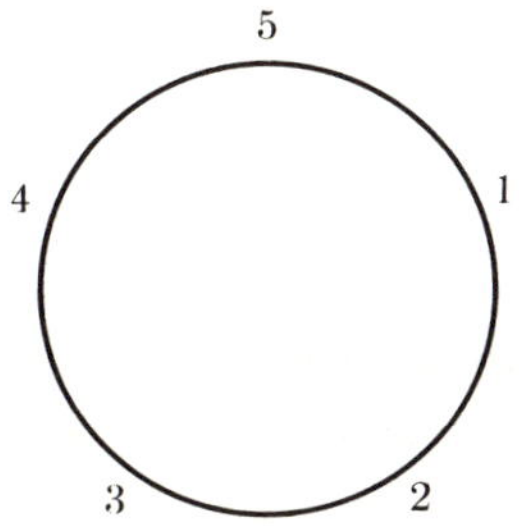

FIGURE 1.1

mind. We see that $3 + 1 = 4$ and $3 + 2 = 5$. Now $3 + 3$ carries us beyond 5 to 1; therefore $3 + 3 = 1$. To find $3 + 4$ we start with 3, count 4 more, and reach 2; hence $3 + 4 = 2$.

Because there are only five elements in our system, it should be possible to find out whether any two of them can be added, and, if so, what the sums would be. How many pairs of elements are there? Any one of the five elements could be the first of two addends; for each of these, any of the five elements could be the second addend; hence, there are five pairs of addends beginning with element 1, five beginning with the element 2, and so on. The total count is 25 pairs of two addends each.

It would be well to organize the addition results so that they will be easy to find and to use. For this purpose an addition table like Table 1.1 may be drawn up, the elements written in order in the column that borders the table on the left and in the row that borders the table across the top.

Table 1.1 Addition Table for the Five-element System

+	1	2	3	4	5
1					
2					
3	4	5	1	2	
4					
5					

Within the body of the table are written the sums found by adding to any element in the left border any element in the top border. Thus, letting any element in the left border be designated by the letter u and any element in the top border by the letter v, the entries in the table represent the sums $u + v$; for example, where the row containing element 3 in its left border crosses the column having 1 in its top border, we write the sum $3 + 1$, which we found to be the element 4.

EXERCISE 1.3

1. Explain why the entries in Table 1.1 appear in the 3 row instead of in the 3 column.

2. Copy and complete the addition table for the five-element system.

3. Use the addition table to find the element e that makes each equality true.

(a) $1 + 2 = e$ (b) $2 + 4 = e$ (c) $5 + 3 = e$
(d) $2 + e = 4$ (e) $3 + e = 2$ (f) $1 + e = 5$
(g) $e + 2 = 2$ (h) $e + 5 = 4$ (i) $e + 3 = 1$

4. Use the addition table to find all elements e that make each statement true.

(a) $1 + 2 \neq e$ (b) $e + 5 = e$ (c) $2 + e \neq 4$
(d) $e + 1 = e$ (e) $e + 2 \neq 2$ (f) $e + e = 1$

5. Make an addition table to show the sums of the first five numbers, 1, 2, 3, 4, 5 (see Section 1.1). Remember that counting with the natural numbers is unlimited.

6. Compare the tables you made for Exercises 2 and 5. The entries show some similarities and some dissimilarities. List two ways in which the entries resemble each other and two ways in which they differ.

1.4 CLOSURE UNDER ADDITION IN THE FIVE-ELEMENT SYSTEM

When we compare the two tables constructed in the preceding exercise, we may notice that every space in the body of each table is filled with one and only one entry. This characteristic shows that each pair of elements has a unique sum and that each pair of the natural numbers 1, 2, 3, 4, 5 has a unique sum.

We may notice further that in the addition table for the five-element system, each entry is one of the original five elements, whereas in the addition table for the first five natural numbers the entries include the new numbers 6, 7, 8, 9, 10. Suppose that the natural numbers 1, 2, 3, 4, and 5 are placed in a fenced enclosure with all the other natural numbers outside. When we add any two of the enclosed numbers, sometimes the sum is to be found within the enclosure, and sometimes we must look outside for it. If, however, the five elements are placed in a fenced enclosure, then the sum of any two of them is always within the enclosure. In mathematical language, the five-element system is said to be *closed* under addition.

In contrast, could it properly be said that the natural numbers are not closed under addition? Certainly the system consisting of the first five of them is not. However, the new numbers 6, 7, 8, 9, 10 are also natural numbers and there are others. It seems reasonable that if we enclosed all the natural numbers, we should find that the sum of any two of them also lies within the enclosure. This conjecture is true.

When a collection of elements possesses at least one operation which, like addition above, combines two elements to produce a unique element of the collection, the terms "number system" or "mathematical system" may be used to indicate the collection of elements, all of the operations defined on it, and the laws concerning the operations. Hence, the term "five-element system" is used correctly even though so far only addition has been defined for it and only one property of addition, namely closure, has been found.

DEFINITION A number system is **closed under addition** if and only if every pair of numbers has a unique sum that is a number in the system.

In the definition, the phrase "if and only if" is used. Mathematicians customarily join two sentences with this phrase when they wish to assert that the sentences mean exactly the same thing. Each sentence is true if the other is true, and neither can be true unless both are true; the first implies the second and the second implies the first. Hence, the definition says that *if* every pair of numbers in a given system has a unique sum which is the same kind of number, *then* the system is closed under addition, and conversely, *if* the system is closed under addition, *then* every pair of numbers in it has a unique sum which is a number in the system. From the discussion in the first two paragraphs of this section we see that the five-element system satisfies the definition of closure under addition.

THEOREM 1.4-1 The five-element system is closed under addition.

When an operation, such as addition, combines two elements of a system in a given order to produce a unique element of the system, the adjective "binary" is applied to it. Theorem 1.4-1 could, therefore, be rephrased to say that addition is a binary operation on the elements of the five-element system.

EXERCISE 1.4

In each of Exercises 1–4 give a reason for your answer.

1. Is the system consisting of only the elements 1, 2, and 3 of the five-element system closed under addition?

2. Is the system consisting of only the element 5 of the five-element system closed under addition?

3. Is the system consisting of all the even natural numbers closed under addition?

4. Is the system consisting of all the odd natural numbers closed under addition?

5. By the law of closure of addition, the sum $1 + 2$ is a single element of the five-element system; hence, the expression $(1 + 2) + 3$ can be considered the sum of exactly two elements, namely, the sum $1 + 2$ and the element 3. To find this sum, first add 1 and 2 as indicated by the parentheses. Then, using the answer just found as first addend and 3 as second addend, reckon the final result. Find the sum.

6. Explain why each of the expressions below can be considered the sum of exactly two elements of the five-element system. Then find each sum.

(a) $1 + (2 + 3)$ (b) $(1 + 2) + (3 + 4)$

7. In Section 1.1 we recalled the fact that the sum of any two natural numbers will be the same regardless of the order in which the numbers are added. From the addition table, determine whether a similar property holds for the five-element system.

1.5 COMMUTATIVITY OF ADDITION IN THE FIVE-ELEMENT SYSTEM

Exercise 1.4.7 asks whether the order of addition of two elements affects the sum. There are several good ways to find the answer. We could, for example, list all 25 pairs of elements, add the elements in each pair in each order, and compare the sums. This method is neither long nor laborious, and it establishes the result that the order of addition does not affect the sum of two elements.

There is a more elegant way of determining this same result that makes us look more critically at the addition table and reason in a more general fashion. Look at the entries in the first row of the body of the table; these are 2, 3, 4, 5, 1. They are, in order, the sums $1 + 1$, $1 + 2$, $1 + 3$, $1 + 4$, and $1 + 5$. Look next at the entries in the first column of the table; these are also 2, 3, 4, 5, 1, but they are, in order, the sums $1 + 1$, $2 + 1$, $3 + 1$, $4 + 1$, and $5 + 1$.

Consider now the significance of our two observations. First, the entries in the first row of the body of the table are the same and in the same order as those in the first column. Second, when the letter e is taken to represent a general element, the sums listed in the first row have the description $1 + e$, while those in the first column have the description $e + 1$. It follows that $1 + e = e + 1$ for all elements e. Making a similar analysis of the entries in the second row and the second column of the body of the table, we find that $2 + e = e + 2$. Further observation of the entries in the remaining rows and the corresponding columns shows that $3 + e = e + 3$, $4 + e = e + 4$, and $5 + e = e + 5$. We conclude that if e and a are any elements of the five-element system, then $a + e = e + a$.

This method of argument has not only provided an answer to the question we asked but has suggested a way of stating our findings in symbols. The definition and theorem below express our conclusion in mathematical language.

DEFINITION In a given number system **addition is commutative** if and only if for any two numbers a and b of the system, $a + b = b + a$.

THEOREM 1.5-1 In the five-element system addition is commutative.

As an application of this theorem, consider the expression $(2 + 3) + 4$. Applying the commutative law of addition to the sum within the parentheses, we find that $(2 + 3) + 4 = (3 + 2) + 4$. If, however, we consider the original expression to be the sum of exactly two elements, namely, the sum $2 + 3$ and the element 4, the commutative law of addition gives the equality $(2 + 3) + 4 = 4 + (2 + 3)$.

EXERCISE 1.5

1. Explain the significance of the phrase "if and only if" in the definition of commutativity.

2. In the discussion above it was shown in detail that $e + 1 = 1 + e$ for all elements e. Show in a similar fashion that $e + 2 = 2 + e$ for all elements e.

3. We have seen that $(2 + 3) + 4 = (3 + 2) + 4 = 4 + (2 + 3)$. Using the commutative law of addition show further that $(2 + 3) + 4 = 4 + (3 + 2)$.

4. By using the laws of closure and commutativity of addition show:

(a) $(3 + 1) + 4 = 4 + (1 + 3)$ (b) $3 + (1 + 4) = (4 + 1) + 3$

5. In the preceding exercise we did not use the actual sums of the elements, but only laws concerning their properties. Hence, if we replace the specific element 3 by the general element a, 1 by b, and 4 by c, the argument establishes the two equalities below. Write out this argument.

(a) $(a + b) + c = c + (b + a)$ (b) $a + (b + c) = (c + b) + a$

Exercises 6–8 deal with systems for which addition is described in Tables 1.2. Each of these tables is to be read in the conventional manner; that is, if u is any element in the left border column and v is any element in the top border row, then the entries in the body of the table are the sums $u + v$. We will make repeated use of these systems. You should keep a record of the growing fund of information about each of them. Note that the same names and symbols are sometimes used for elements in the various systems. The elements themselves in any one system, however, are distinct from those having identical names and symbols in any other system. For example, the sum of the elements called 1 and 4 is the element called 1 in system S and the element called 5 in the five-element system. This use of the same word and symbol to denote items whose identities are determined by the contexts in which they appear is common not only in mathematics but also in the language of ordinary living. Consider, for example, the various interpretations of the words "figure," "solution," and "interest."

Table 1.2 Addition Tables

System Q

+	0	3	6	9
0	0	3	6	9
3	3	0	9	6
6	6	9	0	3
9	9	6	3	0

System M

+	7	8	9
7	8	7	9
8	7	8	9
9	9	8	7

System S

+	1	4	7
1	4	1	7
4		7	
7		4	1

6. Let system Q consist of elements denoted by the symbols 0, 3, 6, 9, and let addition be defined as in Table 1.2. Justify your answer to each question below.

(a) Is Q closed under addition?
(b) Is addition commutative in Q?

7. Let the system M consist of elements denoted by 7, 8, 9, and let addition be defined as in Table 1.2. Justify your answers to the following questions.
(a) Is M closed under addition?
(b) Is addition commutative in M?

8. Let system S consist of elements denoted by 1, 4, 7, and let the system be closed and commutative under addition. Complete its addition chart in Table 1.2.

9. Are the following statements (a) and (b) contradictory? Justify your answer.
(a) Because system S is closed under addition, each pair of elements has a unique sum.
(b) The addition table for S shows that $1 + 7 = 7$ and $4 + 4 = 7$; hence, element 7 does not occur uniquely as a sum.

10. Each of the addition tables with which we have dealt is square in shape because there are as many entries in the top-border row as in the left-border column. The diagonal of the table from upper left to lower right is called the *principal diagonal.* If we fold each table along its principal diagonal, we discover that, in all tables except that for system M, the entries which fall together are exactly alike. For example, in the table for system Q, the entries 3, 6, 9 of the first column fall on corresponding entries 3, 6, 9 of the first row. Each table is said to be *symmetric* with respect to its principal diagonal. This symmetry is due to the identical order of entries in the border row and border column and to the commutativity of addition in the system. Explain why this is so.

1.6 ADDITION OF THREE ELEMENTS IN THE FIVE-ELEMENT SYSTEM

So far we have been concerned with sums of pairs of elements. This emphasis on pairs arises from the fact that addition is a binary operation; that is, it operates on two elements to produce an element. From experience with the natural number system, however, we know that we may find it necessary to add more than two elements. To develop a procedure for doing this, consider the sum of the particular triple of elements 1, 2, 3.

Our accustomed procedure with the natural numbers suggests that we first add 1 and 2 and then add the element 3 to the resulting sum. In this way, calculating the sum of the three elements is reduced to finding successive sums of two elements each. Using parentheses to keep operations distinct, we have

$$1 + 2 + 3 = (1 + 2) + 3$$

where the addition within the parentheses is to be done first. Use of the five-element addition table now gives

$$1 + 2 + 3 = (1 + 2) + 3 = 3 + 3 = 1$$

There is, however, another way of doing this addition which is just as reasonable as the one above. We could associate the elements 2 and 3 and add their sum to the element 1. That is, we could say

$$1 + 2 + 3 = 1 + (2 + 3)$$

Will this method give the same result as the first? Consulting the addition table we find

$$1 + (2 + 3) = 1 + 5 = 1$$

The results of addition by both methods are the same. Either procedure, therefore, could be used as the definition of addition for the triple 1, 2, 3. Let us arbitrarily call the first method the definition and state that the second gives the same result.

DEFINITION For the triple of elements 1, 2, 3 of the five-element system, the sum $1 + 2 + 3 \equiv (1 + 2) + 3$.

In this definition, the symbol $\equiv$ is read "is defined to be" or "is identical with."

STATEMENT For the triple of elements 1, 2, 3 of the five-element system, $(1 + 2) + 3 = 1 + (2 + 3)$.

The definition and statement appear reasonable because they accord with our experience of natural numbers.

Now let us generalize our results. We have no difficulty in constructing a general statement, although we do not know if it is true.

DEFINITION For any three elements a, b, c of the five-element system, the **sum** $a + b + c \equiv (a + b) + c$.

GENERAL STATEMENT For any triple of elements a, b, c of the five-element system, $(a + b) + c = a + (b + c)$.

This statement, which arose from our experience with one triple, asserts a property of all triples of elements. But the fact that it is valid for one triple does not mean that it is valid for all the rest. We would like to know whether the statement is true in its complete generality or whether it fails in some specific cases. That is, we must show that it holds for every set of three elements or else exhibit at least one triple for which it does not hold.

There are in all 125 triples of elements. In the next exercise every one of these triples will be shown to satisfy the equality of the statement; hence, the statement is true. We can now be sure that the sum of any three elements is not affected by the manner in which they are associated into a pair and a single, provided their order is left unchanged. In mathematical language, this means that addition is *associative.*

We may summarize the discussion in a definition and a theorem:

DEFINITION In a given number system **addition is associative** if and only if for any three numbers a, b, and c of the system, $(a + b) + c = a + (b + c)$.

THEOREM 1.6-1 In the five-element system addition is associative.

Finally, the definition of addition of three elements can easily be extended to the addition of four or more. Since the sum of three elements is an element, a fourth can be added to it and, by the law of closure for addition, the sum is an element of the system. To this new sum can be added a fifth element, and again the result is an element. Continuing in this way, as many elements as desired may be added.

EXERCISE 1.6[1]

1. In Table 1.3 the 125 triples of the elements 1, 2, 3, 4, 5 are separated into sets in such a way that each set contains a variety of types of triples. Verify the general statement in the discussion above for the set of triples assigned to you by your instructor. In this way, the truth of the statement can be established by the class for all possible triples of elements without placing a heavy burden on any one student.

2. In the definition of associativity of addition we used the phrase "if and only if." The definition can, therefore, be rewritten as an "if . . . , then . . ." statement and its converse (see Section 1.4). Make this restatement.

3. By the laws of addition in the five-element system verify each of the equalities:

(a) $1 + (2 + 3) = (2 + 1) + 3$ (b) $(3 + 4) + 2 = (3 + 2) + 4$
(c) $(2 + 3) + 4 = 3 + (2 + 4)$ (d) $4 + (3 + 2) = (2 + 3) + 4$

★**4.** Consider the sum of elements a, b, and c. Explain why adding from left to right gives the same result as adding from right to left; that is, explain why $(a + b) + c = (c + b) + a$.

5. From your experience with the natural number system, is addition of natural numbers associative?

[1] The symbol ★ indicates a problem which requires more thought than the unmarked problems do. The symbol ★★ indicates an even harder problem.

Table 1.3 Sets of Ordered Triples of Elements a, b, c

I	II	III	IV	V	VI	VII
1,2,3	1,1,1	1,1,2	1,1,3	1,1,4	1,1,5	1,2,1
3,1,2	1,5,2	1,5,3	1,5,4	1,5,5	2,1,1	2,1,2
2,3,4	2,4,4	2,4,5	2,5,1	2,5,2	2,5,3	2,5,4
3,2,4	3,4,1	3,4,2	3,4,3	3,4,4	3,4,5	3,5,1
2,2,2	4,1,5	4,3,3	4,3,4	4,3,5	4,4,1	4,4,2
	4,3,2	5,2,4	5,2,5	5,3,1	5,3,2	5.3.3

VIII	IX	X	XI	XII	XIII	XIV
1,2,2	1,2,4	1,2,5	1,3,1	1,3,2	1,3,3	1,3,4
2,1,3	2,1,4	2,1,5	2,2,1	2,2,3	2,2,4	2,2,5
2,5,5	3,1,1	3,1,3	3,1,4	3,1,5	3,2,1	3,2,2
3,5,2	3,5,3	3,5,4	3,5,5	4,1,1	4,1,2	4,1,3
4,4,4	4,4,5	4,5,1	4,5,2	4,5,3	4,5,4	4,5,5
5,3,4	5,2,2	5,4,2	5,4,3	5,4,4	5,4,5	5,5,1

XV	XVI	XVII	XVIII	XIX	XX	XXI
1,3,5	1,4,1	1,4,2	1,4,3	1,4,4	1,4,5	1,5,1
2,3,1	2,3,2	2,3,3	2,3,5	2,4,1	2,4,2	2,4,3
3,2,3	3,2,5	3,3,1	3,3,2	3,3,3	3,3,4	3,3,5
4,1,4	4,2,1	4,2,2	4,2,3	4,2,4	4,2,5	4,3,1
5,1,1	5,1,2	5,1,3	5,1,4	5,1,5	5,2,1	4,4,3
5,5,2	5,5,3	5,3,5	5,4,1	5,2,3	5,5,4	5,5,5

6. For system Q of Exercise 1.5.6 addition is associative. For how many triples of elements would the equality of the definition need to be checked to prove that this is true?

7. For system M of Exercise 1.5.7 addition is not commutative. Determine whether it is associative.

8. For system S of Exercise 1.5.8 addition is commutative. Determine whether it is also associative.

****9.** Let system T consist of elements denoted by the symbols a, b, c, d, e, f. Let addition be partially defined by Table 1.4. The system has closure and associativity of addition. Furthermore, when the table is completed, each element appears once and only once in each row and column. The sum $b + d$ can be found by use of the associative law applied to the triple f, b, d. Hence, $(f + b) + d = f + (b + d)$. The table gives sufficient information for reduction of the left member of the equality

Table 1.4 Addition Table for System T

$+$	a	b	c	d	e	f
a	a	b	c	d	e	f
b	b	c				
c	c	a				
d	d			a	b	c
e	e				a	b
f	f	d				a

to a. Hence, a is the sum of f and $(b + d)$ and must lie in the table where the f row and the $(b + d)$ column cross. The only a in the f row is at its intersection with the f column; it follows that $b + d = f$. Employ a similar procedure to find:

(a) The sum $d + b$ by use of the triple d, d, b.

(b) The sum $c + c$ by use of the triple c, c, b.

(c) The sum $c + f$ by use of the triple c, f, f.

10. Find the remaining sums in system T (Table 1.4) by using the property stated for rows and columns of the table.

11. Is addition commutative in system T (Table 1.4)?

1.7 THE LAW OF CANCELLATION FOR ADDITION IN THE FIVE-ELEMENT SYSTEM

Each element appears once and only once in each row and in each column of the body of the addition table. This fact gives rise to several important properties for our number system.

Consider first the sum of two specific elements, say $3 + 4 = 2$. Is there any other element e, different from 4, for which $3 + e = 2$? The answer to this question is "no," because the sum element 2 appears only once in the 3 row of the table, and it appears where this row crosses the 4 column. Hence, whenever $3 + 4 = 3 + e$, we can be sure that e is the element 4.

Let us extend this reasoning to the general case. Given element a, let elements b and c be such that $a + b = a + c$. We ask whether b and c are the same or different. We first find $a + b$ in the body of the table where the a row and the b column cross. Let us designate it by the letter d. Then d is also the sum of a and c. Now let us seek the element c for which the sum $a + c$ equals d. We read across in the a row of the table until we find the entry d and then read up to the heading of the column in which d lies. This heading is the desired element c. Since d appears only once in the a row and since it lies where the a row crosses the b column, the heading gives b as the desired element c. Hence, whenever $a + b = a + c$, b and c are the same element. We say that we have a *left-cancellation law* because we can cancel out the element a which appears on the left-hand side of each sum. By commutativity or by a similar analysis of the columns of the table, we can establish a *right-cancellation law* saying that whenever $b + a = c + a$, then $b = c$. Because addition is commutative, the two laws are equivalent.

THEOREM 1.7-1 *Law of Cancellation for Addition.* For elements a, b, and c of the five-element system, if $a + b = a + c$, then $b = c$.

The left- and right-cancellation laws are not equivalent in all number systems. Consider system M, whose addition table is given in Exercise 1.5. Each element appears once and only once in each row of the table, so that whenever $a + b = a + c$, then $b = c$. In the second column, however, element 8 appears twice, so that $8 + 8 = 9 + 8$. Consequently, when we are given the equality $b + 8 = c + 8$, we cannot assert that $b = c$. Similarly, according to the third column we cannot assert that, when $b + 9 = c + 9$, $b = c$. For system M, therefore, there is a left-cancellation law but no right-cancellation law.

EXERCISE 1.7

In Exercises 1–3 the five-element system is used.

1. Find the element x in each of the following equalities.

(a) $4 + x = 4 + 3$ (b) $x + 1 = 3 + 1$
(c) $2 + x = 3 + 2$ (d) $x + 5 = 5 + 4$

2. Determine whether each statement below is true. Justify your answers.
(a) The converse of the cancellation law: if $b = c$, then $a + b = a + c$.
(b) The inverse of the cancellation law: if $a + b \neq a + c$, then $b \neq c$.
(c) The contrapositive of the cancellation law: if $b \neq c$, then $a + b \neq a + c$.

★**3.** In the discussion above, two methods were suggested for the establishment of a right-cancellation law. Prove that this law holds:
(a) By analysis of the columns of the table.
(b) By use of commutativity of addition applied to the left-cancellation law.

4. Show that the law of cancellation for addition holds for both of systems Q and S defined in Exercise 1.5.

5. The closed and associative but noncommutative system T was introduced in Exercise 1.6. Its complete set of sums is given in Table 1.5, where the letters used to designate its elements are replaced by the symbols 1, 2, 3, 4, 5, 6. Show that this system has (a) a left-cancellation law, (b) a right-cancellation law.

Table 1.5 Addition Table for System T

+	1	2	3	4	5	6
1	1	2	3	4	5	6
2	2	3	1	6	4	5
3	3	1	2	5	6	4
4	4	5	6	1	2	3
5	5	6	4	3	1	2
6	6	4	5	2	3	1

1.8 THE IDENTITY ELEMENT FOR ADDITION IN THE FIVE-ELEMENT SYSTEM

We continue our study of the addition table to find further properties of our number system. In the final column of the table, the elements appear in the same order as in the left-border column. The sums represented are $1 + 5 = 1$, $2 + 5 = 2$, $3 + 5 = 3$, $4 + 5 = 4$, and $5 + 5 = 5$. Expressing these equalities in terms of the general element e, we find $e + 5 = e$; therefore, when 5 is used as a right-hand addend, it produces as a sum the element to which it is added. In mathematical language, 5 is a *right-identity element for addition* or a right additive identity. We can make a similar interpretation of the exact match between the entries in the bottom row of the table and those in the top border row. For every element e, $5 + e = e$; hence, 5 is also a *left identity*. Because 5 has both these properties, let us drop the adjectives and simply call it an *identity element for addition*.

DEFINITION In any number system, a given number i is an **identity for addition** if and only if for every number n of the system, $n + i = i + n = n$.

Return to the addition table and note that for each element e the sum e appears exactly once in the e row and exactly once in the e column. Hence, there is only one addend whose sum with e is e. This addend 5 is unique.

THEOREM 1.8-1 In the five-element system, element 5 is the unique identity element for addition.

EXERCISE 1.8

1. Show that in the addition problem $5 + 2 + 1 + 4 + 3 + 2 + 3 + 1$ the following association of elements is of advantage.

$$5 + 2 + (1 + 4) + (3 + 2) + 3 + 1$$

Complete the addition.

2. In the discussion above, having found that 5 is a right-identity element, we used the addition table to show that 5 is also a left-identity element. Instead, we could have used the commutativity of addition to show this last property. Explain.

3. Investigate the existence of additive identity elements in each of the following systems of Exercise 1.5.

(a) System Q (b) System M (c) System S

4. Investigate the existence of identity elements for addition in system T, whose addition table appears in Exercise 1.7.

5. In Section 1.1 we recalled the natural numbers. Does this system have an additive identity? If so, which natural number is it?

6. Consider the ordinary numbers of arithmetic with which you are familiar. Does this system have an additive identity? If so, which number is it?

1.9 THE ADDITIVE INVERSE IN THE FIVE-ELEMENT SYSTEM

Whenever a system has an additive identity, mathematicians always ask whether each element has a mate such that the sum of the element and its mate is the additive identity. For our system of five elements we phrase the question as follows: given any element e, can we find an element, to be denoted by e^* (read "e star"), such that $e + e^* = 5$?

Searching the addition table for an answer, we find e in the left-border column and the sum 5 in the e row in the body of the table; e^* is then the heading of the column in which the sum 5 lies. Consider, in particular, the element 1. We seek an element denoted by 1^* for which $1 + 1^* = 5$. We locate 1 in the left-border column, look across its row until we find the sum element 5 in the body of the table, and then read up to the heading of the column containing this sum element. Because the column heading is 4, we know that 4 is an acceptable value for 1^*.

A similar investigation of the other equalities $2 + 2^* = 5$, $3 + 3^* = 5$, $4 + 4^* = 5$, and $5 + 5^* = 5$ yields, in order, the results 3, 2, 1, and 5 for e^*. By the commutative law of addition or by an argument similar to the one above in which the words "row" and "column" are interchanged, we know that e^* is effective on the left as well as on the right; that is, $e^* + e = 5$. The element e^* is called an inverse of e under addition or an *additive inverse* of e. Because the sum element 5 appears only once in the e row, there is only one column whose heading gives e^*. The additive inverse is unique.

DEFINITION In any number system having additive identity i, the number n^* is an **inverse under addition** of the number n if and only if $n + n^* = n^* + n = i$.

THEOREM 1.9-1 In the five-element system each element e has a unique additive inverse denoted by e^*. In particular, $1^* = 4$, $2^* = 3$, $3^* = 2$, $4^* = 1$, and $5^* = 5$.

EXERCISE 1.9

1. Explain how to use the addition table to show that 3 is the correct value for 2^* in the equality $2^* + 2 = 5$.

2. In each of the following, find all elements e that make the statement true.

(a) $e^* = 1$	(b) $e^* \neq 4$	(c) $e^* = e$
(d) $e^* \neq e$	(e) $e^* + e \neq 5$	(f) $e^* = 1 + e$
(g) $e^* + 1 = e$	(h) $e^* = 2^*$	(i) $e^* \neq 4^*$

3. Do additive inverses exist in system Q given in Exercise 1.5? If so, what are they?

4. Do additive inverses exist in the noncommutative system T of Exercise 1.7? If so, what are they?

5. Investigate the existence of additive inverses in the ordinary numbers of arithmetic. By what names and by what symbols do we usually indicate such inverses?

Exercises 6–10 investigate properties of the additive inverse in the five-element system.

6. Is every element the inverse of some element? Justify your answer.

7. The element 5 differs from the other elements in one property of its additive inverse. Explain.

8. Denote the additive inverse of the additive inverse of e by the notation $(e^*)^*$ or e^{**}. For example, $1^* = 4$; hence, $1^{**} = (1^*)^* = 4^* = 1$. Find $2^{**}, 3^{**}, 4^{**}, 5^{**}$.

9. We know that $1^* = 4$ and $4^* = 1$. Is it true that for elements a and b, whenever $a^* = b$, then $b^* = a$? Explain.

****10.** Prove or disprove the statement that for elements a and b of the five-element system, $(a + b)^* = a^* + b^*$.

1.10 PROPERTIES OF THE ADDITIVE INVERSE IN THE FIVE-ELEMENT SYSTEM

In the preceding exercise we discovered some properties of the additive inverse. One of the most interesting is the fact that the inverse of a sum is the sum of the inverses. In other words, for elements a and b, $(a + b)^* = a^* + b^*$. This equality can be established, of course, by direct verification: for every one of the 25 pairs of elements, calculate the inverse of the sum and the sum of the inverses and compare answers. This method is long and a little tedious, although it can be shortened considerably by using the commutative property of addition. A better method uses the definition and uniqueness of the additive inverse to justify the equality in completely general terms. It is discussed here to show the power that resides in a mathematical definition as well as the advantage of using theoretical knowledge to handle a theoretical problem.

Begin with the assertion of the definition that $(a + b)^*$ is the element x for which $(a + b) + x = 5$, and find out whether the sum $a^* + b^*$ serves successfully in place of x. Make the substitution and use properties of addition in the five-element system:

$$\begin{aligned}(a + b) + (a^* + b^*) &= (a + b) + (b^* + a^*)\\ &= a + (b + b^*) + a^*\\ &= a + 5 + a^*\\ &= a + a^*\\ &= 5\end{aligned}$$

Now we know that $a^* + b^*$ is indeed an additive inverse of $a + b$. According to Theorem 1.9-1, moreover, it is the only such inverse.

The expression $a^* + b^*$ was presented without any explanation of its origin. Note that it might have been developed by solving the definition $(a + b) + (a + b)^* = 5$ for $(a + b)^*$. The steps in the solution are

$$\begin{aligned}5 &= (a + b) + (a + b)^*\\ a^* + 5 &= a^* + (a + b) + (a + b)^*\\ a^* &= (a^* + a) + b + (a + b)^*\\ &= 5 + b + (a + b)^*\\ a^* &= b + (a + b)^*\\ b^* + a^* &= b^* + b + (a + b)^*\\ &= 5 + (a + b)^*\\ &= (a + b)^*\\ a^* + b^* &= (a + b)^*\end{aligned}$$

whence we know that if the sum $a + b$ has an inverse, then an expression for it will be $a^* + b^*$.

The following properties of the inverse were discovered in Section 1.9. (Two further properties will be established in Section 2.14.)

THEOREM 1.10-1 In the five-element system, the properties listed below are among those which characterize the additive inverse.

(a) If element a is the additive inverse of element b, then b is the additive inverse of a. In symbols: for elements a and b, if $a = b^*$, then $a^* = b$. In particular, $(e^*)^* = e$ for any element e.

(b) The additive inverse of the sum of two elements is the sum of the additive inverses of the elements. In symbols: $(a + b)^* = a^* + b^*$ for elements a and b.

(c) The element 5 is its own additive inverse, and 5 is the only element that is its own additive inverse.

(d) Every element is the additive inverse of some element. In symbols: for every element a there exists an element b such that $a = b^*$.

EXERCISE 1.10

1. Explain why the uniqueness of the additive inverse justifies the assertion that if $a = b$, then $a^* = b^*$.

2. Find element x in each of the following equalities.

(a) $x^* = 4$	(b) $x + 1 = 3^*$
(c) $(x + 1)^* = 3$	(d) $x^* = 1^*$
(e) $x + 1^* = 3$	(f) $(x + 1^*)^* = 3^*$
(g) $x^{**} = 4$	(h) $x^* + 1 = 3^*$
(i) $(x^* + 1)^* = 3$	(j) $x = 4^{**}$
(k) $(x + 1^*)^* = 3$	(l) $(x + 1)^* = 3^*$

3. Prove that for elements a, b, and c,

$$(a + b + c)^* = a^* + b^* + c^*$$

4. Determine whether each statement below is true or false in the five-element system and justify your answer.

(a) If $a = b^*$ and $b = c^*$, then $a = c$.
(b) If $a + b = a^* + b$, then $a = 5$.
(c) If $a + b = c + d$, then $a^* + b^* = c^* + d^*$.

In Exercise 1.9 we discussed the existence of additive inverses for systems Q and T. In the problems below we investigate properties of these inverses.

5. (a) Does Theorem 1.10-1(b) hold for system Q? Justify your answer.
(b) Revise the statement of Theorem 1.10-1(c) to make it valid for system Q.

6. (a) Does Theorem 1.10-1(a) hold for system T? Justify your answer.
(b) Show that in system T, $(4 + 5)^* \neq 4^* + 5^*$.
(c) What does the inequality of part (b) imply concerning the analogue for T of Theorem 1.10-1(b)?
★(d) Reread the derivation of the formula for $(a + b)^*$. Determine what modifications in the formula should be made for a system that is not commutative under addition. State the correct analogue of this property for T.

1.11 USES OF THE ADDITIVE INVERSE IN THE FIVE-ELEMENT SYSTEM

The additive inverse is of use in forming combinations for addition of elements as in the following sums.

$$\begin{aligned}1 + 5 + 3 + 4 + 1 + 4 + 3 + 2 &= 1 + 3 + (4 + 1) + 4 + (3 + 2)\\ &= (1 + 4) + 3\\ &= 3\end{aligned}$$

$$\begin{aligned}&1 + 5 + 3^* + 4 + 1^* + 2^* + 3 + 4 + 2\\ &\quad = (1 + 1^*) + (3^* + 3) + 4 + (2^* + 2) + 4\\ &\quad = 4 + 4\\ &\quad = 3\end{aligned}$$

It is also of use in solving certain types of equations. Let us find element x, if it exists, for which the statement $x + 3 = 1$ is true. Because there are only five elements in our system, we can proceed by trial and error and try each in turn:

$$\begin{aligned}1 + 3 &= 4 \neq 1\\ 2 + 3 &= 5 \neq 1\\ 3 + 3 &= 1\\ 4 + 3 &= 2 \neq 1\\ 5 + 3 &= 3 \neq 1\end{aligned}$$

Hence, $x = 3$ is the one and only solution. Note that we could have arrived at this same result more quickly by looking down the 3 column in our addition table.

We can, however, proceed in a more analytical fashion by saying that

$$\begin{aligned}&\text{if} & x + 3 &= 1\\ &\text{then} & x + 3 + 3^* &= 1 + 3^*\\ &\text{whence} & x + 5 &= 1 + 2\\ &\text{and} & x &= 3\end{aligned}$$

Now we know that if any element makes the statement true, that element is 3. Upon checking we find that $3 + 3 = 1$. Hence, element 3 is a solution.

Let us apply this method to the general equation $a = b + x$, where a and b are any elements. We wish to find an element x, if it exists, for which both members of the equation will have the same value. We use the additive inverse of b to eliminate b from the right member:

$$\begin{aligned}b^* + a &= b^* + (b + x)\\ &= (b^* + b) + x\\ &= 5 + x\\ &= x\end{aligned}$$

If the equation is solvable, the the solution is the sum $b^* + a$. When substituted for x, this sum yields $b + (b^* + a) = (b + b^*) + a = 5 + a = a$. Hence, it does indeed satisfy the equation.

We have shown that if $a = b + x$, then $x = b^* + a$ and that, conversely, if $x = b^* + a$, then $a = b + x$. Recall the use of the phrase "if and only if" in the discussion of Section 1.4. The result of our present argument can be summarized in the statement that $a = b + x$ if and only if $x = b^* + a$.

Is the solution unique? If solution s were found by some other method, would s be different from $b^* + a$, or would s have to be the same as $b^* + a$? We have just seen that $b + (b^* + a) = a$. We know that the new solution also satisfies the equation so that $a = b + s$. Hence, $b + (b^* + a) = b + s$. The law of cancellation now asserts that $b^* + a = s$; therefore, s is identical with the solution already found. The next theorem states this result.

THEOREM 1.11-1 For elements a and b of the five-element system, the equation $a = b + x$ has the unique solution $x = b^* + a$.

Because of the commutative law, this theorem implies that the equation $a = x + b$ has the unique solution $x = a + b^*$. When $a = b$ we obtain a useful corollary. We note that all these equations are of the first degree in x.

Corollary 1.11-1. In the five-element system the equation $a = a + x$ has the unique solution $x = 5$, where 5 is the additive identity.

EXERCISE 1.11

1. Find the element x in each of the following equations.

(a) $2 + 3 + 1^* + 5^* + 4 = x$ (b) $3 + 4 + 5 + 3 + 2 + 4^* = x$
(c) $2 = x + 3$ (d) $3^* = x + 4$
(e) $x + 2 = 3^*$ (f) $4 + x = 5$
(g) $x + 2 = 1$ (h) $3 + x = 3$

2. We obtained a law of cancellation by using the addition table for the five-element system. The additive inverse can be employed in an alternative derivation of this law. Beginning with the relation $a + b = a + c$, add a^* to both members, producing $a^* + (a + b) = a^* + (a + c)$. Now reduce this last equation to $b = c$, the conclusion of the cancellation law.

3. In this exercise use system S of Exercise 1.5.

(a) Find the unknown element x in each of the following equations.

(i) $x = 1 + 4$ (ii) $7 + x = 4$ (iii) $x + 1 = 7$

(b) Explain why we cannot find a formula for the solution of the equation $a = b + x$ in system S.

In Exercises 4–10 use system T of Exercise 1.7.

4. Find the unique element x in each of the following equations.

(a) $x = 3 + 5$ (b) $x = 5 + 3$
(c) $4 = 2 + x$ (d) $4 = x + 2$
(e) $x + 5 = 6$ (f) $5 + x = 6$

5. Why are the answers to parts (a) and (b) of Exercise 4 not identical? To parts (c) and (d)? To parts (e) and (f)?

6. (a) Find a formula for the solution of the equation $a = b + x$.
(b) Does the formula of part (a) give the correct answers for parts (c) and (f) of Exercise 4?

7. (a) Find a formula for the solution of the equation $a = x + b$.
(b) Does the formula of part (a) give the correct answers for parts (d) and (e) of Exercise 4?

8. Write a theorem embodying the results of Exercises 6 and 7.

9. (a) The theorem of Exercise 8 can be called the analogue for system T of Theorem 1.11-1 for the five-element system. Explain.
(b) Write an analogue for system T of Corollary 1.11-1.

10. (a) In Exercise 2 only the use of 5 for the additive identity specialized the derivation of the cancellation law to apply to the five-element system. If 1 had been used for the additive identity, the derivation would have established the law for system T. Make this proof.
(b) In system T, $2^* = 3$. Hence, if we add 3 to each member of the equality $2 + 4 = 2 + 4$, then according to the discussion in part (a) we should produce $4 = 4$. Because $2 + 4 = 6$, the equality $2 + 4 + 3 = 2 + 4 + 3$ gives $6 + 3 = 6 + 3$ and reduces to $5 = 5$. Why do we not get the correct answer?

1.12 TWO GENERALIZATIONS

We have been solving equations of the form $a = b + x$ since we began the study of the five-element system. At first we consulted the addition table to find element x; each equation had to be dealt with as a special case. In Section 1.11, we found that use of the additive inverse could replace consultation of the table. According to the discussion developed there, the general equation has the sum $b^* + a$ as its unique solution. Now we can solve each specific equation in just one step, the evaluation of a simple sum.

Can this analysis be further generalized? Only the use of 5 for the additive identity restricts the discussion in Section 1.11 to the five-element system. If 5 is replaced by the general symbol i, the argument establishes the statement that in any number system possessed of sufficient properties for addition, $a = b + x$ has the unique solution $b^* + a$. The requisite properties are the associative and cancellation laws of addition as well as the existence of an identity and of a unique inverse for each element. If addition is commutative, the solution can be written in the alternative form $a + b^*$.

In similar fashion, we can generalize the first three properties of the additive inverse listed in Theorem 1.10-1. Here, again, only the use of 5 for the additive identity specializes the proofs to the five-element system. If 5 is replaced by the general symbol i, then the equality $e + e^* = e^* + e = i$ defines e^* as an additive inverse of e, but it also defines e as an additive inverse of e^*, whence $(e^*)^* = e$. From this it follows that if $a = b^*$, then $a^* = (b^*)^* = b$. Furthermore, because $i + i = i$, we see that i is an additive inverse of itself. If the additive inverse is unique, then we have established general analogues of Theorem 1.10-1(a) and (c).

We turn next to Theorem 1.10-1(b). For a system having the properties listed above it is easy to show that $(a + b) + (b^* + a^*) = i$. Hence, $b^* + a^*$ is another expression for $(a + b)^*$. If the inverse is unique, then so is this expression.

These proofs, although developed for the five-element system in particular, are really very general. They establish the fact that for any number system which is closed and associative under addition, which has a law of cancellation for addition, which possesses a unique additive identity, and for each of whose elements there exists a unique additive inverse, the analogues of Theorem 1.11-1 and of Theorem 1.10-1(a) and (c) are all valid. If addition is commutative, the actual statement of Theorem 1.10-1(b) is correct.

1.13 THE INTEGERS

The system of natural numbers has no additive identity; hence, it can have no additive inverse for any of its members. In such cases mathematicians frequently seek to extend the limited system by enlarging it to include missing elements. Such an extension creates a new system of which the old one is a part.

We would wish an extension of the natural numbers to include an identity for addition, an additive inverse for each member, and whatever other new numbers are needed to get closure under addition. Let us call the additive identity by the familiar name "zero" and denote it by the symbol 0. Let us designate the additive inverse of any natural number n by the name "the negative of n" and by the symbol $-n$.

The new enlarged system is commonly called the system of *integers*. The former natural numbers are now known as the positive integers, and their additive inverses are known as the negative integers. To display the members of this system we can write

$$\ldots, -4, -3, -2, -1, 0, 1, 2, 3, 4, \ldots$$

where the numbers continue without end in each direction. Investigation reveals that these numbers exhibit closure under addition; hence, no further new numbers must be included to complete the system.

The integers obey all the laws that hold for the natural numbers as well as the laws for identity and inverse in the five-element system. For example, the identity 0 is its own additive inverse and is unique in this property. Moreover, if $-n$ is the additive inverse of positive integer n, then n is the additive inverse of negative integer $-n$. Let us therefore modify our notation to permit the letter n to designate any integer, positive, negative, or zero.

In Figure 1.1 we used a clockface arrangement of elements to assist with addition. To find the sum $a = c + b$ we began at the point marked c and counted b spaces clockwise around the circle to reach a. It follows that if we should begin at point a and count b spaces counterclockwise around the circle we would arrive at point c. By Theorem 1.11-1 we know that $c = a + b^*$. Hence, in counting b spaces counterclockwise, we have performed the operation of adding b^* to a.

The integers can be displayed by taking equally spaced points on a straight line extending infinitely far in each direction. Choose one of these points to correspond to 0; then assign integer 1 to the first point to the right of the zero point and integer -1 to the first point to the left of the zero point. Assign the remaining positive integers in order to the points to the right of point 1 and the remaining negative integers in order to the points to the left of -1, as in Figure 1.2.

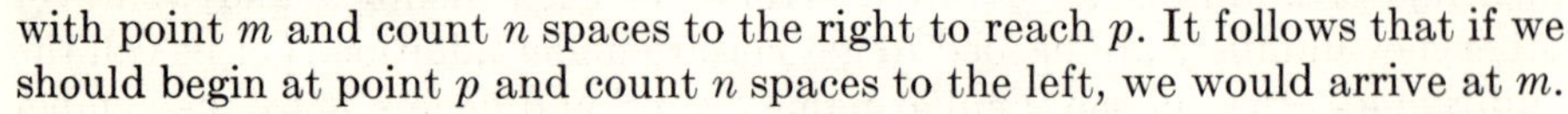

FIGURE 1.2

In order to find the sum $m + n = p$ where n is a positive integer, begin with point m and count n spaces to the right to reach p. It follows that if we should begin at point p and count n spaces to the left, we would arrive at m. Because $m = p + (-n)$, we see that in counting to the left we perform the operation of adding a negative integer to p.

EXERCISE 1.13

1. Write the negative of each of the following integers:

$$2 \qquad -5 \qquad 10 \qquad -6 \qquad 0 \qquad 1 \qquad m \qquad -m$$

2. Construct the integer line from -12 to 12. Then, using this line, find each of the following sums.

(a) $p = 6 + 4$ (b) $p = (-6) + 4$
(c) $p = 2 + (-7)$ (d) $p = (-2) + (-7)$

★3. State the analogue for integers of each of the properties of Theorem 1.10-1. Comment on the laws of negatives which these analogues exhibit.

★4. (a) Write the analogue of Theorem 1.11-1 for the system of integers.
(b) Using the theorem of part (a) solve the following equations for integer x.

(i) $12 = x + 13$ (ii) $6 = x + (-5)$
(iii) $3 + x = -4$ (iv) $(-20) + x = -7$

2
THE FIVE-ELEMENT SYSTEM
MULTIPLICATION

In our study of addition and subtraction in the five-element system, we have seen that some characteristics of the operations arise directly from their definitions and have observed methods whereby other properties may be discovered and justified. We learned the mathematical names for certain concepts and used symbolism in dealing with them. Now we will study the operation of multiplication. The new work will not only extend our knowledge of the five-element system but will also present an opportunity for further development of mathematical skills. The familiar natural numbers are used to open the discussion.

2.1 MULTIPLICATION OF NATURAL NUMBERS

When computing with natural numbers certain additions may be shortened by using a new operation. For example, the indicated sum $7 + 7 + 7$ can be expressed by the notation 3×7 or $3 \cdot 7$, so that $3 \cdot 7 = 7 + 7 + 7$. The operation designated by the symbol $\times$ is called *multiplication*, the numbers on which the operation acts are called *factors*, and the resulting natural number is called the *product*. In the above example, factor 7 is multiplied by factor 3 to give product 21. The first factor, 3, is sometimes called the multiplier, and the second factor, 7, the multiplicand. Using this distinction between factors, we see that $7 \cdot 3$ must be interpreted to mean that as many multiplicands of 3 are to be added as are indicated by the multiplier 7; consequently,

$$7 \cdot 3 = 3 + 3 + 3 + 3 + 3 + 3 + 3$$

It is only because both sums are the same natural number 21 that $3 \cdot 7 = 7 \cdot 3$. The distinction between factors is maintained in the formal description of multiplication.

DEFINITION For any natural number n, **products** are defined as follows:

$$\begin{aligned} 1 \cdot n &= n \\ 2 \cdot n &= (1 \cdot n) + n \\ 3 \cdot n &= (2 \cdot n) + n \qquad \text{and so on} \end{aligned}$$

This definition is said to be *recursive* because each product after the first is found by adding the multiplicand to the preceding product. If n is the number 7,

$$\begin{aligned} 1 \cdot 7 &= 7 \\ 2 \cdot 7 &= (1 \cdot 7) + 7 = 7 + 7 = 14 \\ 3 \cdot 7 &= (2 \cdot 7) + 7 = 14 + 7 = 21 \\ 4 \cdot 7 &= (3 \cdot 7) + 7 = 21 + 7 = 28 \qquad \text{and so on} \end{aligned}$$

Because one of the properties of multiplication in the natural number system is commutativity, it is true that if the multiplication table for 3 is built up to the product $7 \cdot 3$, this product is found to be 21; and that if the multiplication table for 4 is built up to the product $7 \cdot 4$, the answer is found to be 28.

Multiplication need not always be defined by recursive formulas. For some number systems other methods are used, and sometimes a multiplication table is presented without explanation.

EXERCISE 2.1

1. By using the definition of multiplication, construct the multiplication table for the natural number 3 as far as the entry $7 \cdot 3$.

2. For system Q of Section 1.5 let multiplication be defined by Table 2.1. In each of the following, find all elements e that make the statement true.

(a) $6 \cdot 6 = e$ (b) $3 \cdot e = 6$
(c) $9 \cdot e = 3$ (d) $3 \cdot e \neq 3$
(e) $6 \cdot e = 6$ (f) $e \cdot 9 = 9e$
(g) $e \cdot e \neq 0$ (h) $3 \cdot e = e$
(i) $e \cdot 0 = e$

Table 2.1 Multiplication Table for System Q

$\times$	0	3	6	9
0	0	0	0	0
3	0	3	6	9
6	0	0	0	0
9	0	3	6	9

2.2 MULTIPLICATION IN THE FIVE-ELEMENT SYSTEM

To define multiplication in the five-element system let us use a recursive definition similar to that for the natural numbers. The operation so defined will combine two given elements to produce a unique element; and the resulting element will, in general, be different from the one produced by addition. Hence, we are indeed introducing a new binary operation. Let us use customary notations for it.

DEFINITION For elements 1, 2, 3, 4, 5 and for any element e in the five-element system:

$$\begin{aligned} 1 \cdot e &= e \\ 2 \cdot e &= (1 \cdot e) + e \\ 3 \cdot e &= (2 \cdot e) + e \\ 4 \cdot e &= (3 \cdot e) + e \\ 5 \cdot e &= (4 \cdot e) + e \end{aligned}$$

Let us see what these formulas produce when $e = 4$:

$$\begin{aligned} 1 \cdot 4 &= 4 \\ 2 \cdot 4 &= (1 \cdot 4) + 4 = 4 + 4 = 3 \\ 3 \cdot 4 &= (2 \cdot 4) + 4 = 3 + 4 = 2 \\ 4 \cdot 4 &= (3 \cdot 4) + 4 = 2 + 4 = 1 \\ 5 \cdot 4 &= (4 \cdot 4) + 4 = 1 + 4 = 5 \end{aligned}$$

EXERCISE 2.2

1. Using the definition of multiplication, reckon the products $1 \cdot 3$, $2 \cdot 3$, $3 \cdot 3$, $4 \cdot 3$, and $5 \cdot 3$.

2. Construct a multiplication table for the five-element system. (Note that if the table is to be read in the usual way, then the entries in the 3 and 4 columns have already been reckoned.)

3. The product $e \cdot e$ is written e^2 and is called a perfect square. Which elements are perfect squares?

4. In each of the following, find all elements e that make the equality true.

(a) $1 \cdot 2 = e$ (b) $2 \cdot 4 = e$ (c) $5 \cdot 3 = e$
(d) $2e = 4$ (e) $5e = 2$ (f) $3e = 5$
(g) $e \cdot 2 = 2$ (h) $e \cdot 1 = 4$ (i) $e \cdot 5 = 5$

5. In each of the following, find all elements e that make the statement true.

(a) $2 \cdot 3 \neq e$ (b) $e = e \cdot 1$ (c) $2e \neq 3$
(d) $e \cdot 2 = e$ (e) $e \cdot 5 \neq e$ (f) $e \cdot 3 = 3 \cdot e$

6. Copy and complete Table 2.2. The asterisk denotes "additive inverse."

Table 2.2

a	b	a^*	b^*	a^*b^*	a^*b	ab^*	ab	$(ab)^*$
1	2							
3	3							
2	4							

7. From the table prepared in Exercise 6, determine which of the following equalities appear to be true and which are false.

(a) $(ab)^* = a^*b^*$ (b) $(ab)^* = ab^*$
(c) $a^*b^* = ab$ (d) $ab^* = a^*b$

8. In Section 2.14 it will be proved that equalities (b), (c), and (d) of Exercise 7 are true for all elements. Show, therefore, that each of the following statements is true.

(a) If $ab = cd$, then $a^*b^* = c^*d^*$.
(b) If $a^*b = c^*d$, then $ab^* = cd^*$.
(c) If $ab = cd$, then $a^*b = cd^*$.
(d) If $a^*b^* = c^*d^*$, then $ab = cd$.
(e) If $a^*b = c^*d$, then $a^*b^* = c^*d^*$.

9. In each of the following, find all elements e for which the equality holds.

(a) $4^*e^* = 3$ (b) $(4e)^* = 3$ (c) $4e^* = 3$

10. Compare the tables for multiplication and for addition in the five-element system. List two ways in which the entries resemble each other and two ways in which they differ. What is the significance of the likenesses?

2.3 CLOSURE AND COMMUTATIVITY OF MULTIPLICATION IN THE FIVE-ELEMENT SYSTEM

Because there are similarities in the addition and multiplication tables for the five-element system, there should be common properties of the two operations. Two of these are explored in Exercise 2.3, using definitions and proofs similar to those employed for addition. For general statements let us assume that we know what is meant by the words "number system" and "multiplication" in the system. Let us begin with closure.

EXERCISE 2.3

1. Explain the statement: a given number system is closed under multiplication.

2. Explain how the multiplication table shows that the five-element system is closed under multiplication.

3. Which of the following systems are closed under multiplication? Justify your answer in each case.
- (a) The natural number system.
- (b) The even natural numbers.
- (c) The odd natural numbers.
- (d) The system consisting of only the natural numbers 1 and 2.
- (e) The system containing only elements 1 and 2 of the five-element system.
- (f) The system consisting of elements 1, 2, 3, and 4 of the five-element system.
- (g) The system containing only elements 1, 4, and 5 of the five-element system.

4. Explain the statement: in a given number system multiplication is commutative.

5. Prove that multiplication is commutative in the five-element system.

6. By using the laws of closure and commutativity of multiplication in the five-element system, justify the following equalities.
- (a) $(2 \cdot 3) \cdot 4 = 4 \cdot (2 \cdot 3)$
- (b) For elements a, b, and c, $(ab)c = c(ba)$.

7. Is multiplication commutative for system Q as described in Exercise 2.1.2?

2.4 ASSOCIATIVITY OF MULTIPLICATION IN THE FIVE-ELEMENT SYSTEM

Not all the characteristics of addition (for example, associativity) were immediately obvious from a careful reading of the addition table. This property is developed for multiplication in the exercise below, where you are asked to adapt the discussion in Section 1.6. The following theorem summarizes the results.

THEOREM 2.4-1 The five-element system is closed, commutative, and associative under multiplication.

EXERCISE 2.4

1. Define the product of a triple of elements in the five-element system. Model your definition on that given for addition of three elements in Section 1.6.

2. Explain the statement: in a given number system, multiplication is associative.

3. Verify the formula defining associativity of multiplication for the set of triples in Exercise 1.6.1 assigned to you by your instructor.

4. By using the laws of multiplication in the five-element system, show that each of the following equalities is true.

(a) $(3 \cdot 4) \cdot 2 = 3 \cdot (4 \cdot 2)$
(b) $(3 \cdot 4) \cdot 2 = 4 \cdot (3 \cdot 2)$
(c) $(3 \cdot 4) \cdot 2 = 2 \cdot (4 \cdot 3)$

★5. Given elements a, b, and c of the five-element system, explain why multiplying from left to right gives the same result as multiplying from right to left. That is, explain why $(ab)c = (cb)a$.

6. By rules analogous to those for addition, the definitions and laws can be extended to the product of more than three elements. Applying associative and commutative laws, elements can be multiplied in any order and in any grouping. For example, in the five-element system,

$$\begin{aligned} 4 \cdot 3 \cdot 2 \cdot 4 \cdot 1 \cdot 3 &\equiv ((((4 \cdot 3) \cdot 2) \cdot 4) \cdot 1) \cdot 3 = 3 \\ &= (4 \cdot 4) \cdot (3 \cdot 3) \cdot (2 \cdot 1) = 3 \end{aligned}$$

Multiply these elements together in some other order or grouping.

7. Find the following products.

(a) $2 \cdot 1 \cdot 4 \cdot 3 \cdot 3 \cdot 2$ (b) $3 \cdot 4 \cdot 5 \cdot 2 \cdot 1 \cdot 4$

2.5 IDENTITY AND INVERSE UNDER MULTIPLICATION IN THE FIVE-ELEMENT SYSTEM

One further similarity between the addition and multiplication tables is of interest. In the body of each table one row repeats the top-border row and one column has elements identical with those in the left-border column. In the addition table these identical entries are the last row and column, and as a result, element 5 is the additive identity. In the multiplication table these corresponding elements appear in the first row and column, and only there. Hence, element 1 has uniquely the property that $e \cdot 1 = 1 \cdot e = e$ for all elements e, and it is the unique *identity for multiplication.*

THEOREM 2.5-1 In the five-element system, element 1 is the unique identity element for multiplication.

Having discovered an identity, let us proceed to the problem of finding for each element e a multiplicative inverse, to be denoted by e', such that $ee' = e'e = 1$. We investigate all appearances of the element 1 in the body of the multiplication table because the mates e and e' whose products are 1 are the captions of the row and column at whose intersection the element 1 appears. The first four rows and columns yield $1' = 1$ and $4' = 4$, whereas $2' = 3$ and $3' = 2$. Because element 1 appears only once in each of these rows and columns, the inverses are unique. In the fifth row and column the identity element for multiplication does not appear. Hence, element 5 has no multiplicative inverse.

THEOREM 2.5-2 In the five-element system each element $e \neq 5$ has a unique multiplicative inverse denoted by e'. In particular, $1' = 1$, $2' = 3$, $3' = 2$, and $4' = 4$.

EXERCISE 2.5

In this exercise properties of the multiplicative inverse in the five-element system are developed and then extended to other systems. Justify each answer.

1. Is every element the inverse of some element?

2. Denote the inverse of the inverse of element e by $(e')'$ or e''; for example, $2'' = (2')' = 3' = 2$. Is it true that for every element e different from 5, $e'' = e$?

3. Explain how the definition of multiplicative inverse would lead you to expect that the two relations $2' = 3$ and $3' = 2$ should occur together. Is it true that for any elements a and b, if $a = b'$, then $a' = b$?

4. When investigating additive inverses, we discovered that the additive inverse of a sum of elements is the sum of the additive inverses. Does a corresponding relation hold for multiplicative inverses? That is, for a and b different from 5, does $(ab)' = a'b'$?

5. Under addition in the five-element system, only the identity is its own inverse. Under multiplication, however, not only does $1' = 1$, but also $4' = 4$. Explain this dissimilarity in the number of elements that are identical with their inverses.

6. Consider the system of natural numbers.

(a) Is there a multiplicative identity in the natural numbers? If so, which number is it?

(b) Which, if any, natural numbers have multiplicative inverses?

7. Answer the questions of Exercise 6 for the system of integers.

2.6 PROPERTIES OF THE MULTIPLICATIVE INVERSE IN THE FIVE-ELEMENT SYSTEM

In the preceding section we discovered that each of the properties of the additive inverse listed in Theorem 1.10-1 has an analogue for the multiplicative inverse. Let us write these as a theorem along with a further statement giving the special property of the element 5.

THEOREM 2.6-1 In the five-element system the properties listed below are among those that characterize the multiplicative inverse.

(a) If a is the multiplicative inverse of b, then b is the multiplicative inverse of a. In symbols: for elements a and b, if $a = b'$, then $a' = b$. In particular, for $e \neq 5$, $(e')' = e$.

(b) For elements different from 5 the multiplicative inverse of the product of two elements is the product of their multiplicative inverses. In symbols: $(ab)' = a'b'$ for elements a and b different from 5.

(c) The elements 1 and 4 are their own multiplicative inverses and are the only elements with this property.

(d) Every element except 5 is the multiplicative inverse of some element. In symbols: for every element $a \neq 5$, there exists an element b such that $a = b'$.

(e) The element 5 is the only element that has no multiplicative inverse.

EXERCISE 2.6

1. Find element x in each of the following equalities.

(a) $x' = 4$	(b) $2x = 3'$	(c) $(2x)' = 3$
(d) $x' = 1'$	(e) $3'x = 4$	(f) $(3'x)' = 4'$
(g) $x'' = 4$	(h) $2x' = 3'$	(i) $(2x')' = 3$
(j) $x = 1''$	(k) $(3'x)' = 4$	(l) $(3x)' = 4$

2. By direct verification show that if $a^* = b'$, then $a' = b^*$.

3. For what elements a and b is $ab = ab'$?

4. Determine whether each statement below is true or false in the five-element system and justify your answer.

(a) If $a = b'$ and $b = c'$, then $a = c$.

(b) For elements a, b, c, and d all different from 5, if $ab = cd$, then $a'b' = c'd'$.

5. Prove that for elements a, b, and c all different from 5, $(abc)' = a'b'c'$.

2.7 USES OF THE MULTIPLICATIVE INVERSE IN THE FIVE-ELEMENT SYSTEM

The multiplicative inverse has many uses. One of them is to establish a law of cancellation for multiplication. Assume that $ab = ac$ for elements a, b, and c. We seek to prove that $b = c$. We can eliminate a from both members of the first equality by using the multiplicative inverse of a. Thus, $a'(ab) = a'(ac)$; whence $(a'a)b = (a'a)c$ and $(1)b = (1)c$ so that $b = c$. The existence of a' is essential to the success of this procedure and is guaranteed whenever $a \neq 5$. (The theorem below includes the required restriction.) Note that because of the commutative law the conclusion remains valid when $ba = ca$.

THEOREM 2.7-1 *Law of Cancellation for Multiplication.* For elements $a \neq 5$, b, and c of the five-element system, if $ab = ac$, then $b = c$.

A further use of the multiplicative inverse produces a formula for the solution of the equation $a = bx$, where a and $b \neq 5$ are known and element x is to be found. If we eliminate b from the right member, the resulting equality will give an expression for x. Multiplication of both members of the equation by b' yields

$$\begin{aligned} b'a &= b'(bx) \\ &= (b'b)x \\ &= (1)x \\ &= x \end{aligned}$$

If there is a solution for the given equation, it must be the product $b'a$. Because substitution of this product for x gives

$$\begin{aligned} a &= b(b'a) \\ &= (bb')a \\ &= (1)a \\ &= a \end{aligned}$$

we know that $b'a$ is a solution for this first-degree equation.

Is this solution unique? One of the best ways to answer this question is to use the method that dealt successfully with a similar situation in Section 1.11. Assume that we are presented with another solution s and are asked to determine whether s is the same as $b'a$. We know that $b(b'a) = a$. Furthermore, because s is a solution, $a = bs$. Hence, $bs = b(b'a)$. Since $b \neq 5$, the law of cancellation is applicable. It asserts that $s = b'a$. Consequently, the "new" solution s is the same as $b'a$.

THEOREM 2.7-2 For elements a and b of the five-element system, when $b \neq 5$, the equation $a = bx$ has the unique solution $x = b'a$.

Here again, the commutative law causes the theorem to imply more than is explicitly stated, namely that $a = xb$ has the unique solution ab' for x. When $a = b$, a useful corollary is obtained.

Corollary 2.7-2 For $a \neq 5$, the equation $a = ax$ has the unique solution $x = 1$.

EXERCISE 2.7

1. Find all elements x that make each of the following equalities true.

(a) $3x = 4$	(b) $2x = 2$	(c) $4x = 4 \cdot 3$
(d) $4x = 3$	(e) $2x = 1$	(f) $3x = 2 \cdot 3$
(g) $5x = 5 \cdot 2$	(h) $5x = 4x$	(i) $2x = 2 \cdot 5$

2. Criticize the following "proof" that element 3 equals element 4: we have $5 \cdot 3 = 5$ and $5 \cdot 4 = 5$; hence, $5 \cdot 3 = 5 \cdot 4$; whence, by cancellation, $3 = 4$.

2.8 THE REDUCED FIVE-ELEMENT SYSTEM

Let us delete from the multiplication table the final row and column, both headed by element 5, to create a new table containing only 1, 2, 3, and 4. It resembles the addition table in that each element appears once and only once in each row and column. Hence, the four elements are closed under multiplication, satisfy a general law of cancellation, and have a unique solution for the equation $a = bx$. They form a new algebraic system. Let us give it the name the "reduced system" of elements. The new system inherits commutativity and associativity of multiplication from the old one. The identity 1 is a member of the system. Because all the remaining elements have inverses, the theorems on the inverse already stated still hold. Thus, the reduced system possesses under multiplication all the properties that characterize the five-element system under addition.

2.9 SOME GENERALIZATIONS

Some of the properties of the reduced system of elements can be established by proofs in which only changes in terminology and symbolism are required to effect the transformation from addition in the whole system to multiplication in the reduced system.

The law of cancellation is an example. In Exercise 1.11.2 the additive inverse was used to show that when $a + b = a + c$, then $b = c$. In Section 2.7 a similar use of the multiplicative inverse reduced the relation $ab = ac$ to $b = c$. A second example deals with the solution of equations. In Section 1.11, $b^* + a$ was established as the unique solution of equation $a = b + x$. In Section 2.7 similar methods were used to develop $b'a$ as the unique solution of equation $a = bx$ in the reduced system.

Let us consider a general number system with a general operation and then construct and prove corresponding theorems for it. Call the system N. Instead of a specific operation on its elements, such as addition or multiplication, use an unspecified binary operation. Call the operation "circle" and designate it by the symbol $\circ$. Hence circle may represent addition, subtraction, multiplication, or any other process that operates on two elements to produce a unique element of N. Impose the following conditions:

(1) Circle is associative. (No assumption is made about its commutativity.)

(2) Let i represent an identity element of N under circle. For example, if circle is addition, and if N is the five-element system, i is 5; or if N is the system of integers, i is 0.

(3) Let the symbol $e^\dagger$ (read "e dagger") denote an inverse of e under circle; thus if circle is addition and N is the integer system, $e^\dagger$ is a name for $-e$. On the other hand, if circle is multiplication and N is the reduced five-element system, $e^\dagger$ represents e'.

Arguments patterned on those above allow us to derive analogous properties.

THEOREM N.1 For elements a, b, and c of system N, if $a \circ b = a \circ c$ or if $b \circ a = c \circ b$, then $b = c$.

THEOREM N.2 For elements a and b the equations $a = b \circ x$ and $a = x \circ b$ have unique solutions for element x given, respectively, by the formulas $x = b^\dagger \circ a$ and $x = a \circ b^\dagger$.

Next consider the uniqueness properties of the identity and inverse. For the five-element system we proved these by use of the addition and multiplication tables. We have no such tables for system N. Can we derive the desired properties on the basis of our theoretical knowledge? Yes, if we use the method that successfully established uniqueness for the solutions of the equations of Section 1.11.

THEOREM N.3 The identity element is unique, and for each element there is a unique inverse.

Corresponding properties of the additive and multiplicative inverse had derivations that differed only in terminology and notation. They have analogues in system N.

THEOREM N.4 For elements a and b, $a^{\dagger\dagger} = a$, $(a \circ b)^\dagger = b^\dagger \circ a^\dagger$, and if $a = b^\dagger$, then $a^\dagger = b$.

The theorems for N are established in the optional discussion that follows. You are encouraged to work through the proofs and understand them. It should then be clear that many of the theorems and derivations that we have developed for the five-element system are special cases of more general ones which hold for many different number systems.

Discussion (Optional)

In the proof of the law of cancellation, the inverse is used to show that if $a \circ b = a \circ c$, then $b = c$. Operate with $a^\dagger$ on the left side of both members of the first equality and get $a^\dagger \circ (a \circ b) = a^\dagger \circ (a \circ c)$. By the associative law this equation can be written $(a^\dagger \circ a) \circ b = (a^\dagger \circ a) \circ c$; whence $i \circ b = i \circ c$, or $b = c$. A similar use of the inverse establishes the solution $b^\dagger \circ a$ for the equation $a = b \circ x$. Uniqueness can be proved by demonstrating that any element which acts like a solution must coincide with the solution. We first used these methods in Section 1.11, where we showed that $a = b + x$ has the unique solution $b^* + a$ for x. If in that discussion we replace $+$ by $\circ$ and $*$ by $\dagger$, the argument is adapted to our present use.

We now apply the uniqueness-proof technique to show that any element which acts like identity i must coincide with i. Element i is defined by the property that $n \circ i = i \circ n = n$ for every element n. Now let element j also have the property that $n \circ j = j \circ n = n$ for every element n. It follows that $n \circ j = n \circ i$; whence, by the cancellation law, $j = i$. In like manner, we can show that any element which acts like inverse $e^\dagger$ coincides with $e^\dagger$.

What about other important properties of the inverse? Is it true that $e^{\dagger\dagger} = e$ and that if $a = b^\dagger$, then $a^\dagger = b$? The equality $e \circ e^\dagger = e^\dagger \circ e = i$ defines not only $e^\dagger$ as an inverse of e but also e as an inverse of $e^\dagger$. Similarly, the equality $a \circ b = b \circ a = i$ defines b as an inverse of a and a as an inverse of b. Since the inverse is unique, the word "an" can properly be changed to "the" in the two sentences above.

Finally, let us inquire about the inverse of the expression $a \circ b$. To derive its formula, we can use the method employed in Section 1.10 for addition in the five-element system. If, instead, we simply wish to verify the formula suggested by previous study, namely, $(a \circ b)^\dagger = b^\dagger \circ a^\dagger$, then we can show that $(a \circ b) \circ (b^\dagger \circ a^\dagger) = i$. In the following exercise you are asked to make this verification.

EXERCISE 2.9 (OPTIONAL)

1. Prove that for elements a, b, and c of system N, if $b \circ a = c \circ a$, then $b = c$.

2. Prove that for the elements a and b of system N, $a = b \circ x$ has the unique solution $x = b^\dagger \circ a$.

3. Prove that the inverse $e^\dagger$ of element e is unique.

4. Derive the formula for the inverse of $a \circ b$.

5. Show that $(a \circ b) \circ (b^\dagger \circ a^\dagger) = i$.

★6. For elements a, b, and c of system N find a correct formula for the inverse of $a \circ b \circ c$.

2.10 GROUPS

We have seen that when an algebraic system has a few simple properties, it also has certain other properties. Recognizing the importance of this situation, mathematicians have developed the concept of a group. As this new concept is defined and discussed, it will become clear that the system N is actually a general group.

DEFINITION A nonempty set of elements on which a binary operation is defined forms a **group** with respect to the operation provided the following conditions are satisfied:

(a) The set of elements is closed under the operation.
(b) The operation is associative.
(c) There is an identity element for the operation.
(d) Each element has an inverse under the operation.

If the operation is commutative, the group is commutative. In order to state the properties that hold for any group, we use the notation developed in the preceding section. Let us designate the system of elements by N and denote the binary operation by $\circ$ (circle). And let us represent an identity element by i and an inverse of element e by $e^{\dagger}$ (e dagger). In any group, the following theorems hold.

THEOREM G.1 The identity element is unique.

THEOREM G.2 For each element the inverse is unique.

THEOREM G.3 For elements a, b, and c, if $a \circ b = a \circ c$, or if $b \circ a = c \circ a$, then $b = c$.

THEOREM G.4 For elements a and b the equations $a = b \circ x$ and $a = x \circ b$ have unique solutions given, respectively, by the formulas $x = b^{\dagger} \circ a$ and $x = a \circ b^{\dagger}$.

Furthermore, the inverse has the properties stated in the following corollary.

Corollary G.2 For elements a and b of a group, $a^{\dagger\dagger} = a$; $(a \circ b)^{\dagger} = b^{\dagger} \circ a^{\dagger}$; and if $a = b^{\dagger}$, then $a^{\dagger} = b$.

As an application of these properties, consider system W whose elements are denoted by symbols 1, 2, 3, 7, 8, 9. Let a binary operation $\circ$ be defined on this set by Table 2.3. It can be shown that circle is associative. We see from the table that W is closed but not commutative under circle. The entries in the 7 row and 7 column show that 7 is an identity element. Furthermore, this identity appears in every row and column, and it is easy to verify that each element has an inverse under circle. All conditions set forth in the definition are satisfied; hence, system W forms a group with respect to

Table 2.3 Operation Table for System W

$\circ$	1	2	3	7	8	9
1	7	9	8	1	3	2
2	8	3	7	2	9	1
3	9	7	2	3	1	8
7	1	2	3	7	8	9
8	2	1	9	8	7	3
9	3	8	1	9	2	7

circle. It can now be asserted, without further discussion, that the following statements are true:

(1) Element 7 is the unique identity element.

(2) Each element e has a unique inverse $e^\dagger$ under circle. Table 2.3 shows that $2^\dagger = 3$ and $3^\dagger = 2$, while all other elements are their own inverses. Furthermore, $(a \circ b)^\dagger = b^\dagger \circ a^\dagger$. For example, $(2 \circ 9)^\dagger = 1^\dagger = 1$ and $9^\dagger \circ 2^\dagger = 9 \circ 3 = 1$.

(3) If $a \circ b = a \circ c$, or if $b \circ a = c \circ a$, then $b = c$.

(4) The equation $a = b \circ x$ has the unique solution $x = b^\dagger \circ a$, and the equation $a = x \circ b$ has the unique solution $x = a \circ b^\dagger$. The lack of commutativity makes these solutions differ as the equations differ. Thus $8 = 3 \circ x$ has solution $x = 9$, whereas $8 = x \circ 3$ has solution $x = 1$.

It is clear that the establishment of the group property has powerful consequences for any number system. It was relatively easy to read the table and get the simple requirements for the existence of the group, after associativity was assured. Then group theorems relieved us of the necessity of making long existence and uniqueness proofs to establish statements (1)–(4). It will be well worth our while to look for the group property and make use of its consequences in all later work.

The concept of a group has been introduced as an effective means of generalizing algebraic systems and their properties. Such generalization is only one of the many uses which have been found for this important concept since its formulation near the beginning of the nineteenth century. One of its first applications was to the problem of solving algebraic equations. Another application was to the field of geometry. Here the elements of the groups are not numbers but are equations that describe ways in which geometric objects may be transformed. For example, in euclidean geometry a figure may be moved from one position in space to another; in some other geometries its shape may be changed, as a balloon is deformed by squeezing.

The expression of many of the concepts of quantum physics in terms of groups is another example of the multitude of uses of this mathematically young idea.

EXERCISE 2.10

Determine whether each of the following sets of elements forms a group under the operation indicated. Justify your answer.

1. The natural numbers under addition.
2. The integers under addition.
3. The even integers under addition. (Zero is considered an even integer.)
4. The odd integers under addition.
5. System Q of Exercise 1.5 under addition.
6. System T of Exercise 1.8 under addition.
7. Elements 1 and 4 of the five-element system under multiplication.
8. Elements 3 and 4 of the five-element system under multiplication.
9. Elements 1, 3, and 4 of the five-element system under multiplication.
10. Elements 2, 3, and 7 of system W under circle.
11. The natural numbers under multiplication.
12. The odd integers under multiplication.

2.11 COMMENT ON GROUP CRITERIA

The preceding exercise not only provided an acquaintance with some interesting groups, but also gave experience in the techniques of testing for group requirements. Exercise 2.10.2 demonstrated that the integers are a group under addition. In Exercise 2.10.3 you were asked about the even integers. Since these numbers are a subset of the integers and since the addition operation is the same, there was no need to repeat all the tests. The test for associativity could be omitted because associativity of addition for all integers implies the property for even integers. The test for closure could not be omitted: although the sum of two even integers is an integer, it cannot be assumed without verification that the sum is even. When we discover that it is, then we can assert closure.

We know that every even integer has an additive inverse because every integer possesses one. Until we test, however, we do not know that the inverse of an even integer is even and belongs to the system. Finally, we are assured without any further testing that the additive identity is a member of the set because the sum of each even integer and its inverse is the identity. Thus, the answer to Exercise 2.10.3 required verification only of closure and the fact that every member of the set has an inverse in the set. The even integers are a *subgroup* of the additive group of all integers.

To generalize the discussion for a general system N that is a group under the operation circle, let C be a subset of elements of N. Whenever C satisfies all the criteria for a group under circle, C is said to be a *subgroup* of N. What tests are required to determine whether a subset of N is a subgroup of N? If the operation is associative or commutative for the totality of numbers in N, it remains so for any subset of N. On the other hand, even if N is closed under the operation circle, a subset of N may not be; further testing is required. When we consider the inverse, we find that every element of C possesses an inverse in N, but for any given element the inverse may or may not be an element of C. Here again, further testing is required. If the inverse condition is met, the identity condition is also met, because when the subset C contains e and $e^{\dagger}$, it must also contain $e \circ e^{\dagger} = i$. We thus have the following definition and theorem:

DEFINITION Let N be a group under operation circle. Then a subset C of N is a **subgroup** of N if and only if C is a group under circle.

THEOREM 2.11-1 Let N be a group under circle. Then a subset C of N is a subgroup of N if and only if C is closed under circle and contains an inverse for each of its elements.

EXERCISE 2.11

In each part of Exercises 1 and 2, C is a subset of N where N forms a group under the operation indicated. Determine whether C is a subgroup of N.

1. N is the system of integers under addition. C is the set of multiples of 3.

2. N is the system W of Section 2.10 under circle.
(a) C is the set of elements 1, 7.
(b) C is the set of elements 7, 8.
(c) C is the set of elements 1, 7, 8.

★3. In any group the identity is its own inverse; in the group consisting of the reduced system of elements under multiplication, both the identity and element 4 are their own inverses. Prove that any group containing an even number of elements always possesses at least one element different from the identity which is its own inverse.

2.12 THE MULTIPLICATIVE PROPERTY OF THE ELEMENT 5

In this section we return to the five-element system to discuss a consequence of one of the dissimilarities between the addition and multiplication tables. The addition table presents each element once and only once in each row and column. The multiplication table displays this characteristic in the first four rows and columns of the table but not in the fifth. The final row and column are composed entirely of the element 5, and this element appears in no other position. It follows that a product is 5 whenever one of the factors is 5; and conversely, in order to yield a product of 5, one of the factors must be 5. These observations are stated in a theorem.

THEOREM 2.12-1 *Multiplicative Property of the Element 5.* For elements a and b of the five-element system, $ab = 5$ if and only if $a = 5$ or $b = 5$.

The word "or" in the theorem is the conjunctive "or"; it has the meaning and/or. Thus, the final statement should be interpreted as "a is 5, or b is 5, or both a and b are 5." The word "or" is customarily used in this sense in mathematical writing.

EXERCISE 2.12

1. Find all elements e that make the following statements true.

(a) $3e = 5$	(b) $3e \neq 5$	(c) $3 \cdot 5 = 3e$
(d) $5e = 3$	(e) $5e \neq 3$	(f) $5 \cdot 4 = 5e$
(g) $5e = 5$	(h) $5 \cdot 4 \neq 5e$	(i) $2 \cdot 5 = 2e$

2. If a product of three or more elements is 5, what can be said about at least one of the factors of the product?

3. If one of the three or more factors of a product of elements is 5, what can be said about the product?

4. If a product of elements is different from 5, what can be said about each factor of the product?

5. If none of the factors of a product of elements is 5, what can be said about the product?

6. For elements a and b, if $ab = 5$ and $a \neq 5$, what can be said about element b?

7. Consider the statement: for elements a and b, if $ab \neq 5$, then $a \neq 5$ and $b \neq 5$; conversely, if $a \neq 5$ and $b \neq 5$, then $ab \neq 5$.

(a) Explain why the statement is true.

(b) Rephrase the statement using the expression "if and only if."

8. Consider the statement: for elements a and b, each of a and b has a multiplicative inverse if and only if $ab \neq 5$.

(a) Rephrase the statement as two "if . . . , then . . ." sentences.

(b) Explain why each of the sentences of part (a) is true.

9. It is true that $a^* = 4a$ for every element a, where a^* is the additive inverse of a.

(a) Verify this statement by comparing the values of a^* and $4a$ for every a.

(b) Prove the statement by use of the definitions of additive inverse and of the product $5a$.

10. Does the integer 0 have a multiplicative property for the system of integers similar to the multiplicative property of the element 5 in the five-element system? Explain.

2.13 COMBINING OPERATIONS IN THE FIVE-ELEMENT SYSTEM

Experience with the natural numbers indicates that it is sometimes convenient to combine the operations of addition and multiplication. Let us, therefore, consider two obvious combinations: first, the product of a sum of elements and a single element, and second, the sum of a product of elements and a single element.

The expression $(b + c)a$, where a, b, and c are elements, indicates the product of a sum and a single element. By laws of closure, the sum $b + c$ is an element, and the product of the sum $b + c$ and the element a is an element. Hence, a unique element represented by such an expression can always be found. For example, $(2 + 4) \cdot 3 = 1 \cdot 3 = 3$.

The expression $(bc) + a$, on the other hand, indicates the sum of a product and a single element. By laws of closure, the product bc is an element, and the sum of the product bc and the element a is an element. Consequently, a unique element represented by such an expression can always be found. For instance, $(2 \cdot 4) + 3 = 3 + 3 = 1$.

Might there be different ways of finding the single element that such an indicated product or indicated sum represents? For example, can $(2 + 4) \cdot 3$ be written as $(2 \cdot 3) + (4 \cdot 3)$? Does $(2 \cdot 4) + 3$ equal $(2 + 3) \cdot (4 + 3)$? A brief calculation shows that the answer to the first question is "yes," but to the second it is "no." Because multiplication is defined in terms of addition (but not vice versa), these answers are not surprising.

Consider the general formula $(b + c)a = ba + ca$. To determine whether it holds for all elements a, b, and c in the five-element system, for each of the 125 triples of elements each member of the equality must be calculated, and it must then be noted whether the two answers are the same. The result is that in every case the product $(b + c)a$ equals the sum $ba + ca$. In mathematical language, the factor a can be *distributed from the right* over the two summands b and c. We thus have the following three-part definition and a theorem which the law of commutativity of multiplication permits us to state in strong form.

DEFINITIONS

In a given number system multiplication is **right distributive** with respect to addition if and only if for any three numbers a, b, and c of the system, $(b + c)a = ba + ca$.

In a given number system multiplication is **left distributive** with respect to addition if and only if for any three numbers a, b, and c of the system, $a(b + c) = ab + ac$.

In a given number system multiplication is **distributive** with respect to addition if and only if it is both right and left distributive.

THEOREM 2.13-1 In the five-element system, multiplication is distributive with respect to addition.

A counterexample showed that the indicated sum $(bc) + a$ cannot be written as an indicated product. Hence, addition is not right distributive with respect to multiplication. The law of commutativity now prevents addition from being left distributive with respect to multiplication.

By agreement within the mathematical community, the indicated sums $(bc) + a$ and $a + (bc)$ are written in the simpler form $bc + a$ and $a + bc$, respectively, with the understanding that multiplication is to precede addition in determining their value. So firmly is this convention established that computers are frequently programmed to handle mathematical expressions by performing indicated multiplications and divisions before indicated additions and subtractions.

Let us now apply the distributive law to the product $(a + b)(c + d)$. By closure of addition, the sum $a + b$ can be considered a single element. Hence, by the left-distributive law the product is $(a + b)c + (a + b)d$. Now applying the right-distributive law, we get the expansion in the corollary below.

Corollary 2.13-1 For elements a, b, c, and d of the five-element system, $(a + b)(c + d) = ac + bc + ad + bd$.

EXERCISE 2.13

1. Verify the right-distributive law of multiplication with respect to addition for your set of triples (see Exercise 1.6.1). In this way the truth of the law will be established by the class for all possible triples of elements.

2. Write each of the following indicated sums as an indicated product.

   (a) $4x + 3x$ (b) $3x + 3 \cdot 2$
   (c) $a \cdot 4 + ab$ (d) $3(a + b) + 4(a + b)$

3. Find element x in each of the following equations.

   (a) $3(x + 2) = 4$ (b) $4(1 + 2) = 4(1 + x)$

4. By use of Corollary 2.13-1 verify the equalities below. (Note that $a^2 = a \cdot a$.)

   (a) $(2a + 1)(3b + 2) = ab + 3b + 4a + 2$
   (b) $(2a + 1)(3a + 2) = a^2 + 2a + 2$

5. Show that the equality in Corollary 2.13-1 is in accord with the usual method for finding the product of two binomials.

★6. In Exercise 1 you proved that multiplication is right distributive with respect to addition. Use the commutative law to show that multiplication is also left distributive with respect to addition. Do not make any calculations.

★7. Using the distributive law on the left member, show that the following equality is true for elements a and b having multiplicative inverses a' and b':

$$a'b'(a + b) = a' + b'$$

★8. Consider the formula of Exercise 7.

(a) Explain why both addition and multiplication must be commutative for the formula to hold.

(b) Consider a system in which multiplication is distributive with respect to addition.

(i) What correction should be made in the right member of the formula if multiplication is commutative but addition is not?

(ii) What correction should be made in the left member of the formula if addition is commutative but multiplication is not?

★9. We studied addition in system S of Exercise 1.5. In this system multiplication is closed, commutative, and distributive with respect to addition. Use these facts to complete the entries in Table 2.4. [Suggestion: find the product $1 \cdot 4$ by using the distributive law applied to the triple 1, 1, 1. That is, $1(1 + 1) = 1 \cdot 1 + 1 \cdot 1$; whence $1 \cdot 4 = 4 + 4 = 7$.]

Table 2.4 Multiplication Table for System S

×	1	4	7
1	4		
4		1	
7			7

★10. Find x in each of the following equations, where all elements belong to system S.

(a) $x = (4 + 7) \cdot 1$ (b) $4x + 7x = 7$
(c) $(x + 1) \cdot 4 = 7$ (d) $1 \cdot (7 + x) = 7$

2.14 APPLICATIONS OF THE DISTRIBUTIVE LAW IN THE FIVE-ELEMENT SYSTEM

In this section we study three of the many useful applications of the distributive law. The first shows that the multiplicative property of the element 5 is an inevitable consequence of the five-element system rather than a capricious characteristic. We begin by confirming the statement that $5a = 5$ for every element a. Corollary 1.11-1 asserts that in every equality of the form $Y + X = X$, where X and Y are expressions in terms of elements and Y is the element 5. Therefore we seek an expression X for which the equality $5a + X = X$ holds. Such an expression is ba, where b is an element; for, using the distributive law, $5a + ba = (5 + b)a = (b)a = ba$. It follows that $5a$ is the additive identity, or, in symbols, $5a = 5$ for any a.

We next verify the statement that if $ab = 5$, then $a = 5$ or $b = 5$. Since for elements $a = b = 5$ the statement is trivially true, let one of the elements, say a, be different from 5. Then the equation $ab = 5$ has the unique solution $b = 5a'$, so that $b = 5$.

Let us now develop a theoretical justification of the equalities used in Exercise 2.2.6.

THEOREM 2.14-1 For elements a and b of the five-element system, $(ab)^* = a^*b = ab^*$ and $a^*b^* = ab$.

We begin with the definition $a + a^* = 5$; whence, $(a + a^*)b = 5b$. Both members of the equation can be changed in form to obtain $ab + a^*b = 5$. It follows that $(ab)^* = a^*b$. (Why?) If we had started with the relation $b + b^* = 5$ and had multiplied both members on the left by a, we would have found $(ab)^* = ab^*$. Laws of the additive inverse now justify the continued equality $a^*b^* = (ab^*)^* = ((ab)^*)^* = ab$.

EXERCISE 2.14

1. Derive the continued equality for a^*b^*.

2. An alternative proof of the final equality in Theorem 2.14-1 can be made by applying the associative and distributive laws in order to the sum $ab + a^*b + a^*b^*$. Make this proof.

2.15 THE RATIONAL NUMBERS

When the integers are considered under the operation of multiplication, it is found that closure, commutativity, and associativity hold, that the integer 1 serves as identity, but that only the integers 1 and -1 have inverses. The additive identity 0 has the property that for integers m and n, $mn = 0$ if and only if $m = 0$ or $n = 0$. As in the five-element system there is a restricted law of cancellation saying that for $k \neq 0$, if $km = kn$, then $m = n$. There is, however, no general solution for the equation $m = nx$.

In Section 1.13 we extended the natural numbers to the integers in order to obtain properties lacking in the former system. Now that the integers are seen to be subject to severe restrictions, a further extension might be profitable. So let us define a new system containing all the integers, a multiplicative inverse for each nonzero integer, and whatever other new numbers are needed for closure under multiplication. Designate this extended set as the *rational number system*. Numbers 0 and 1 remain the additive and multiplicative identities, respectively. For any nonzero rational number r, the multiplicative inverse is indicated by the symbol $1/r$ and by the name "reciprocal of r." Every property of the five-element system under addition and multiplication has an analogue in the system of rationals. In particular, reciprocals of rational numbers r and s satisfy the relations $1/(1/r) = r$, $1/(rs) = (1/r)(1/s)$, and if $r = 1/s$, then $s = 1/r$.

Consider the following examples. The new system includes the integer 2 and its reciprocal $\frac{1}{2}$. It also contains the product of the integer 3 and the fraction $\frac{1}{2}$. This product is the fraction $\frac{3}{2}$. Then, of course, it must contain the reciprocal of $\frac{3}{2}$, which is written $1/(\frac{3}{2})$ and which has the value $\frac{2}{3}$. The negative of $\frac{2}{3}$ is written $-\frac{2}{3}$.

EXERCISE 2.15

1. Write the reciprocal of each of the following rational numbers, where $mn \neq 0$.

$$\begin{array}{ccccc} 14 & \frac{1}{27} & \frac{3}{4} & \frac{11}{9} & n \\ 1/n & m/n & -6 & -\frac{4}{5} & -n \end{array}$$

2. Find the rational number x that makes each of the following equalities true.

(a) $x = \frac{2}{3} \cdot 17$	(b) $x = 4 \cdot 0$	(c) $x = -\frac{1}{6} \cdot 0$
(d) $12x = 1$	(e) $\frac{7}{4}x = 1$	(f) $2x = 3$
(g) $x = \frac{4}{5}(-\frac{13}{14})$	(h) $7x = -21$	(i) $3x = \frac{5}{2}$
(j) $\frac{1}{3}x = 11$	(k) $4x = 4 \cdot \frac{2}{3}$	(l) $\frac{2}{7}x = \frac{3}{4} \cdot \frac{2}{7}$

3. Find all rational numbers x that make each of the following statements true. (Sometimes your answer will be "none" or "all.")

(a) $x = 0 \cdot 0$	(b) $0x = 4$	(c) $4x = 0$
(d) $0x = 0$	(e) $9x = 37x$	(f) $0x = 0 \cdot 2$
(g) $0x = (-\frac{2}{3})x$	(h) $\frac{2}{5}x = \frac{2}{5} \cdot 0$	

4. Criticize the following "proof" that $1 = 0$ in the rational number system.

Let $x = 1$; then $x^2 = 1$. Furthermore, $x - 1 = 0$ and $x^2 - 1 = 0$; whence, $x - 1 = x^2 - 1$. Both members of the equality can be factored to obtain

$$1 \cdot (x - 1) = (x + 1)(x - 1)$$

Cancellation of the factor $x - 1$ from both sides gives $1 = x + 1$; whence, $x = 0$. But it is known that $x = 1$; hence, $0 = 1$.

5. (a) Explain why the continued equality $ab = a(0 + b) = a \cdot 0 + ab$ holds for all rational numbers a and b.

(b) Explain why the equality of part (a) yields the relation $a \cdot 0 = 0$ for all rational numbers a. (Suggestion: see the discussion in the first paragraph of Section 2.14.)

6. Theorem 2.14-1 states multiplicative properties of the additive inverse. Restate this theorem for the rational number system.

2.16 THE REAL NUMBERS

In Section 1.13 the integers were placed on a number line. Now the rational number $\frac{1}{2}$ is placed on this same line by assigning it to the point halfway between 0 and 1. Similarly, $-\frac{1}{2}$ is assigned to the point halfway between 0 and -1, and the number $\frac{9}{4}$ to the point one-fourth of the way between 2 and 3. By such a scheme every rational number can be associated with a unique point on the line. When, conversely, it is asked whether each point on the line is associated with some rational number, the answer is "no."

Consider, for example, a line segment AB which represents the hypotenuse of an isosceles right triangle each of whose legs has length 1 (see Figure 2.1).

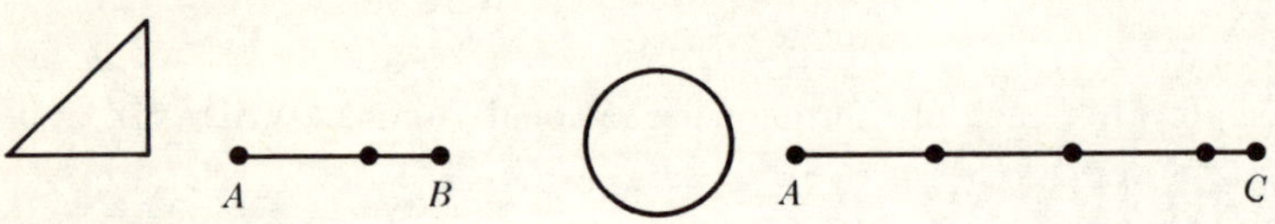

FIGURE 2.1

By the Pythagorean theorem it has length $\sqrt{2}$. Or consider a segment AC equal to the circumference of a circle of diameter 1 (see Figure 2.1). It has length π. When A is placed at the zero point of the number line, as in Figure 2.2, then B and C fall at points to which should be assigned the numbers $\sqrt{2}$ and π, respectively. These numbers, however, are not rational. Hence we have found points on the number line that are not associated with rational numbers. They correspond instead to numbers that are called irrational (nonrational). There are many such numbers, for it can be shown that between any two rationals there lie irrationals. Together the two kinds of numbers compose the *real number system*, the one that serves us best in the calculations required in our daily lives. Its members completely fill up the number line. It forms a group under addition with 0 as its additive identity.

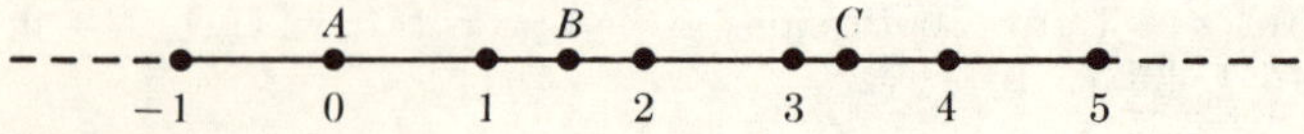

FIGURE 2.2

With the exception of 0 it forms a group under multiplication with 1 as its multiplicative identity. Its notation resembles that for the rationals. Thus the additive inverse of real number g is called its negative and is written $-g$, and the multiplicative inverse of g (when $g \neq 0$) is called its reciprocal and is written $1/g$. Furthermore, in this system multiplication is distributive with respect to addition. Hence, its laws of operation are the same as those found for the five-element system and for the rationals.

EXERCISE 2.16

1. Place each of the following rational numbers on the number line.

(a) $2/3$ (b) $-3/2$ (c) $11/6$ (d) $-7/3$

2. The number $\sqrt{3}$ is irrational. It lies between $3/2$ and $7/4$. Place it approximately on the number line.

3. Place the number $-\sqrt{3}$ approximately on the number line.

4. In Section 1.13 the sum of two integers was placed on the integer line. Follow the method of that section to locate the following sums on the real number line.

(a) $2/3 + 2$ (b) $2/3 + (-2)$
(c) $-2/3 + 2$ (d) $-2/3 + (-2)$

5. Place the following sums approximately on the real number line:

(a) $\sqrt{3} + 4$ (b) $\sqrt{3} + (-4)$
(c) $(-\sqrt{3}) + 4$ (d) $(-\sqrt{3}) + (-4)$

★6. For positive real numbers g and h, tell how to place the following sums on the real number line:

(a) $g + h$ (b) $g + (-h)$
(c) $(-g) + h$ (d) $(-g) + (-h)$

3

THE FIVE-ELEMENT SYSTEM FURTHER PROPERTIES

We have observed the five-element system under addition and multiplication. Let us now investigate the behavior of the elements under the inverse operations of subtraction and division, and discuss two special topics. One of these is a "practical" use for the system based on its limited number of elements. The other, with which the chapter begins, carries forward our work on the solution of linear equations.

3.1 ON THE SOLUTION OF EQUATIONS IN THE FIVE-ELEMENT SYSTEM

The solution of equations is an interesting topic because of the algebraic manipulation involved and also because of the general results that can sometimes be obtained. In this section, we recall types of equations that have already been solved and treat others of more elaborate form.

We easily find the solution for $x + 3 = 2$, namely, $x = 2 + 3^* = 4$, and the solution for $3x = 4$, namely, $x = 3' \cdot 4 = 3$. Now consider the equation

$$3x + 3 = 2 \tag{1}$$

It is clear that if $3x$ is assumed to be a simple variable, say y, then Equation (1) reduces to $y + 3 = 2$, which, as we have seen, has solution $y = 4$.

The replacement of y by its value $3x$ gives $3x = 4$, whose solution is $x = 3$. Hence the solution of Equation (1) is $x = 3$. In the next exercise you are asked to solve equations of this more complicated form.

EXERCISE 3.1-A

1. Solve each of the following equations.

(a) $3x + 4 = 2$ (b) $4x + 1 = 3$ (c) $2x + 5 = 1$
(d) $4x + 3 = 4$ (e) $3x + 2 = 5$ (f) $4x + 3 = 5$
(g) $2x + 4 = 5$ ★(h) $ax + b = 5$ where $a \neq 5$

2. Equation (a) in Exercise 1 can be reduced to Equation (e) by the addition of element 3 to both members, or (e) can be reduced to (a) by the addition of element 2 to both members. We can, therefore, say that Equations (a) and (e) are two expressions for the same equality.

(a) Show that Equations (b) and (f) are two expressions for the same equality.
(b) Are Equations (c) and (g) two expressions for the same equality?
(c) Rewrite Equation (d) in the form $ax + b = 5$.

From Exercise 3.1-A.2 we see that the general linear equation in the five-element system has the form

$$ax + b = 5 \qquad a \neq 5 \tag{2}$$

The discussion above suggests an effective procedure for finding its solution. By Theorem 1.11-1, Equation (2) yields uniquely $ax = 5 + b^*$ or $ax = b^*$. By Theorem 2.7-2, $x = a'b^*$. It follows that the unique solution of Equation (2) is $x = a'b^*$.

We need not "check" our answer in the original equation because the theorems we have used are strong ones, asserting "this is the unique solution" and not saying merely, "if there is a solution, then this is it." Of course, in numerical problems it is always wise to check the accuracy of the arithmetic.

THEOREM 3.1-1 For elements $a \neq 5$ and b of the five-element system, the equation $ax + b = 5$ has the unique solution $x = a'b^*$.

Consider next the second-degree equation produced by setting a product of two linear factors equal to 5. The general form would be

$$(ax + b)(cx + d) = 5 \qquad ac \neq 5 \tag{3}$$

By the multiplicative property of the additive identity, the product in the left member is 5 if and only if at least one of its factors is 5; hence, $ax + b = 5$ or $cx + d = 5$. These two equations have unique solutions $a'b^*$ and $c'd^*$, respectively. Each solution reduces one of the factors in the left member of Equation (3) to 5 and therefore reduces the product of the factors to 5. Furthermore, an element different from $a'b^*$ or $c'd^*$ fails to reduce either factor to 5; hence, it cannot reduce the product of factors to 5. We have found the two unique solutions for Equation (3).

THEOREM 3.1-2 For elements a, b, c, d of the five-element system, when $ac \neq 5$, the equation $(ax + b)(cx + d) = 5$ has the unique solutions $x = a'b^*$ and $x = c'd^*$.

In the real number system the analogues of these theorems are important. There are an infinite number of reals; hence, it is impossible to solve linear equations by the trial-and-error process of direct substitution. The theorem below not only tells how many solutions there are, but also gives a formula for writing each one.

THEOREM For real numbers a, b, c, and d, with $ac \neq 0$, the equation $ax + b = 0$ has the unique solution $x = (-b)/a$, and the equation

$$(ax + b)(cx + d) = 0$$

has unique solutions $x = (-b)/a$ and $x = (-d)/c$.

EXERCISE 3.1-B

3. What is the significance of the condition $ac \neq 5$ in describing Equation (3) and in the hypothesis of Theorem 3.1-2?

4. Use Theorems 3.1-1 and 3.1-2 to solve the following equations in the five-element system.

(a) $3x + 1 = 5$ (b) $(x + 1)(2x + 3) = 5$
(c) $(3x + 5)(x + 4^*) = 5$ (d) $(3x + 3)(4x + 1) = 5$
(e) $(x + 4)(x + 2^*) = 5$ (f) $(4x + 3^*)(3x + 4) = 5$

5. Find all solutions of each of the following equations in the five-element system.

(a) $5x + 3 = 5$ (b) $5x + 4 = 4$
(c) $5(x + 1) = 5$ (d) $(5x + 1)(5x + 2) = 5$
(e) $(2x + 4)(5x + 5) = 5$ (f) $(2x + 5)(5x + 2) = 5$

6. Consider the equation $(x + 1)(x + 2) = 3$. If each factor is set equal to 3, then two equations are produced, namely, $x + 1 = 3$ with solution $x = 2$ and $x + 2 = 3$ with solution $x = 1$.

(a) Show that elements 2 and 1 do not satisfy the original equation.
(b) Explain why this method does not produce correct solutions.

7. The following equations are stated for the real number system. Find all solutions for each equation. (Sometimes your answer may be "none" or "all real numbers.")

(a) $3x + 7 = 0$ (b) $9x + \pi = 0$
(c) $0x + 7 = 0$ (d) $9x + 0 = 0$
(e) $0x + 0 = 0$ (f) $\frac{1}{5}x + 7 = 0$
(g) $(-8)x + \frac{2}{3} = 0$ (h) $\pi x + \sqrt{2} = 0$
(i) $(17x + 24)(5x + \pi) = 0$ (j) $[\frac{3}{4}x + 5][\frac{2}{7}x + (-\frac{3}{14})] = 0$

3.2 SUBTRACTION IN THE FIVE-ELEMENT SYSTEM

In Chapter 1 we studied addition of elements and developed many of its properties. Let us now define the operation of subtraction and then study it in the five-element system.

DEFINITION In any number system, given the numbers a and b, the **difference** $a - b$ is the number c, if it exists uniquely, for which $a = b + c$. The process of finding the difference is called **subtraction.**

This definition says that the difference of the two specific elements 2 and 3, written $2 - 3$, is c whenever c is the unique solution of the equation $2 = 3 + c$. Theorem 1.11-1 not only asserts that c does indeed exist uniquely, but also gives a formula for finding it, namely, $c = 3^* + 2 = 4$. It follows that $2 - 3 = 4$.

Let us now find the difference $2 - 4$. We seek a unique element c for which $2 = 4 + c$. The equation has the unique solution $c = 4^* + 2 = 3$, so we know that $2 - 4 = 3$. This answer could have been anticipated. The difference $2 - 3$ established the equality $2 = 3 + 4$, and commutativity of addition gives $2 = 4 + 3$, which defines 3 as the difference $2 - 4$.

Let us generalize our argument. Given two elements a and b, we seek an element c for which $a = b + c$ uniquely. By the properties of addition in the five-element system, c is given uniquely by the formula $b^* + a$, and this sum has one and only one value. Hence, for every pair of elements a and b, the conditions of the definition are met, and the difference $a - b$ can be found by the formula $b^* + a$. We have the two theorems below. The corollary follows when a is replaced by 5.

THEOREM 3.2-1 The five-element system is closed under subtraction.

THEOREM 3.2-2 For elements a and b of the five-element system, $a - b = a + b^*$.

Corollary 3.2-2 For element b of the five-element system, $5 - b = b^*$.

Returning to the definition, we may note that the equations $a = b + c$ and $a = c + b$ define, respectively, c as the difference $a - b$ and b as the difference $a - c$. By commutativity of addition, $b + c = c + b$. Hence, the two relations $c = a - b$ and $b = a - c$ always occur together.

THEOREM 3.2-3 *For elements a, b, and c of the five-element system, $a - b = c$ if and only if $a - c = b$.*

Because the difference c of a and b is defined by the equality $a = b + c$, we can use the addition table instead of Theorem 1.11-1 to find c. For example, to find $2 - 3$ we seek an element c for which $3 + c = 2$. We enter the addition table at 3 in the left-border column and look across in the 3 row until we reach the sum 2; then the desired element c is the heading of the column in which the sum 2 lies. Hence, $c = 4$. Since the sum 2 is found once and only once in the 3 row, the difference 4 surely exists and is unique. By a similar argument we can show that the difference of any two elements exists and is unique, and thus give a second proof of Theorem 3.2-1.

EXERCISE 3.2

1. Using Theorem 1.11-1, find the differences $1 - 4$ and $1 - 2$. Without using Theorem 3.2-3, show that the second of these results could be anticipated from the first.

2. Using the addition table find the differences $3 - 1$ and $3 - 2$. Do your results illustrate Theorem 3.2-3?

3. (a) Find the difference $3 - 4$.
(b) Explain why element 3 could be a candidate for a left identity for subtraction.
(c) Find the difference $3 - 5$.
(d) Does element 3 remain a candidate for left identity? Explain.

4. Make a subtraction table for the five-element system. Explain how the table is to be read.

5. In each of the following equalities, find the element x.

(a) $x - 1 = 3$ (b) $2 - x = 1$ (c) $x - 2 = 5$
(d) $4 - x = 2$ (e) $x - 5 = 4$ (f) $a - x = b$

6. In this exercise solutions are found for linear equations.
(a) By using the definition of subtraction, show that equation $a = x - b$ has solution $x = a + b$.
(b) By using Theorem 3.2-3, show that equation $a = b - x$ has solution $x = b - a$.

7. This exercise concerns subtraction of natural numbers.

(a) Let the difference of natural numbers be defined in the same manner as the difference of elements. When possible, find the differences below.

(i)	$15 - 3$	(ii)	$15 - 12$
(iii)	$6 - 1$	(iv)	$6 - 5$
(v)	$14 - 14$	(vi)	$3 - 10$
(vii)	$8 - 18$	(viii)	$4 - 2$

(b) Explain how you know that the natural numbers are not closed under subtraction.

(c) Under what circumstances is subtraction possible in the natural numbers?

(d) When subtraction is possible, is the following statement true? For natural numbers k, m, and n, if $k - m = n$, then $k - n = m$. (Justify your answer.)

8. This exercise concerns subtraction of integers. Let the difference of integers be defined in the same manner as the difference of elements.

(a) Show that for integers k and m, $k - m = k + (-m)$.

(b) Why is it true that for any two integers k and m, the sum $k + (-m)$ always exists and is unique? It follows that the integers are closed under subtraction. Explain.

(c) Is it true that for integers k, m, and n, if $k - m = n$, then $k - n = m$? Justify your answer.

9. Answer Exercise 8 for real numbers instead of integers.

3.3 PROPERTIES OF SUBTRACTION IN THE FIVE-ELEMENT SYSTEM

Let us now investigate associativity for subtraction. The property would hold if and only if for each triple of elements a, b, c,

$$(a - b) - c = a - (b - c) \tag{4}$$

Forming these differences with elements 1, 2, 3, respectively, we find $(1 - 2) - 3 = 1$, while $1 - (2 - 3) = 2$. For the triple 1, 2, 3, therefore, equality (4) does not hold. Since it fails to be true for one triple of elements, it cannot be true for all triples. Hence, we cannot establish the theorem that in the five-element system subtraction is associative.

A particular example for which a suggested theorem is shown false is called a *counterexample* to the proposed theorem. It prevents the assertion from being true for all possible cases and so warns that the general statement cannot be proved. Sometimes counterexamples suggest modifications that can transform false statements into true ones. For instance, it is true that in the five-element system subtraction is associative for all triples whose third element is 5; that is, $(a - b) - 5 = a - (b - 5)$ for every triple a, b, 5.

In Section 1.8 we defined an identity element for addition. The following analogous definition may be made for subtraction: in any number system, a given number i is an identity for subtraction if and only if for every number n of the system, $n - i = i - n = n$. But is there an identity element for subtraction? For elements 1 and 2 we find $1 - 2 = 4$. Because 4 is different from both 1 and 2, we see that neither 1 nor 2 can be the desired identity. Similarly, the relations $4 - 3 = 1$ and $5 - 4 = 1$ prevent any of elements 3, 4, or 5 from qualifying. It follows that there is no identity for subtraction and that there can be no subtractive inverses.

EXERCISE 3.3

1. Is subtraction commutative in the five-element system? Justify your answer.

2. By use of Theorem 3.2-2 and properties of addition in the five-element system, the equality $b - a = c - a$ yields $b = c$, thus establishing a law of right cancellation for subtraction. Make the complete argument.

3. Establish a law of left cancellation for subtraction.

4. For all elements a, b, and c, $(a + b^*) + c^* = a + (b^* + c^*)$. Show how to translate this equality into a law for subtraction: $(a - b) - c = a - (b + c)$.

5. By methods similar to those used in Exercise 4 establish the laws below.

(a) $(a + b) - c = a + (b - c)$ (b) $(a - b) + c = a - (b - c)$

6. By the distributive law, $a(b + c^*) = ab + ac^*$. Using Theorems 2.14-1 and 3.2-2 transform this relation into a law of left distributivity of multiplication with respect to subtraction.

7. Is there a law of right distributivity of multiplication with respect to subtraction? If so, write it.

3.4 COMMENT ON SUBTRACTION

The operation of subtraction possesses few of the properties found for addition. Fortunately, the relation $a - b = a + b^*$ permits subtraction problems to be restated in terms of addition. In many systems such a restatement is customarily made. Of course, when no additive inverse exists, then subtraction must be performed by use of its definition. Recall, for example, the strongest statement that can be made for the natural numbers, namely, that if $k = m + n$, then $k - m = n$ and $k - n = m$.

When making the subtraction table for the five-element system, we could have found the differences by counting on the clockface of Section 1.3. In Section 1.13 we found that when we start at a and move b spaces around the face in a counterclockwise direction, we reach the sum $a + b^*$; hence, we arrive at the element representing the difference $a - b$. For example, to find the particular difference $2 - 3$ we start at element 2, move counterclockwise three spaces, and find ourselves at element 4; hence, $2 - 3 = 4$.

To find the difference of two real numbers r and s we use the relation $r - s = r + (-s)$ on the real number line of Section 2.16 together with the answers to Exercise 2.16.6. If s is positive, then we use part (b) or (d) according as r is positive or negative. If s is negative, then $-s$ is positive, and we proceed as directed in part (a) or (c).

EXERCISE 3.4

1. Use the clockface to find the following differences in the five-element system.

(a) $4 - 2$ (b) $2 - 4$
(c) $3 - 5$ (d) $1 - 1$

2. Use the real number line to find the following differences of real numbers. (Suggestion: see answers to Exercises 2.16.4 and 2.16.5.)

(a) $2/3 - 2$ (b) $(-2/3) - 2$
(c) $2/3 - (-2)$ (d) $(-2/3) - (-2)$
(e) $\sqrt{3} - 4$ (f) $\sqrt{3} - (-4)$
(g) $(-\sqrt{3}) - 4$ (h) $(-\sqrt{3}) - (-4)$

In the remaining exercises well-known relations between addition and subtraction are established for the five-element system.

3. This exercise shows that for elements a, b, c, and d, if $a + d = c + b$, then $a - b = c - d$.

(a) Add the sum $b^* + d^*$ to both members of the equality in the hypothesis and use laws of addition to get $a + b^* + (d + d^*) = c + d^* + (b + b^*)$.

(b) Reduce the equality of part (a) to the form $a + b^* = c + d^*$, whence $a - b = c - d$.

4. This exercise shows that for elements a, b, c, and d, if $a - b = c - d$, then $a + d = c + b$.

(a) Change the equality in the hypothesis into the form $a + b^* = c + d^*$, then add the sum $b + d$ to both members.

(b) Use laws of addition to obtain the desired conclusion.

5. Combine the statements of Exercises 3 and 4 into an "if and only if" theorem.

★6. Show that $(a + b) - (c + d) = (a - c) + (b - d)$ for elements a, b, c, and d.

★7. Using distributive laws as well as the laws of addition employed in previous exercises, establish the equalities below for elements a, b, c, and d.

$$\begin{aligned}(a - b)(c - d) &= ac + b^*d^* + ad^* + b^*c \\ &= (ac + bd) + (ad + bc)^* \\ &= (ac + bd) - (ad + bc)\end{aligned}$$

3.5 INTRODUCTION TO QUOTIENTS

Let us now return to multiplication of elements and seek the element c in the equation $a = bc$ when a and b are known.

EXERCISE 3.5

1. Given 4 as the first element and 3 as the second, use the multiplication table to find a third element c such that $4 = 3c$.

2. Repeat Exercise 1 taking the first element as 3 and the second as 4.

3. Repeat Exercise 1 using 1 as the first element and 2 as the second.

4. Consider the following problem: given a first element a and a second element b, can we find a third element c such that $a = bc$?

(a) Explain how this question is answered by use of the multiplication table.
(b) Can the desired element c always be found? Explain.
(c) When c can be found, is it unique? Explain.

5. Find element c such that $a = bc$ when

$$\text{(a)}\quad a = 4,\ b = 5 \qquad \text{(b)}\quad a = b = 5$$

6. Complete Table 3.1. The entries already listed are the elements c found in Exercises 1, 2, 3, and 5.

Table 3.1 Element c for which $a = bc$

	Element b				
	1	2	3	4	5
Element a					
1		3			
2					
3				2	
4			3		None
5					All

3.6 DIVISION IN THE FIVE-ELEMENT SYSTEM

There have been occasions in the preceding work when it would have been convenient to use the operation of division. Let us now define and study this operation.

DEFINITION In any number system, given the numbers a and b, the **quotient** $a \div b$ is the number c, if it exists uniquely, for which $a = bc$. The process of finding the quotient is called **division.**

This definition says that the quotient of two specific elements, say 4 and 3 (written $4 \div 3$), is element c, if it exists and is unique, for which $4 = 3c$. In Exercise 3.5.1, c was found to be the unique element 3; hence,

$4 \div 3 = 3$. In Exercise 3.5.2, c took the unique value 2 when the equation $3 = 4c$ was solved. By the definition above, 2 is the quotient $3 \div 4$.

It is now clear that the table prepared in Exercise 3.5.6 presents a complete set of possible quotients. Note that each of the first four columns contains a unique entry for c. In the final column there are no entries in the first four places and five entries in the fifth. Translated into the terminology of division, the table shows that a proper quotient exists when any element is divided by 1, 2, 3, or 4. Division by 5 is not possible, because either there is no answer at all for c in the equation $a = 5c$, or else the answer fails to be unique. The five-element system is closed under division, except by 5.

Instead of using Table 3.1, we can use the known properties of multiplication. When $b \neq 5$, the equation $a = bc$ has the unique solution ab' for element c. In this case $a \div b = ab'$. The multiplicative property of 5, however, indicates that no element provides a solution for c when $b = 5$ but $a \neq 5$, while all elements are solutions when $a = b = 5$. Hence, in both cases c fails to meet the requirements of a quotient. These results accord with the earlier ones and justify the second theorem below.

THEOREM 3.6-1 With the omission of division by 5, the five-element system is closed under division.

THEOREM 3.6-2 For elements $b \neq 5$ and a of the five-element system, $a \div b = ab'$.

Assume that $b \neq 5$. Then the equation $a = bc$ defines c as the quotient of a and b. Similarly, when $c \neq 5$ the equation $a = cb$ defines b as the quotient of a and c. By commutativity of multiplication, $bc = cb$, so that the two relations always occur together.

THEOREM 3.6-3 For non-5 elements a, b, and c of the five-element system, $a \div b = c$ if and only if $a \div c = b$.

EXERCISE 3.6

1. Find the quotients $3 \div 2$ and $3 \div 4$. Without using Theorem 3.6-3, show how the second result could have been anticipated from the first.

2. Make a division table for the five-element system. (Note that this table has five rows but only four columns.)

3. For each equation below find all solutions x.

(a) $x \div 2 = 3$ (b) $4 \div x = 5$ (c) $x \div 4 = 5$

(d) $x \div 3 = 1$ (e) $4 \div x = 3$ (f) $5 \div x = 5$

(g) $4 \div x = 4 \div 3$ (h) $5 \div x = 5 \div 3$ (i) $x \div 2 = 3 \div 2$

4. Since $5 = 2 \cdot 5$ is a true equality, the definition of quotient permits us to write $5 \div 2 = 5$. The equality $5 = 5 \cdot 2$ is also true, but the definition of quotient does not permit us to write $5 \div 5 = 2$. Explain.

5. For which element a is it true that $a \div b = a \div d$ even when $b \neq d$?

6. A student is asked why he cannot divide by 5 in the five-element system. He replies, "Because $a \div b = ab'$, and 5 has no multiplicative inverse." Explain why his answer is incomplete.

7. Is the restriction $a \neq 5$ introduced arbitrarily into the hypothesis of Theorem 3.6-3, or is it an inevitable consequence of the preceding discussion? Explain.

8. Let the quotient of rational numbers be defined in the same manner as the quotient of elements. Define quotient and division of rational numbers.

9. State the analogue of Theorem 3.6-1 for the rational number system. Why is it true?

10. (a) State the analogue of Theorem 3.6-2 for the rational number system.
(b) The theorem of part (a) yields the rule: to divide two fractions, multiply the dividend by the reciprocal of the divisor. Illustrate this rule by finding the quotients $a \div b$ for the rational numbers below.

a:	3	7	$4/5$	$3/4$
b:	5	$2/3$	9	$5/7$

11. (a) State the analogue of Theorem 3.6-3 for the rational number system.
(b) Illustrate this analogue using for a and b the numbers of Exercise 10(b).

12. Answer Exercises 8 and 9 for the real instead of the rational number system.

3.7 PROPERTIES OF DIVISION IN THE FIVE-ELEMENT SYSTEM

In continuing the discussion begun in the preceding section, the restriction against division by 5 must be kept in mind, and statements must be modified to agree with this limitation. For example, it would clearly be unprofitable to propose the theorem: in the five-element system, division is commutative; that is, for elements a and b, $a \div b = b \div a$. The proposition is obviously false because, if a is 5, the quotient in the right member of the equation does not exist, and if b is 5, the quotient in the left member has no meaning. In discussing commutativity, sufficient restrictions should be included so that the resulting statement might possibly be true. A more tenable proposition would be: with the omission of element 5, the five-element system is commutative under division; that is, for elements a and b both different from 5, $a \div b = b \div a$. Further consideration may show that this statement also is false, but its truth or falsity is not immediately apparent.

Similarly, concerning associativity, our statement should be modified to: in the five-element system, division except by 5 is associative; that is, for elements a, b, and c, with b and c different from 5, $(a \div b) \div c = a \div (b \div c)$.

If we should attempt to check the truth of this equality, the only triples of elements that would qualify for testing are those in which the second and third elements are different from 5. There are 80 such triples because there are five choices for a and four each for b and c. If we use Theorem 3.6-2, however, we can completely avoid the test by calculation.

EXERCISE 3.7

1. (a) By using Theorem 3.6-2 and properties of multiplication, prove or disprove the modified statement on associativity of division.

(b) For appropriate triples of your set, test the truth of the modified statement on associativity of division. Can you reach a conclusion for the whole system?

2. Prove or disprove the statement: the elements 1, 2, 3, and 4 of the five-element system are commutative under division.

3. (a) In Section 1.8 there is a definition of identity element for addition. Modify this to give a definition of identity element for division.

(b) Is there an identity element for division? Justify your answer.

4. Consider elements a, b, and c, with $abc \neq 5$.

(a) By the law of cancellation for multiplication, if $ab' = ac'$, then $b' = c'$. From this statement derive the following law of cancellation for division: if $a \div b = a \div c$, then $b = c$.

(b) By the law of cancellation for multiplication, if $ba' = ca'$, then $b = c$. Convert this statement into a law of cancellation for division.

5. For $a \neq 5$ there is the relation $(b + c)a' = ba' + ca'$. Transform this relation into a law of right distributivity of division with respect to addition.

6. Is there a law of left distributivity of division with respect to addition? If so, write it.

***7.** By the associative law of multiplication, the relation $(ab')c' = a(b'c')$ is valid when $bc \neq 5$. Show that this statement can be translated into the following law for division: for elements a, b, and c with $bc \neq 5$, $(a \div b) \div c = a \div (bc)$.

3.8 COMMENT ON DIVISION

The operation of division possesses few of the properties of multiplication. Fortunately, the relation $a \div b = ab'$ permits every valid division problem to be restated in terms of multiplication. It is reasonable to make such a restatement because we were well acquainted with multiplication before we began the study of division.

Since the same properties are true of the two operations in the real number system, why do we often use division instead of multiplication by reciprocal? The answer probably is that we were taught to divide long before being introduced to the concept of the multiplicative inverse. Now, however, because we understand that, with the exception of 0, all real numbers have reciprocals and that these reciprocals are inverses under multiplication, we should be able to use whichever process promises simpler arithmetic. Recalling that division of real numbers introduces fractions into our reckoning, we now may write three familiar laws. (It is understood, of course, that no denominators are zero.)

$$\frac{a}{b} \cdot \frac{c}{d} = \frac{ac}{bd} \tag{5}$$

$$\frac{a}{b} + \frac{c}{d} = \frac{ad + bc}{bd} \tag{6}$$

$$\frac{a}{b} = \frac{c}{d} \quad \text{if and only if } ad = bc \tag{7}$$

In the optional exercise below analogues of these formulas are considered for the five-element system.

Finally, it may be noted that some systems contain few multiplicative inverses, so that the relation $a \div b = ab'$ has little significance. For such systems all quotients must be found by using the definition of Section 3.6. For example, the strongest statement possible for integers asserts that for numbers k, m, and n all different from 0, $0 \div k = 0$, and if $k = mn$, then $k \div m = n$ and $k \div n = m$.

EXERCISE 3.8 (OPTIONAL)

1. In this exercise the five-element formula $(a \div b)(c \div d) = (ac) \div (bd)$ when $bd \neq 5$ is justified. Explain why each of expressions (b) through (e) follows from the preceding one.

(a) $(a \div b)(c \div d)$ (b) $(ab')(cd')$
(c) $(ac)(b'd')$ (d) $(ac)(bd)'$
(e) $(ac) \div (bd)$

2. Show that when the formula of Exercise 1 is interpreted for real numbers, the resulting equality (5) justifies the rule that to multiply fractions we multiply the numerators and multiply the denominators.

3. This exercise shows that for elements a, b, c, d, with $bd \neq 5$, if $ad = cb$, then $a \div b = c \div d$. Explain why each of equalities (b) through (e) follows from the preceding one.

(a) $ad = cb$ (b) $(ad)(b'd') = (cb)(b'd')$
(c) $(ab')(dd') = (cd')(bb')$ (d) $ab' = cd'$
(e) $a \div b = c \div d$

4. The converse of Exercise 3, namely, if $a \div b = c \div d$, then $ad = cb$, is now proved. Reverse the steps in Exercise 3 and show that each equality, in turn, from (d) to (a), is a consequence of the following one.

5. State the "if and only if" theorem that has been proved in Exercises 3 and 4.

6. (a) Show that when the theorem of Exercise 5 is interpreted for the real number system it yields statement (7) above.

(b) If the statement of part (a) is translated into the language of proportion, there results the familiar law that in any proportion the product of the extremes equals the product of the means. Explain.

7. In the five-element system the analogue of formula (6) is the following statement: for elements a, b, c, d with $bd \neq 5$, $(a \div b) + (c \div d) = (ad + bc) \div (bd)$. To show that the right member of the formula can be transformed into the left member, explain why each of expressions (b) through (f) follows from the preceding one.

(a)	$(ad + bc) \div (bd)$	(b)	$(ad + bc)(bd)'$
(c)	$(ad + bc)b'd'$	(d)	$ab'dd' + cd'bb'$
(e)	$ab' + cd'$	(f)	$(a \div b) + (c \div d)$

3.9 A NOTE ON SYMBOLISM

The natural numbers, the integers, the rationals, the real numbers, and the elements of the five-element system have some common properties under the four operations we have studied. In discussing them we employed identical symbols to indicate certain members of all the systems. In particular, 1, 2, 3, 4, and 5 denoted elements or numbers, according to context. When we introduced the elements, we might have used the symbol 0 instead of 5, thus representing them by 0, 1, 2, 3, and 4. The additive identity would have had the same name and shape for numbers and elements, and we might have anticipated some of its properties. This very familiarity, however, could have been a disadvantage because we might have relied too heavily upon it. Since we had no preconceived ideas about the symbol 5, we were free and unprejudiced in determining the characteristics of the element it represents. We found its properties and then, as a second step, recognized them and gave them familiar names. Because the choice of a symbol was arbitrary, we used 5 in order to gain a degree of freedom from connotation.

3.10 THE FIVE-ELEMENT SYSTEM IN CRYPTOGRAPHY

Many of us have probably enjoyed the experience of sending messages in code. In this section, we shall use elements to construct ciphers for an alphabet of four letters and one punctuation mark; the method can be adapted to the English alphabet of 26 letters and 3 of its symbols of punctuation. The design of such ciphers is one of the practical uses of systems such as the five-element system.

One of the simplest schemes for disguising a message is, of course, the substitution of numbers for letters. We make the following rule of correspondence, using the letters *a*, *h*, *n*, *s*: for the letter *a* use element 1, for *h* use 2, for *n* use 3, for *s* use 4, and for the full stop use element 5. This rule can be written in tabular form:

$$\begin{matrix} a & h & n & s & . \\ 1 & 2 & 3 & 4 & 5 \end{matrix} \tag{8}$$

Now we encipher the message *Nan has a sash.* We write the letters of the message in order without separation into words and insert occasional periods where they would assist a "friendly" decoder or deter an "enemy" decoder. We get the following sequence of letters.

n a n h a s . a s a s h

Replace each letter by its element equivalent:

3 1 3 2 1 4 5 1 4 1 4 2

The message will probably be transmitted more accurately if it is associated into sets of four elements each. Additional elements 5 may be used in order to complete the four-element sets. The transmitted message is 3132 1451 4142. A friendly decoder can easily read it.

Now consider the message *Hans has Ann's sash.* We have no way of indicating the apostrophe, so the decoder will have to be clever enough to recognize the word *Anns* as *Ann's*. There are 15 letters in the message, so we should insert a period to bring the number to 16, a multiple of four. Where would the period best clarify the message for the decoder? Probably between the words *has* and *Anns*. Of course, we could insert five periods, one between each two words of the message, one at the beginning, and one at the end; in this way, the number of letters is brought to 20, also a multiple of four. Before enciphering this message we write it in either of two ways:

h a n s h a s . a n n s s a s h

or . h a n s . h a s . a n n s . s a s h .

EXERCISE 3.10-A

1. Encipher the message *Hans has . Ann's sash*
2. Decipher the message 5133 1521 4531 3454 1425

The decoder could be expected to read the message with ease, but it is probable that the enemy would guess such a simple scheme and would also read it with ease. Let us construct a less obvious cipher by multiplying by 2 each element in key (8), obtaining the following association of letters and elements:

a h n s .
2 4 1 3 5

If we translate the sequence 2 4 1 3 5 back into letters using (8), we will have a scheme for translating letters into letters. Hence, our new cipher can take either of the forms below:

a h n s .	a h n s .	
2 4 1 3 5	h s a n .	(9)

EXERCISE 3.10-B

3. Show that in key (9) *Nan has . a sash* translates into 1214 2352 3234 or into ahas hn.h nhns.

4. Using key (9), encipher the message *Hans has . Ann's sash* into elements and into letters.

5. Decipher the following messages using key (9).

(a) 4213 5542 3554 2345 (b) ahas hn.h aahn nhns

6. Show that if the multiplier 3 is used for each element in key (8), we get the new cipher:

a h n s .	a h n s .	
3 1 4 2 5	n a s h .	(10)

7. Using key (10), decipher the messages below.

(a) 5345 3215 2321 5132 5134 4315 (b) hnha n.an h.ns .nhh

8. Form the following new cipher by using the multiplier 4 for each element in key (8).

a h n s .	a h n s .	
4 3 2 1 5	s n h a .	(11)

9. Using key (11), decipher the following messages.

(a) 4224 3415 3421 3413 (b) ansn .nsa ash. nsas asan

Our cipher can be made more difficult to break if we agree that the first four letters of our message should be enciphered by key (9), the next four by key (10), and the third set of four by key (11), and that we should repeat this process as often as necessary. Let us also agree not to consider the period as a letter for the purpose of choosing the key. To use this scheme on the message *Nan has . a sash,* first write it so that the successive sets of four letters, exclusive of periods, are made evident: *n a n h   a s . a s   a s h;* then encode it:

Using (9)	Using (10)	Using (11)
n a n h	a s . a s	a s h
1 2 1 4	3 2 5 3 2	4 1 3
a h a s	n h . n h	s a n

Therefore, the message finally reads

1214 3253 2413 or ahas nh.n hsan

EXERCISE 3.10-C

10. Using the three-part enciphering method described above, encipher the message

Hans has . Ann's sash

11. Using the three-part method decipher each of the following messages.

(a) 4213 1325 1413 (b) aha. snh. ns.s an.a hns.

★**12.** You are the "enemy." Try to decipher the following messages for which a four-part method has been used.

(a) 5342 4513 2242 5344 5131 3453 2215
(b) .sha nanh .ash hss. ashn hns.

3.11 RETROSPECT AND PROSPECT

In studying the five-element system we learned how mathematicians create and develop some of the concepts with which they work. We constructed a new number system, then defined and studied the four usual operations for it. There are only 25 sums and 25 differences; the multiplication tables extend only to 5×5; and there are 20 quotients. There are no fractions, no decimals, no need for long division. As a figment of the mathematical imagination, this system might be likened to a court minuet in dance. It is dainty and small, formal and smooth; yet it has both symmetry and variation in its development. By interpreting some of our results in relation to traditional systems we gained new understanding of certain rules of operation.

In our investigation we set up patterns by means of which the structure of other algebraic systems might be explored. We freed ourselves from consideration of any particular numbers and asked about general systems, introducing the important concept of a mathematical group. We used various methods of proof, thus increasing our ability to deal with mathematical material.

Our study raises further questions. For example, in forming our system we counted to 5. Suppose we had counted only to 4 or had counted further to 6, would the four- or six-element system have behaved in all respects like that of five elements? As a second example, could this system be used as a basis for constructing a more complicated system? These problems are dealt with in later chapters.

4

A SIX-ELEMENT SYSTEM

In this chapter we will study a system of six elements. The results will be as inclusive and detailed as those for the five-element system but will be developed more easily because we shall use many of the important mathematical concepts treated in the first three chapters.

4.1 FORMATION OF THE SYSTEM

Let us assume we can count to 6 and then must begin again with 1. We thus create a system of exactly six elements for which we use the familiar names and symbols 1, 2, 3, 4, 5, 6. Further, two elements are equal if and only if they are the same element.

4.2 ADDITION IN THE SIX-ELEMENT SYSTEM

Addition is defined as for the five-element system except, of course, that 6 must be reached before starting again with 1. A clockface arrangement like Figure 4.1 may help. To find the sum $4 + 3$, first count to 4, then proceed 3 spaces farther around the clockface. Because the counting terminates at

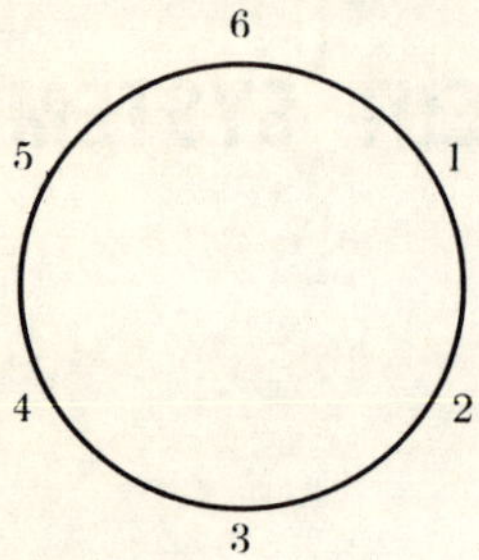

FIGURE 4.1

element 1, in this system $4 + 3 = 1$. A partial addition table is presented in Table 4.1.

Table 4.1 Partial Addition Table for the Six-element System

+	1	2	3	4	5	6
1		3		5	6	1
2	3	4	5		1	2
3	4	5	6	1		3
4	5		1	2	3	
5	6	1		3		5
6		2	3	4	5	6

EXERCISE 4.2

1. Copy and complete the addition table for the six-element system.

In each of Exercises 2–16 find all elements x that make the statement true.

2. $1 + 4 = x$

3. $2 + x = 5$

4. $4 + x = 4$

5. $x + 3 = 1$

6. $x + 2 = 6$

7. $4 + x = 4 + 5$

8. $2 + 3 = x + 3$

9. $3 + x = x + 3$

10. $x + 5 = 5 + x$

11. $x + 6 = x$

12. $2 + x \neq 6$

13. $x + 6 \neq x$

14. $3 + x \neq x$

15. $6 + x \neq 5$

16. $1 + x \neq x + 1$

17. Which property of addition is suggested by the answers to each of the following sets of exercises?

(a) 9, 10, 16 (b) 11, 13 (c) 7, 8

18. (a) Does the addition table show commutativity of addition?
(b) Why was your answer to Exercise 17(a) not sufficient to establish this property?

19. (a) Does the addition table show the existence of an identity for addition? If so, which element is it?
(b) Does this result agree with your findings in Exercise 17(b)?

20. If an additive identity exists, determine whether each element has an inverse with respect to it.

21. Does the addition table show closure under addition? Explain.

22. Have the answers to Exercises 18–21 established the six-element system as a group under addition? Explain.

4.3 PROPERTIES OF ADDITION IN THE SIX-ELEMENT SYSTEM

In the preceding exercise you discovered that addition in the six-element system exhibits closure and commutativity. Element 6 is an identity for addition, and each element has an inverse. All group requirements have been met except associativity. A rather long process was required to establish this property for the five-element system, and the same process would establish it for the new system. For each of the $6 \cdot 6 \cdot 6 = 216$ triples of elements we would find that the associative formula holds. The number of detailed verifications could be reduced by use of the additive identity, because whenever one or more of the elements concerned is 6, the formula is satisfied.

The new system meets all the criteria for a commutative group under addition. Certain consequences are immediate. The identity 6 is unique, and each element e has a unique inverse e^*. In particular, $1^* = 5$, $2^* = 4$, $4^* = 2$, and $5^* = 1$, while elements 3 and 6 are their own inverses. There is a law of cancellation for addition, and equations of the form $a = b + x$ have unique solutions $x = b^* + a$. Under addition, the six-element system behaves essentially as did that of five elements.

EXERCISE 4.3

The following problems refer to the six-element system.

1. Show that the associative formula $(a + b) + c = a + (b + c)$ is immediately satisfied when

(a) $a = 6$ (b) $b = 6$ (c) $c = 6$

2. For how many triples of elements would the formula in Exercise 1 still need verification in order to establish associativity of addition for the six-element system?

3. Let addition of three elements be defined in the customary manner; that is, $a + b + c \equiv (a + b) + c$. Show that $(a + b) + c = (c + b) + a$.

4. Theorem 1.10-1 states properties of the additive inverse. For each property write an analogue for the six-element system and give an illustration.

5. Find all elements e that make each of the following equalities true.

(a) $e^* = 1$ (b) $e^* = 4^*$ (c) $e^* = e$
(d) $e^* + 2 = 6$ (e) $e^* = e + 4$ (f) $e = e^* + 4$

6. In each equation below find element x.

(a) $x + 3 = 5$ (b) $x + 4 = 1$ (c) $x^* + 4 = 3$
(d) $(x + 1)^* = 2$ (e) $x + 3 = 4 + 3$ (f) $x^* + 2 = 2 + 5$

7. Let addition of more than three elements be defined in the customary manner (see Section 1.6). Find each of the following sums.

(a) $2 + 3 + 5 + 4 + 1 + 6 + 3$ (b) $3 + 6 + 4^* + 5 + 3^* + 2 + 4$

4.4 MULTIPLICATION IN THE SIX-ELEMENT SYSTEM

The second customary operation of an algebraic system is multiplication. Our discussion begins with a definition reminiscent of that for the five-element system.

DEFINITION For elements 1, 2, 3, 4, 5, 6 and for any element e of the six-element system,

$$\begin{aligned} 1 \cdot e &= e \\ 2 \cdot e &= (1 \cdot e) + e \\ 3 \cdot e &= (2 \cdot e) + e \\ 4 \cdot e &= (3 \cdot e) + e \\ 5 \cdot e &= (4 \cdot e) + e \\ 6 \cdot e &= (5 \cdot e) + e \end{aligned}$$

When we specialize the definition for $e = 5$, we get

$$\begin{aligned} 1 \cdot 5 &= 5 \\ 2 \cdot 5 &= (1 \cdot 5) + 5 = 5 + 5 = 4 \\ 3 \cdot 5 &= (2 \cdot 5) + 5 = 4 + 5 = 3 \\ 4 \cdot 5 &= (3 \cdot 5) + 5 = 3 + 5 = 2 \\ 5 \cdot 5 &= (4 \cdot 5) + 5 = 2 + 5 = 1 \\ 6 \cdot 5 &= (5 \cdot 5) + 5 = 1 + 5 = 6 \end{aligned}$$

In the exercises below you are asked to construct a multiplication table and to investigate some of its characteristics.

EXERCISE 4.4

1. Use the definition above to find the products $1 \cdot 2$, $2 \cdot 2$, $3 \cdot 2$, $4 \cdot 2$, $5 \cdot 2$, and $6 \cdot 2$.

2. Construct a multiplication table for the six-element system. (Note that the 2 and 5 columns have already been calculated.)

3. Why does the note in Exercise 2 specify the 2 and 5 columns rather than the 2 and 5 rows?

4. Compare the multiplication tables for the five-element and six-element systems. The entries in the tables show some similarities and some dissimilarities.

(a) List three ways in which the tables resemble each other. What is the significance of each of these similarities?

(b) List two ways in which the tables differ from each other. For each of these dissimilarities identify a property of multiplication in the five-element system that is lacking in the six-element system.

In each of Exercises 5–16 find all elements x that make the equality true in the six-element system.

5. $5x = 2$

6. $3x = 3$

7. $x \cdot 4 = 2$

8. $3x = x \cdot 3$

9. $x \cdot 4 = x$

10. $2x = 1$

11. $1 \cdot x = x$

12. $5x = x \cdot 5$

13. $x \cdot 1 = x$

14. $3x = 4$

15. $x \cdot 5 = 1$

16. $4x = x \cdot 2$

17. What do the answers to Exercises 5, 6, 7, 10, 14, and 15 suggest about the number of solutions of an equation of the form $ax = b$ or $xa = b$?

18. In Exercise 9, $2 \cdot 4 = 2$, $4 \cdot 4 = 4$, and $6 \cdot 4 = 6$. Hence, element 4 is a candidate for identity element for multiplication. Does it have the other necessary qualifications? Explain.

19. (a) What do the answers to Exercises 11 and 13 suggest about the identity for multiplication?

(b) Is there an identity element for multiplication in the six-element system? If so, is it unique?

20. If you found an identity element in Exercise 19, determine whether each element has an inverse under multiplication.

4.5 PROPERTIES OF MULTIPLICATION IN THE SIX-ELEMENT SYSTEM

The preceding exercise shows that the six-element system has closure and commutativity under multiplication. Element 1 is the unique multiplicative identity, and both it and 5 are their own inverses. None of 2, 3, 4, and 6, however, has an inverse; so the six-element system does not form a multiplicative group, and we cannot call on group theorems to establish further properties. The next three theorems record the results obtained so far, along with associativity and distributivity, which can also be established.

THEOREM 4.5-1 Under multiplication the six-element system is closed, commutative, associative, and distributive with respect to addition.

THEOREM 4.5-2 In the six-element system, element 1 is the unique identity element for multiplication.

THEOREM 4.5-3 Each of the elements 1 and 5 of the six-element system is its own unique inverse under multiplication.

Further study of the multiplication table reveals that in the 1 and 5 rows and columns each element appears once and only once. For these two elements, therefore, there should be a cancellation law and a solution for the equation $a = bx$. For element 1 the theorems concerned are trivial; hence, only those pertaining to element 5 are presented.

THEOREM 4.5-4 *Law of Cancellation for Multiplication.* For elements b and c of the six-element system, if $5b = 5c$, then $b = c$.

THEOREM 4.5-5 For element a of the six-element system, the equation $a = 5x$ has the unique solution $x = 5a$.

EXERCISE 4.5

1. Find all elements x that make each of the following equalities true.

(a) $5x = 1$	(b) $5x' = 1$	(c) $4x = 6$
(d) $6x = 6$	(e) $5x = 6$	(f) $2x = 5$
(g) $5x = 2$	(h) $5x' = 2$	(i) $3x = 2$
(j) $3x = 6$	(k) $2x = 2 \cdot 3$	(l) $2x = 5 \cdot 2$
(m) $5x = 5 \cdot 4$	(n) $5x + 2 = 6$	(o) $5x + 3 = 1$

2. Prove Theorem 4.5-5.

3. Explain why associativity of multiplication could be established by verifying the associative formula for only the 125 triples that omit element 6.

4. If addition is left distributive with respect to multiplication, then $a + (bc) = (a + b)(a + c)$ for all triples of elements a, b, and c. Test the equality for the triples below.

(a) 3, 2, 4 (b) 4, 3, 3 (c) 5, 3, 4 (d) 3, 2, 5

5. In Exercise 4, which triples of elements present counterexamples to the left distributivity of addition with respect to multiplication?

6. If multiplication is left distributive with respect to addition, then $a(b + c) = ab + ac$ for elements a, b, and c. Show that the formula is immediately satisfied when:

(a) $a = 6$ (b) $b = 6$ (c) $c = 6$

7. For how many triples of elements would the formula in Exercise 6 still require verification in order to establish left distributivity for the six-element system?

8. The complete six-element system *does not* form a group under multiplication. Nevertheless, some subset of its elements may do so. Consider each of the sets below and determine whether it is a multiplicative group:

(a) 1 and 6 (b) 1 and 5 (c) 1, 3, and 5

4.6 DIVISORS OF THE ADDITIVE IDENTITY IN THE SIX-ELEMENT SYSTEM

In the five-element system the additive identity possessed special properties under multiplication. Corresponding properties for element 6 would assert that a product of elements is 6 if one of the elements is 6; that if a product is 6, one of the factors is 6; and that 6 is the only element which lacks a multiplicative inverse. The first part of this statement is true, but the last two parts are false. In this system the additive identity 6 can have a pair of factors or divisors neither of which is the identity itself. When such a situation occurs, each of the factors is called a *divisor of the additive identity*. The equalities $2 \cdot 3 = 6$ and $3 \cdot 4 = 6$ define elements 2, 3, and 4 as divisors of the additive identity 6.

THEOREM 4.6-1 *Multiplicative Property of the Element 6.* For elements a and b of the six-element system, if $a = 6$ or $b = 6$, then $ab = 6$.

DEFINITION In any number system let i be the identity for addition. Then numbers a and b are **divisors of the additive identity** i if and only if $ab = i$ but $a \neq i$ and $b \neq i$.

THEOREM 4.6-2 In the six-element system, each of elements 2, 3, and 4 is a divisor of the additive identity 6, and these elements are the only such divisors.

We have already discovered two consequences of the existence of these divisors in the severe restrictions placed on the cancellation law in Theorem 4.5-4 and the solutions of the equations in Theorem 4.5-5. An example will illustrate a third consequence: when an equation is formed by setting element 6 equal to a product of linear factors, the number of solutions is no longer predictable.

Let us find all solutions for the equation

$$(x + 1)(5x + 4) = 6 \qquad (1)$$

By the multiplicative property of 6, if either of the factors $x + 1$ or $5x + 4$ is 6, the product will be 6. Setting each in turn equal to 6 produces the linear equations

$$x + 1 = 6 \text{ with unique solution } x = 1^* = 5$$

and

$$5x + 4 = 6 \text{ with unique solution } x = 5 \cdot 4^* = 4$$

Elements 5 and 4 reduce (1) to $6 \cdot 5 = 6$ and $5 \cdot 6 = 6$, respectively; hence, they are solutions. Now, however, we seek other elements x that can reduce the left member of Equation (1) to the product $2 \cdot 3$ or $3 \cdot 2$ and so give it the value 6. The first factor, $x + 1$, has value 2 when $x = 1$. Furthermore, when $x = 1$, the second factor, $5x + 4$, has value 3. Hence, $x = 1$ is a solution of the equation. Again, the first factor yields 3 when $x = 2$, and the second yields 2; consequently, $x = 2$ is a solution.

Finally, we must seek elements x for which the left member of the equation reduces to the product $3 \cdot 4$ or $4 \cdot 3$. We have already seen that when the first factor becomes 3, the second becomes 2; so there is no element which produces the combination $3 \cdot 4$. Will any element produce $4 \cdot 3$? When $x = 3$ the first factor takes the value 4, but the second factor takes the value 1. Since $4 \cdot 1 \neq 6$, element 3 is not a solution. It follows that only 1, 2, 4, and 5 are solutions of this equation.

In Exercise 4.6.2 below we discover that the number of solutions cannot be predicted for an equation of the form $(ax + b)(cx + d) = 6$. In this respect, the six-element system differs from that of five elements with its firm assertion in Theorem 3.1-2.

EXERCISE 4.6

In this exercise the six-element system is used.

1. Why are elements 1 and 5 not divisors of additive identity 6?

2. Find all solutions of each of the following equations.

(a) $(5x + 4)(x + 3) = 6$ (b) $(5x + 2)(5x + 1) = 6$
(c) $(x + 1)(x + 5) = 6$ (d) $(2x + 6)(3x + 6) = 6$
(e) $(4x + 5)(2x + 3) = 6$ (f) $(3x + 2)(4x + 1) = 6$

3. How many solutions has the equation $ax = 6$

(a) When $a = 2$? (b) When $a = 3$? (c) When $a = 4$?

4. What do the answers to Exercise 3 imply concerning the number of solutions of the equation $ax = 6$ when a is a divisor of the additive identity 6?

5. (a) Show that $2b = 2(b + d)$ when b is any element and d is 3 or 6.
(b) Show that $2b \neq 2(b + d)$ when b is any element and d is 1, 2, 4, or 5.
(c) Hence, show that if $2b = 2c$, then $c = b$ or $c = b + 3$.

6. Show that the results of Exercise 5 hold when $2b$, $2c$, and $2(b + d)$ are replaced by $4b$, $4c$, and $4(b + d)$, respectively.

7. (a) Show that $3b = 3(b + d)$ when b is any element and d is 2, 4, or 6.
(b) Show that $3b \neq 3(b + d)$ when b is any element and d is 1, 3, or 5.
(c) Hence, show that if $3b = 3c$, then $c = b$, or $c = b + 2$, or $c = b + 4$.

★8. Combine the results in Exercises 5, 6, and 7 to complete the statement beginning "for elements a, b, and c where a is a divisor of 6, if $ab = ac$, then . . . ."

4.7 STRUCTURE OF THE SIX-ELEMENT SYSTEM

The six-element system forms a commutative group under addition, has closure, commutativity, and associativity under multiplication, and possesses a multiplicative identity. Furthermore, multiplication is distributive with respect to addition.

For the five-element system we used the distributive law to show that $(ab)^* = ab^* = a^*b$ and $a^*b^* = ab$. The verifications of these formulas in Section 2.14 apply to the six-element system with the simple replacement of element 5 by element 6. A similar replacement in the derivation of the formula $5a = 5$ establishes the relation $6a = 6$ of Theorem 4.6-1.

Note, however, that the specification $a \neq 6$ is not sufficient to guarantee that the equation $ab = 6$ will have the unique solution $b = 6a' = 6$. The strongest statement that can be made is that if $ab = 6$ and a is 1 or 5, then $b = 6$.

EXERCISE 4.7 (OPTIONAL)

Verify the following formulas for elements a and b of the six-element system.

1. $6a = 6$

2. $(ab)^* = ab^* = a^*b$

3. $a^*b^* = ab$

4.8 OCCURRENCE OF DIVISORS OF THE ADDITIVE IDENTITY

Because of the similarities in the construction of the five- and six-element systems, you should be curious as to why one system has familiar multiplication properties while the other has not. One explanation is based on the unlike multiplicative characteristics of the natural numbers 5 and 6. The number 5 is *prime;* that is, it has as divisors only itself and 1. The number 6, on the other hand, is *composite;* besides 6 and 1 it also has divisors 2 and 3. Dissimilarities in the factorability of the numbers produce dissimilarities in the multiplicative properties of the element systems. In order to understand how this comes about, let us investigate the relationship between the numbers and the elements.

In the five-element system, elements correspond to natural numbers in such a way that a product of elements can be 5 when and only when the corresponding product of natural numbers is 5, 10, 15, or some other multiple of 5. As an illustration, let us interpret $a = 3$ and $b = 4$ either as natural numbers or as elements. Because the natural-number product ab is 12 and hence is not a multiple of 5, the element product ab is not 5. Conversely, because the element product 2 is different from 5, we know that the natural-number product is not a multiple of 5. On the other hand, for $a = 2$ and $b = 5$ the natural number $2 \cdot 5 = 10$ is a multiple of 5 and corresponds to the element product $2 \cdot 5 = 5$. Furthermore, when elements a and b are interpreted as natural numbers, they do not exceed 5. Because 5 is prime, a product of two such numbers is a multiple of 5 if and only if at least one of the numbers is 5. Interpretation of this statement for elements gives the multiplicative property of 5 in Theorem 2.12-1.

In the six-element system we find, as expected, that a product of elements can be 6 when and only when the corresponding product of natural numbers is a multiple of 6. This situation occurs not only when one of the factors is such a multiple but also, since 6 is composite, when the factors are suitable mates. For example, $5 \cdot 6$ is surely a multiple of 6 because it contains the factor 6. The product $3 \cdot 4 = 12$ is also a multiple of 6. When there are exactly two factors and neither is greater than 6, their product is a multiple of 6 if and only if they are associated as in one of the following equalities:

$$
\begin{array}{ll}
1 \cdot 6 = 2 \cdot 3 = 6 & \quad 2 \cdot 6 = 3 \cdot 4 = 12 \\
3 \cdot 6 = 18 & \quad 4 \cdot 6 = 24 \\
5 \cdot 6 = 30 & \quad 6 \cdot 6 = 36
\end{array}
$$

Hence, as a product of elements, $ab = 6$ if and only if a and b are one of the following couples: 6 and 1, 2, 3, 4, 5, or 6; 2 and 3; or 3 and 4. The first set of couples justifies the multiplicative property of 6 given in Theorem 4.6-1, and the last two couples yield Theorem 4.6-2 on divisors of the additive identity.

By similar reasoning we can show that for a system of n elements constructed in the same manner as our five- and six-element systems, no divisors of the additive identity exist when n is prime, but such divisors do exist when n is composite. For example, there are none in a system of 29 elements. (Note that this system would be suitable for the enciphering of messages involving the 26 letters of the English alphabet and three marks of punctuation, as suggested in Section 3.10.) A system of ten elements, however, has divisors of the additive identity. We can determine which elements they are by considering all distinct products of numbers not exceeding 10 that give multiples of 10, namely,

$$\begin{array}{ll} 1 \cdot 10 = 2 \cdot 5 = 10 & 2 \cdot 10 = 4 \cdot 5 = 20 \\ 3 \cdot 10 = 6 \cdot 5 = 30 & 4 \cdot 10 = 8 \cdot 5 = 40 \\ 5 \cdot 10 = 50 & 6 \cdot 10 = 60 \\ 7 \cdot 10 = 70 & 8 \cdot 10 = 80 \\ 9 \cdot 10 = 90 & 10 \cdot 10 = 100 \end{array}$$

Obviously, 2, 4, 5, 6, and 8 are divisors of the additive identity 10, while each of 1, 3, 7, and 9 must be multiplied by 10 in order to give a product of 10.

EXERCISE 4.8

1. For a system of $n = 9$ elements constructed similarly to the five- and six-element systems:

(a) Determine whether or not there are divisors of the additive identity.

(b) If there are divisors of the additive identity, find them.

2. Repeat Exercise 1 using $n = 12$.

3. Repeat Exercise 1 using $n = 13$.

4. Repeat Exercise 1 using $n = 43$.

4.9 RETROSPECT AND PROSPECT

The six-element system has been worth investigation because of its unexpected multiplicative characteristics. We have answered one of the questions raised at the end of Chapter 3: a system of six elements does not behave in all respects like one of five elements. Furthermore, our work has demonstrated the need for inquiry into the behavior of general n-element systems, but this investigation is left to professional number theorists.

The second question of Section 3.11 concerns the use of the five elements as a tool in the formation of a more elaborate system. In the next chapter we shall construct ordered pairs of elements, consider each pair to be a number, and discover the characteristics of the system that they compose.

5
A SYSTEM OF PAIRS OF ELEMENTS

Mathematicians, like artists and composers, are accustomed to developing simple concepts into more interesting forms. In this chapter the five-element system is extended to a new system whose properties are determined, not only by the characteristics of the original simpler system, but also by the manner in which the extension is made.

5.1 DEFINITION AND NOTATION

This extended system is created by forming ordered pairs of elements. The ordered pair having first element 2 and second element 3 is written (2,3); it *differs* from the pair (3,2), which obviously has first entry 3 and second entry 2. A pair consisting of two general elements, for example, (e_1,e_2), may be designated by a single capital letter; thus, E would be a symbol for (e_1,e_2). On occasion, it is more convenient to use two separate letters for the elements of a pair. When no confusion or ambiguity could arise, we may write an ordered pair as (x,y) in place of the more traditional form.

If its name is to describe it completely, the new system should be designated as a *system of ordered pairs of elements of the five-element system,* but let us call it simply a *system of pairs.* In finding its properties, however, we must use the rules of formation of its members.

5.2 EQUALITY OF PAIRS

As a first step, we need a definition of equality of pairs, that is, an *equivalence relation* between pairs. Let us use the simplest one, namely, that equality means identity. In Chapter 6 a different concept of equivalence of pairs will be discussed.

DEFINITION **Two pairs are equal** if and only if their corresponding elements are equal. In symbols, for pairs $P = (p_1,p_2)$ and $Q = (q_1,q_2)$, $P = Q$ if and only if $p_1 = q_1$ and $p_2 = q_2$.

For example, $(2^*,1^*) = (3,4)$ because $2^* = 3$ and $1^* = 4$. On the other hand, $(3,4) \neq (4,3)$, although they are composed of the same elements.

According to the definition, any two nonidentical pairs are distinct from each other. As a consequence, there are five different choices of element for the first entry in any pair and five for the second; hence, there are a total of 25 pairs in the system.

The definition of equality of pairs ensures three properties which mathematicians always require of a good equivalence relation. First, equality of pairs is *reflexive;* that is, each pair is equal to itself. Second, equality of pairs is *symmetric;* that is, for pairs P and Q, if $P = Q$, then $Q = P$. Third, equality of pairs is *transitive;* that is, pairs equal to equal pairs are equal to each other. Frequent use will be made of this last property.

EXERCISE 5.2

1. Find p_1, q_1, p_2, q_2 in the equalities below.

(a)	$(p_1,p_2) = (4,2)$	(b)	$(2,p_2) = (p_1,3)$
(c)	$(2',p_2) = (p_1,5)$	(d)	$(q_1,q_2) = (5,5)$
(e)	$(q_1,1) = (4,q_2)$	(f)	$(q_1,4^*) = (3',q_2)$

2. Give at least three other names for the pair (2,3).

3. Which, if any, of the following pairs are equal?

$P = (1,2)$	$Q = (2,2)$
$R = (3,1)$	$S = (1,3)$
$T = (1,3^*)$	$U = (3',3^*)$
$V = (2',4')$	$W = (4,5)$

4. List the 25 distinct pairs in the system of pairs.

5. List the pairs that satisfy each of the following conditions.
(a) The order of the two elements is immaterial.
(b) The second entry is the additive inverse of the first.
(c) The second entry is three times the first.
(d) The sum of the two entries is 2.

5.3 ADDITION OF PAIRS

How shall two pairs be added? There are many ways, of which the simplest is to add corresponding entries: for example, (2,3) + (4,2) = (1,5). This is the method to be used here.

DEFINITION The **sum of two pairs** is a pair whose elements are the sums of the corresponding elements of the given pairs. In symbols, for pairs $P = (p_1,p_2)$ and $Q = (q_1,q_2)$, $P + Q = (p_1 + q_1,\ p_2 + q_2)$.

To find the properties of addition, we could form an addition table and employ it as we did with the five-element system. Such a procedure would be long and tedious, because with 25 pairs in the u column and 25 in the v row there would be 625 entries in the body of the table. The difficulty of finding the common properties of so many entries is evident. Clearly, we must seek a better method. Let us try to reason from properties of the five-element system to properties of the system of pairs.

Consider first the fact that the five-element system is closed under addition. What does this imply for addition in the pair system? The use of symbols will help. Given $P = (p_1,p_2)$ and $Q = (q_1,q_2)$, where p_1, q_1, p_2, and q_2 are any elements whatever, so that P and Q are any pairs whatever; let us investigate the entries $p_1 + q_1$ and $p_2 + q_2$ in the pair representing $P + Q$. Because the five-element system is closed under addition, the sum $p_1 + q_1$ always exists, is an element, and is unique. It follows that there is one and only one element that may be written in place of the sum $p_1 + q_1$. Similarly, there is one and only one element that may be written in place of the sum $p_2 + q_2$. Hence, the symbol $(p_1 + q_1,\ p_2 + q_2)$ defines a unique member of the system of pairs; that is, the sum $P + Q$ is a unique pair. The property of closure carries over to become a property of the new system.

THEOREM 5.3-1 The system of pairs is closed under addition.

EXERCISE 5.3

1. Find the following sums of pairs.

(a) (1,2) + (3,4) (b) (3,1) + (3,1) (c) (2,4) + (2*,4*)
(d) (2,3) + (5,5) (e) (4,5) + (3,4) (f) (5,1) + (2,5)

2. Each combination of symbols below purports to be a statement about pairs, and each fails of its purpose. Explain.

(a) (2,1) + (3,4) = 5,5
(b) (2,1) + (3,4) = (2 + 3)(1 + 4)
(c) (2,1) + (3,4) = (2 + 3) + (1 + 4)

3. Find all elements x and y for which each of the following statements is true.

(a) $(2,3) + (x,4) = (1,2)$ (b) $(4,1) + (3,y) = (2,5)$
(c) $(x,1) + (3,2) = (4,4)$ (d) $(3,4) + (x,5) = (3,4)$
(e) $(x,y) + (5,5) = (x,y)$ (f) $(5,5) + (x,y) \neq (x,y)$
(g) $(3,1) + (1,y) = (5,5)$ (h) $(2,4) + (x,y) = (5,5)$
(i) $(x,y) + (x^*,y^*) = (5,5)$ (j) $(x^*,y^*) + (x,y) \neq (5,5)$

4. Is there an identity for addition in the system of pairs? If so, what is it? Justify your answer.

5. If you found an identity in Exercise 4, determine whether each pair has an inverse. If an inverse exists, give a formula for it.

6. Find all elements x and y for which each of the following statements is true.

(a) $(2,1) + (x,y) = (x,y) + (2,1)$
(b) $(x,y) + (3,4) = (3,4) + (x,y)$
(c) $(1,3) + (x,y) \neq (x,y) + (1,3)$

7. What property of addition of pairs is suggested by your answers to Exercise 6?

5.4 COMMUTATIVITY OF ADDITION OF PAIRS

The preceding exercise suggests that addition of pairs may be commutative. Let us determine whether all pairs P and Q satisfy the commutative relation

$$P + Q = Q + P \tag{1}$$

Using customary notation,

$$P + Q = (p_1 + q_1, p_2 + q_2) \qquad Q + P = (q_1 + p_1, q_2 + p_2)$$

By definition, equality (1) holds if and only if

$$p_1 + q_1 = q_1 + p_1 \qquad \text{and} \qquad p_2 + q_2 = q_2 + p_2 \tag{2}$$

Commutativity of addition in the five-element system establishes relations (2); these, in turn, justify formula (1).

THEOREM 5.4-1 In the system of pairs, addition is commutative.

The discussion preceding the theorem forms what the ancient Greeks called the *analysis* of its proof. It began with the conclusion that we wished to establish [formula (1)] and worked back to relations [formulas (2)] that are known to hold because pairs are composed of elements of the five-element system. Now let us seek to write what the Greeks called the *synthesis* of the proof, that is, an argument which proceeds smoothly from hypothesis to conclusion. A good synthesis convinces one of the truth of the theorem but may give no clue as to the way in which the proof was originally conceived. There are various ways of making a synthesis, one of them being by good expository prose. Here a formal method already familiar to you from your work in secondary school geometry will be used. Write the theorem as a conditional sentence, namely, "if P and Q are pairs, then $P + Q = Q + P$." Next present the proof as a series of such conditionals, the first of which begins with the clause "if P and Q are pairs" and the last of which ends with the clause "then $P + Q = Q + P$." For each two contiguous conditionals, the assertion in the "then" clause of the first is the assertion in the "if" clause of the second. Each conditional in the series is justified by a definition or by a previous theorem or observation. Let us proceed with the argument.

THEOREM 5.4-1 Restated If P and Q are pairs of the system of pairs, then $P + Q = Q + P$.

Proof 1. If $P = (p_1,p_2)$ and $Q = (q_1,q_2)$ are pairs of the system of pairs, then p_1, q_1, p_2, and q_2 are elements of the five-element system. (True by definition of pairs.)

2. If p_1, q_1, p_2, and q_2 are elements of the five-element system, then $p_1 + q_1 = q_1 + p_1$ and $p_2 + q_2 = q_2 + p_2$. (True by law of commutativity of addition in the five-element system.)

3. If $p_1 + q_1 = q_1 + p_1$ and $p_2 + q_2 = q_2 + p_2$, then $(p_1 + q_1, p_2 + q_2) = (q_1 + p_1, q_2 + p_2)$. (True by definition of equality of pairs.)

4. If $(p_1 + q_1, p_2 + q_2) = (q_1 + p_1, q_2 + p_2)$, then $P + Q = Q + P$. (True by definition of addition of pairs.)

We have constructed a chain of conditional sentences beginning with the clause "if P and Q are pairs" and ending with the clause "then $P + Q = Q + P$." Such proofs are customarily presented with less repetition of words and symbols in the following "argument–reason" form, which includes an argument that makes evident the transition from step 3 to step 4.

THEOREM 5.4-1 Restated If P and Q are pairs of the system of pairs, then $P + Q = Q + P$.

Given Pairs $P = (p_1,p_2)$ and $Q = (q_1,q_2)$.

To prove $P + Q = Q + P$.

Proof

Arguments	*Reasons*
1. p_1, q_1, p_2, and q_2 are elements of the five-element system.	1. Definition of pairs.
2. $p_1 + q_1 = q_1 + p_1$ and $p_2 + q_2 = q_2 + p_2$	2. In the five-element system, addition is commutative.

Arguments	*Reasons*
3. $(p_1 + q_1, p_2 + q_2) = (q_1 + p_1, q_2 + p_2)$	3. Definition of equality of pairs.
4. $(p_1 + q_1, p_2 + q_2) = P + Q$ and $(q_1 + p_1, q_2 + p_2) = Q + P$	4. Definition of addition of pairs.
5. $P + Q = Q + P$	5. Steps 3 and 4 and transitive property of equality of pairs.

In this final finished form, the proof proceeds logically, by properly justified steps, from hypothesis to conclusion.

EXERCISE 5.4

1. A system of T-pairs is constructed in the same way (by analogous definitions for pair, equality, and addition) as the system of pairs, except that the elements are drawn from system T of Exercise 1.7. Find the following sums of T-pairs.

(a) (1,2) + (3,4) (b) (3,4) + (1,2) (c) (1,1) + (4,6)
(d) (4,6) + (1,1) (e) (2,3) + (5,4) (f) (5,4) + (2,3)

2. (a) Draw a conclusion about commutativity of addition in the system of T-pairs.

(b) For which of the following reasons is the conclusion of part (a) different from the corresponding conclusion for the system of pairs? Explain.

(i) Construction of the systems of pairs.

(ii) Characteristics of the systems from which the elements of the pairs are drawn.

3. In the argument–reason proof of this section, which argument would fail if the elements of the pairs were drawn from system T instead of from the five-element system? Explain.

In the remaining exercises the system of pairs of elements of the five-element system is used.

4. By closure of addition, the sum of (2,3) and (4,2) is a single pair. Hence, the sum [(2,3) + (4,2)] + (5,1) can be considered the sum of exactly two pairs, namely, the sum pair for (2,3) and (4,2) and the pair (5,1). Find this final sum.

5. Explain how the sum (2,3) + [(4,2) + (5,1)] can be considered the sum of exactly two pairs. Find this sum.

6. What property of addition of pairs is suggested by the numerical answers to Exercises 4 and 5?

5.5 ASSOCIATIVITY OF ADDITION OF PAIRS

Experience with the five-element system suggests that we next address ourselves to the problem of adding many pairs. It also indicates that such an addition should be reduced to successive additions of two pairs each. For three pairs, the process is given in the definition below.

DEFINITION For pairs P, Q, and R of the system of pairs, the sum

$$P + Q + R \equiv (P + Q) + R$$

It would be convenient if an associative law should hold. Let us investigate the general statement that for pairs P, Q, and R,

$$(P + Q) + R = P + (Q + R) \tag{3}$$

Using customary notation, we wish to know whether

$$[(p_1 + q_1) + r_1, (p_2 + q_2) + r_2] = [p_1 + (q_1 + r_1), p_2 + (q_2 + r_2)] \tag{4}$$

Relation (4) is equivalent to the couple of equalities

$$\begin{aligned} &(p_1 + q_1) + r_1 = p_1 + (q_1 + r_1) \\ \text{and} \quad &(p_2 + q_2) + r_2 = p_2 + (q_2 + r_2) \end{aligned} \tag{5}$$

Because p_1, q_1, r_1, p_2, q_2, and r_2 are elements of the five-element system in which addition is associative, the equalities in (5) hold; hence, relation (4) is true and so is its interpretation in the equivalent statement (3). We have completed the analysis which shows that addition of pairs is associative. In the exercises that follow, you are asked to write a synthesis of the proof in argument–reason form.

THEOREM 5.5-1 In the system of pairs addition is associative.

As noted in Section 1.6, the definition of addition can be extended to the sum of four or more pairs. Then, associative and commutative laws together permit us to add pairs in any order and in any grouping.

EXERCISE 5.5

1. Give two numerical illustrations of Theorem 5.5-1.
2. Make an argument–reason proof of Theorem 5.5-1.
3. In each of the following equalities find elements x and y.

   (a) $(3,4) + (4,y) + (4,5) = (x,1)$
   (b) $(x,1) + (1,3) + (3,5) = (5,y)$
   (c) $(3,4) + (2,1) + (2,y^*) = (x^*,4)$
   (d) $(2,4) + (3,1) + (4,3) + (5,1) = (x,y)$
   (e) $(4,4^*) + (3^*,1) + (3,4) + (2,2) = (x',y')$
   (f) $(1,y) + (4,3') + (x^*,y) + (5,4) = (1,1^*)$

In the remaining exercises consider a system of "cross-pairs" for which the pairs and their equality are defined exactly like those in the system we have been using. Addition, however, has the following definition. For pairs $P = (p_1,p_2)$ and $Q = (q_1,q_2)$, $P + Q = (p_1 + q_2, p_2 + q_1)$.

4. Find the following sums in the system of cross-pairs.

(a)	$(1,2) + (2,4)$	(b)	$(2,4) + (1,2)$
(c)	$(3,4) + (5,5)$	(d)	$(5,5) + (3,4)$
(e)	$(4,1) + (2,3)$	(f)	$(2,3) + (4,1)$
(g)	$[(1,2) + (5,4)] + (3,4)$	(h)	$(1,2) + [(5,4) + (3,4)]$

5. The above sums show that addition is not commutative in the system of cross-pairs. Explain how it is possible for this statement to be correct when addition is commutative in the five-element system from which the entries of the pairs are drawn.

6. The sums in Exercise 4 also show that addition is not associative in the system of cross-pairs. Explain how this result can be consistent with the fact that addition is associative in the five-element system.

5.6 THE ADDITIVE GROUP OF PAIRS

According to Exercises 5.3.4 and 5.3.5, the pair (5,5) is an identity pair for addition. Let us call it the *five-pair* and designate it by the latter F. Each pair $P = (p_1,p_2)$ has an additive inverse given by $P^* = (p_1^*,p_2^*)$. Since we have shown closure, commutativity, and associativity of addition of pairs, the system satisfies all requirements for a commutative group.

THEOREM 5.6-1 The system of pairs forms a commutative group under addition.

Theorems and corollaries developed in Section 2.10 set forth the long list of properties possessed by the elements and operation of a group. Interpreted for pairs they yield the following assertions.

THEOREM 5.6-2 The five-pair $F = (5,5)$ is the unique identity for addition.

THEOREM 5.6-3 Every pair $P = (p_1,p_2)$ has a unique inverse under addition called P^* and given by the formula (p_1^*,p_2^*).

Corollary 5.6-3 For pairs P and Q, $P^{**} = P$; $(P + Q)^* = P^* + Q^*$; and if $P = Q^*$, then $P^* = Q$.

THEOREM 5.6-4 For pairs P and Q, the equation $P + X = Q$ has the unique solution $X = P^* + Q$.

THEOREM 5.6-5 *Law of Cancellation for Addition.* For pairs P, Q, and R, if $P + Q = P + R$, then $Q = R$.

EXERCISE 5.6

1. Find pair X in each of the following equations.

(a) $X + (2,5) = (3,1)$
(b) $X + (4,3)^* = (1,2)$
(c) $X^* + (1,2) = (4,5)$
(d) $X^* + (2,3)^* = (4,1)$
(e) $X + (3,4) = (3,4)$
(f) $X + (3,4) = F$
(g) $X + (1,5) = (3,4) + (1,5)$
(h) $X + (3,5) = (1,2)^* + (3,5)$
(i) $X^* + (1,1) = (4,2) + (1,1)$
(j) $X^* + (3,2) = (1,4)^* + (3,2)$

2. There are five pairs which have first element 5.
(a) List the pairs.
(b) Show that this small system of pairs forms a commutative group under addition, a subgroup of the additive group of pairs. (Suggestion: note Theorem 2.11-1.)

3. Four sets of pairs are listed in Exercise 5.2.5. For each set determine whether it is a subgroup of the additive group of pairs.

5.7 ASSOCIATED PAIRS-AND-ELEMENTS RELATIONS

Every property of addition of pairs gives rise to two associated properties of addition of elements, one concerning the first entries and one concerning the second entries in the pairs. Not only did we justify properties of pairs by establishing equivalent statements about elements, but we also solved pair equations by considering separately the two corresponding element equations. Pair notation thus permits us to deal with two simultaneous, although distinct, element relations without difficulty or confusion.

Here is a practical application to a problem in physics (the use of the five elements as components of the pairs makes the numbers in it absurd, but it is nonetheless a real problem). Assume a world in which the five-element system is the usual and accustomed number system. A boat sails across a river. The current is carrying it downstream at the rate of 2 miles per hour and its engines are pushing it directly across at 3 miles per hour. Then the pair (2,3) is a convenient way of recording the motion of the boat. If, now, a wind blows downstream at 2 miles per hour and if the auxiliary engine is used to give an additional speed across stream of 1 mile per hour, the pair (2,1) records the additional velocities in the directions down and across the river, respectively. The total effect of current and wind and of engine and auxiliary is $(2,3) + (2,1) = (4,4)$; consequently, the boat sails with the combined velocities of 4 miles per hour downstream and 4 miles per hour across stream.

5.8 MULTIPLICATION OF PAIRS

An investigation of multiplication for pairs must begin with a definition. Let us choose a simple one stating that to multiply two pairs we multiply corresponding elements. Let us further agree to denote the operation in any of the usual ways; then, for two specific pairs,

$$(2,3) \times (4,1) = (2,3) \cdot (4,1) = (2,3)(4,1) = (2 \cdot 4,\ 3 \cdot 1) = (3,3)$$

DEFINITION The **product of two pairs** is a pair whose elements are the products of the corresponding elements of the given pairs.

The definition should imply closure for the operation. That it does so can be confirmed by an argument similar to that which established closure for addition.

THEOREM 5.8-1 The system of pairs is closed under multiplication.

EXERCISE 5.8

1. Why is it not necessary to construct and use a table for multiplication of pairs?

2. Why is Theorem 5.8-1 true?

Find the products of pairs indicated in Exercises 3–8.

3. $(1,2)(3,4)$

4. $(3,4)(1,2)$

5. $(1,1)(4,5)$

6. $(2,3)F$

7. $(5,2)(3,5)$

8. $(2',3')(2,3)$

In Exercises 9–19, find all elements x and y for which each statement is true.

9. $(2,3)(x,y) = (2,3)$

10. $(x,4)(3,4) = (1,1)$

11. $(5,3)(x,y) = (2,1)$

12. $(3,5)(x,y) = F$

13. $(5,y)(x,5) = F$

14. $(1,1)(x,y) \neq (x,y)$

15. $(x',y')(x,y) = (1,1)$

16. $(x,y)F \neq F$

17. $(3,2)(x,y) = (x,y)(3,2)$

18. $(x,y)(1,4) \neq (1,4)(x,y)$

19. $(x,5)(5,y) \neq F$

20. Which property of multiplication of pairs is suggested by each of the following sets of problems chosen from Exercises 3–19 above?

(a) 3, 4, 17, 18 (b) 5, 9, 14
(c) 8, 10, 15 (d) 6, 16
(e) 7, 12, 13, 19

21. By closure of multiplication the product of (2,3) and (4,2) is a single pair. Hence, the product [(2,3)(4,2)](3,1) can be considered the product of exactly two pairs, namely, the product pair for (2,3) and (4,2) and the pair (3,1). Find this final product.

22. Explain how the product (2,3)[(4,2)(3,1)] can be considered the product of exactly two pairs; then find the product.

23. What property of multiplication is suggested by the numerical answers to Exercises 21 and 22?

In Exercises 24–27 find pair P.

24. $P = (1,4)[(3,2) + (4,2)]$

25. $P = (1,4)(3,2) + (1,4)(4,2)$

26. $P = [(3,4) + (1,3)](2,2)$

27. $P = (3,4)(2,2) + (1,3)(2,2)$

28. What property of multiplication is suggested by the numerical answers to Exercises 24 and 25? Exercises 26 and 27?

5.9 PROPERTIES OF MULTIPLICATION OF PAIRS

The similarity of the processes by which multiplication and addition are performed suggests that these operations have in common not only closure but also commutativity and associativity. The preceding exercise supports this conjecture, and arguments analogous to those in Sections 5.4 and 5.5 confirm it.

THEOREM 5.9-1 In the system of pairs, multiplication is commutative.

DEFINITION For pairs P, Q, and R, the product $PQR \equiv (PQ)R$.

THEOREM 5.9-2 In the system of pairs, multiplication is associative.

According to Exercise 5.8.28 multiplication of pairs may be distributive with respect to addition. By calling on the corresponding characteristic of the five-element system, it can be shown that this property holds. Some immediate consequences, stated in the corollary below, follow by arguments analogous to those for elements in Section 2.14.

THEOREM 5.9-3 In the system of pairs, multiplication is distributive with respect to addition.

Corollary 5.9-3 For pairs P and Q, $PF = F$; $(PQ)^* = P^*Q = PQ^*$; and $P^*Q^* = PQ$.

EXERCISE 5.9

1. Find x and y in the following products.

(a) $(2,3) \cdot (4,1) \cdot (1,1) = (x,y)$

(b) $(3,4) \cdot (2,y) \cdot (4,3) = (x,2)$

(c) $(x,1) \cdot (4,2) \cdot (3,5) = (5,y)$

(d) $(3,4)[(2,y) + (4,3)] = (x,2)$

(e) $[(x,1) + (4,2)](2,3) = (5,y)$

2. Does the system of pairs have the property that when a product of pairs is F, then at least one of the pairs must be F? Justify your answer.

3. Write an argument–reason proof of Theorem 5.9-1.

4. Write an argument–reason proof of Theorem 5.9-2.

5. Using symbols state the distributive law for pairs.

★6. Prove the left-distributive law for pairs.

★7. Establish the results stated in Corollary 5.9-3.

5.10 DIVISORS OF THE ADDITIVE IDENTITY IN THE SYSTEM OF PAIRS

Exercise 5.9.2 shows that the system of pairs has the property that a product of two pairs may be F without either pair being F. Whenever two factors, neither of which is the identity for addition, multiply together to give that identity, we call each of the factors a divisor of the additive identity. (See the definition of divisor of the additive identity in Section 4.6.) For example, pairs (3,5) and (5,4) are divisors of the additive identity F because $(3,5)(5,4) = F$, although $(3,5) \neq F$ and $(5,4) \neq F$. Why does the five-element system, in which there are no such divisors, produce a system of pairs in which such divisors exist? The explanation is suggested in Exercises 5.8.7 and 5.8.13. Whenever any pair having 5 as its first entry is multiplied with any pair having 5 as its second entry, the product pair has 5 for both entries. In symbols, $(5,p)(q,5) = (5,5) = F$ for all p and q. Now, when p and q are both different from 5, neither factor $(5,p)$ nor $(q,5)$ is F, and each is, by definition, a divisor of F. The existence of these divisors is, therefore, primarily a consequence of the method of construction of the pairs and the method chosen for their multiplication.

How many such divisors are there? In the pair $(5,p)$, the second entry p can be any of the elements 1, 2, 3, or 4. In the pair $(q,5)$, q can range over this same set of elements. Hence, 8 of the 25 pairs are divisors of F. We therefore have the theorem below in which the condition p and q both different from 5 is replaced by the shorter equivalent condition $pq \neq 5$.

THEOREM 5.10-1 The eight pairs $(5,p)$ and $(q,5)$ with $pq \neq 5$ are divisors of the additive identity F, and these pairs are the only such divisors. If $PQ = F$, but both elements of Q are different from 5, then $P = F$.

This theorem must be taken into account in seeking solutions of an equation of the form

$$[X + (2,3)][X + (1,4)] = F \qquad (6)$$

In each square bracket the sum of pairs is a pair. Hence, the equation represents a product of two pairs set equal to F. Because the product will surely be F if either factor is F, two obvious solutions are $(2,3)^*$ and $(1,4)^*$. We must now determine whether any pairs reduce the left member of Equation (6) to a product of associated divisors of F whose product is F, because all such pairs produce the equality $F = F$ and are also solutions. If $X = (4,2)$, the left member of Equation (6) becomes $(1,5)(5,1)$, and if $X = (3,1)$ it becomes $(5,4)(4,5)$. Equation (6) thus has four distinct solutions: $(2,3)^* = (3,2)$; $(1,4)^* = (4,1)$; $(4,2)$; and $(3,1)$.

EXERCISE 5.10

1. List the eight pairs that are divisors of F.

2. Find all solutions for each of the following equations.

(a) $[X + (1,2)][X + (3,4)] = F$ (b) $[X + (3,1)][X + (1,5)] = F$
(c) $[X + (4,2)][X + (1,3)] = F$ (d) $[X + (5,1)][X + (4,2)] = F$
(e) $[X + (2,3)][X + (2,4)] = F$ (f) $[X + (4,3)][X + F] = F$

3. Show that each of the following equations has five solutions for pair X.

(a) $(2,5)X = F$ (b) $(5,4)X = F$

4. What can be said of the number of solutions of the equation $PX = F$ when P is a divisor of F?

5. Let $PQ = PR$ where $Q = (q_1,q_2)$ and $R = (r_1,r_2)$.

(a) If $P = (2,5)$, show that q_1 must equal r_1 but that q_2 and r_2 may be any elements.

(b) If $P = (5,4)$, then what relation, if any, holds between q_1 and r_1, and between q_2 and r_2?

5.11 MULTIPLICATIVE IDENTITY AND INVERSE IN THE SYSTEM OF PAIRS

In Exercise 5.8.20(b) we found that the pair (1,1) serves as a multiplicative identity. In Exercise 5.8.20(c) we discovered that every pair $P = (p_1,p_2)$ for which $p_1p_2 \neq 5$ has an inverse of the form (p_1', p_2'). There are 16 such pairs because there is a choice of four elements for the first entry and four for the second. They are exactly the pairs that are not F or divisors of F. Moreover, they form a commutative group under multiplication. Some of the consequences of their group character are listed below, stated, when possible, for the whole system of pairs, rather than restricted to the multiplicative group. In the exercises you are asked to justify these extended statements.

THEOREM 5.11-1 In the system of pairs, pair $I = (1,1)$ is the unique identity pair for multiplication.

THEOREM 5.11-2 Every pair that is not F or a divisor of F has a unique multiplicative inverse. In symbols, every pair $P = (p_1,p_2)$ with $p_1p_2 \neq 5$ has a unique multiplicative inverse $P' = (p_1',p_2')$.

Corollary 5.11-2 For pairs P and Q having inverses, $(P')' = P$; $(PQ)' = P'Q'$; and if $P = Q'$, then $P' = Q$.

THEOREM 5.11-3 For pairs P and Q where P is not F or a divisor of F, the equation $PX = Q$ has the unique solution $P'Q$ for pair X.

THEOREM 5.11-4 *Law of Cancellation for Addition.* For pairs P, Q, and R where P is not F or a divisor of F, if $PQ = PR$, then $Q = R$.

The system of pairs closely resembles the system of six elements. Cancellation laws and rules for the solution of equations are of more significance for the 25 pairs, 16 of which have multiplicative inverses, than for the six elements, only two of which have such inverses. Nevertheless, the structure of the pair system is no more sophisticated than that of the elements.

EXERCISE 5.11

1. Find pair X in each product below.

(a) $(2,1) \cdot (3,4) \cdot I \cdot (2,1)' = X$ (b) $(1,3) \cdot (2,4) \cdot F \cdot (3,2)' \cdot (1,3) = X$

2. Find all pairs X that satisfy each of the following equations.

(a) $(4,1) \cdot (2,3)' = X$ (b) $(2,3)X = (3,1)$
(c) $(2,3)X' = (3,1)$ (d) $[(4,1)X]' = (2,3)$
(e) $(1,2)X = F$ (f) $X^* = X'$
(g) $X = X \cdot X$ (h) $(2,3)X = (4,4)(2,3)$
(i) $(1,4)X' = (3,2)(1,4)$ (j) $(2,1)X' = (2,1)(3,4)'$

3. Justify Theorem 5.11-1.

4. Justify Theorem 5.11-3.

5. Justify Theorem 5.11-4.

6. Explain why the 16 pairs lacking element 5 form a group under multiplication.

7. Four pairs of the multiplicative group have element 1 as first element.
(a) List the four pairs.
(b) Show that the four pairs form a subgroup under multiplication.

8. Show that the pairs of type (p,p) with $p \neq 5$ form a subgroup of the multiplicative group.

★9. The multiplicative group contains the identity I, which is its own multiplicative inverse. Show that at least one other pair must also be its own multiplicative inverse. Are there more than two such pairs? Explain.

5.12 COMMENT ON THE FIRST SYSTEM OF PAIRS

In this chapter we have discovered that under addition the system of pairs closely resembles the five-element system from which it is derived. Under multiplication, however, it is more like the six-element system. It is clear that when one system is constructed from another, the properties of the parent system do not automatically characterize the new one. The behavior of the derived system is determined by the definitions governing its construction as well as by the properties of its parent.

In the next chapter another system of pairs of elements will be developed whose dissimilarity from the present one begins with a new definition of equivalence.

6

A SECOND SYSTEM OF PAIRS OF ELEMENTS

The definitions of equality, addition, and multiplication by which the first system of pairs is constructed are all arbitrary. It is reasonable to believe that a change in one or more of them would result in a system with different properties. In this chapter the assumption that new definitions will produce new systems is tested.

6.1 EQUIVALENCE OF PAIRS

The members of the new system are formed and written in a manner identical with those of the first: they are ordered pairs of elements of the five-element system. Instead of interpreting equivalence as identity, however, a new definition is chosen here which arises by means of a seemingly unusual association.

Mathematicians generally reserve the word *equality* for identity and use the word *equivalence* for a relation such as that about to be defined. This custom will be followed here, the symbol $=$ being used only for identity; for equivalence one of the standard symbols, namely, $\sim$, is adopted. Hence, $A = (a_1,a_2)$ means that pair A is identical with (a_1,a_2), and $A \sim B$ means that pair A is equivalent to pair B.

DEFINITION For pairs $A = (a_1,a_2)$ and $B = (b_1,b_2)$, A **is equivalent to** B (in symbols, $A \sim B$) if and only if $a_1 + b_2 = a_2 + b_1$.

According to this definition, $(3,4) \sim (3,4)$. Also, $(3,4) \sim (5,1)$ because $3 + 1 = 4 + 5$, but $(3,4) \nsim (2,1)$ because $3 + 1 \neq 4 + 2$. For any two pairs we can distinguish between equivalence and nonequivalence.

In Section 5.2 we discussed three properties of an equivalence relation. It is easy to show that the relation just defined has the first two of these properties. It is reflexive; that is, each pair is equivalent to itself. It is symmetric; that is, for pairs A and B, if $A \sim B$, then $B \sim A$. Let us now show that it is also transitive; that is, for pairs A, B, and C, if $A \sim B$ and $B \sim C$, then $A \sim C$. We must begin by translating the statement we wish to prove into the language of elements.

STATEMENT For pairs $A = (a_1,a_2)$, $B = (b_1,b_2)$, and $C = (c_1,c_2)$, if $a_1 + b_2 = a_2 + b_1$ and $b_1 + c_2 = b_2 + c_1$, then $a_1 + c_2 = a_2 + c_1$.

Note that the left members of all three equalities contain elements a_1 and c_2 and that the right members contain a_2 and c_1. Hence, it may be profitable to add the two equations in the "if" clause. Making this addition and using the commutative and associative laws for elements, we find

$$(a_1 + c_2) + (b_1 + b_2) = (a_2 + c_1) + (b_1 + b_2)$$

The law of cancellation for elements permits us to eliminate the unwanted sums $(b_1 + b_2)$. Hence, we obtain the desired result $a_1 + c_2 = a_2 + c_1$.

EXERCISE 6.1

1. Which of the following pairs are equivalent?

$A = (3,3)$	$B = (2,3)$	$C = (4,2)$	$D = (5,1)$
$G = (3,4)$	$H = (3,1)$	$J = (2,2)$	$K = (3,5)$

6.2 A SECOND SYSTEM OF PAIRS

In the new system, pairs with dissimilar elements may be equivalent to each other. In this respect it differs from the one investigated in Chapter 5. In order to distinguish it from the earlier one, let us name it "the second system of pairs."

There are five special pairs whose equivalence is evident, namely, $(1,1) \sim (2,2) \sim (3,3) \sim (4,4) \sim (5,5)$. In Table 6.1 the 25 sets of ordered couples of elements are listed, and each of the five pairs equivalent to (5,5) is circled.

Table 6.1 Pairs of Elements

(1,1)	(2,1)	(3,1)	(4,1)	(5,1)
(1,2)	(2,2)	(3,2)	(4,2)	(5,2)
(1,3)	(2,3)	(3,3)	(4,3)	(5,3)
(1,4)	(2,4)	(3,4)	(4,4)	(5,4)
(1,5)	(2,5)	(3,5)	(4,5)	(5,5)

In the exercise below you are asked to determine all pairs that are equivalent to (1,5). These can be discovered by trial and error. In the preceding exercise, however, this method was found to be slow and tedious. Let us try to make an algebraic argument that will be quicker and more efficient. Note that a general pair (x,y) is equivalent to (1,5) if and only if its entries satisfy the definition

$$x + 5 = y + 1 \tag{1}$$

This equation reduces to $x + 4 = y$. If for x we substitute, in order, the elements 1, 2, 3, 4, and 5, then for y we get the corresponding elements 5, 1, 2, 3, and 4. All pairs composed of the associated x and y entries are equivalent to (1,5) because their elements meet criterion (1). Hence,

$$(1,5) \sim (2,1) \sim (3,2) \sim (4,3) \sim (5,4)$$

EXERCISE 6.2

1. (a) Copy the table of pairs of elements.
   (b) Circle all those that are equivalent to (5,5).
   (c) Cross with a diagonal line \ all those that are equivalent to (1,5).
2. Draw a line above those that are equivalent to (2,5).
3. Cross with a diagonal line / all those that are equivalent to (3,5).
4. Underline all those that are equivalent to (4,5).
5. Are any pairs left unmarked?
6. Justify the statement that there are only five nonequivalent pairs in this system.

6.3 EQUIVALANCE SETS OF PAIRS

Since there are only five nonequivalent pairs in the second system of pairs, why should we deal with all 25? Why not let one pair represent all the pairs equivalent to it? For example, why not let (5,5) represent the whole set of five pairs that were circled in Table 6.1? Of course, they could be represented just as well by (1,1) or (2,2) or any other one of them.

Each set of five pairs is called an *equivalence set* because it is composed of all and only those pairs that are equivalent to each other. Customary notation for the equivalence set of pair A is $[A]$ (read "bracket A"); thus, $[(5,5)]$—or, better, $[5,5]$—denotes the set circled in Table 6.1. Since $[1,1]$ denotes the same set, as does also $[2,2]$, we may say that all three equivalence sets are equal. Clearly,

$$[1,1] = [2,2] = [3,3] = [4,4] = [5,5]$$

Note that the pairs themselves are not identical, but the equivalence sets they represent are equal.

DEFINITION The aggregate of all and only those pairs that are equivalent to a given pair A is called the **equivalence set** of pair A and is denoted by $[A]$. Two equivalence sets are *equal* if and only if they contain exactly the same pairs.

Equality for equivalence sets is used in its customary sense of *identity*. Furthermore, when $[A] = [B]$, then pair A of the first symbol is equivalent to pair B of the second. The definition for equivalence of pairs now justifies the theorem below.

THEOREM 6.3-1 Let $A = (a_1,a_2)$ and $B = (b_1,b_2)$. Then $[A] = [B]$ or $[a_1,a_2] = [b_1,b_2]$ if and only if $a_1 + b_2 = a_2 + b_1$.

When the members of any set are listed, the list is generally enclosed in braces. Thus,

$$[5,5] = \{(1,1), (2,2), (3,3), (4,4), (5,5)\}$$

This enumeration tells us that $(4,4)$ is a pair of the equivalence set $[5.5]$. It also tells us that $(3,4)$ is not a member of this set because it is not listed within the braces. The symbol $\in$ is used to express membership; thus, $(4,4) \in [5,5]$ and $(3,4) \notin [5,5]$.

The seemingly unusual association that produced our definition of equivalence of pairs becomes clear when the pairs that belong to a given equivalence set are examined critically. Associate the general pair (a_1,a_2) with the difference of its entries $a_1 - a_2$; then all pairs in its equivalence set are associated with the same difference. For example, $(5,5)$ and $(4,4)$, corresponding to $5 - 5$ and $4 - 4$, respectively, are both associated with element 5. Pairs $(3,2)$ and $(5,4)$ have differences $3 - 2$ and $5 - 4$, respectively; both yield difference 1. According to this association, the criterion for two general pairs (a_1,a_2) and (b_1,b_2) to be in the same equivalence set is the equality $a_1 - a_2 = b_1 - b_2$. By Exercise 3.4.5, this relation can be rewritten as $a_1 + b_2 = a_2 + b_1$, which is the definition given in Section 6.1. It follows that, as an alternative definition of equivalence of pairs, we could assert that $(a_1,a_2) \sim (b_1,b_2)$ if and only if $a_1 - a_2 = b_1 - b_2$. We see that the "unusual association" has a reasonable explanation. It places in the same equivalence set all those pairs that share a specific difference for their first and second entries.

Let us return briefly to the first system of pairs and explain why in dealing with it the concept of equivalence sets was not introduced. Because pairs with dissimilar entries are unequal in that system, no two distinct pairs can be equivalent. Hence, each set would contain exactly one member. No new characteristics of the system would emerge from a study of the sets. In the second system, however, investigations of pairs and equivalence sets are mutually profitable.

EXERCISE 6.3

1. Let $A = (1,2)$ and $B = (2,3)$.
   (a) Why is $A \neq B$?
   (b) Why is $A \sim B$?
   (c) Why is $[A] = [B]$?
2. Write in words the meaning of each of the following statements.
   (a) $[3,1] = \{(3,1), (4,2), (5,3), (1,4), (2,5)\}$
   (b) $[4,1] = [1,3]$
   (c) $[2,3] \neq [5,2]$
   (d) $(1,5) \in [3,2]$
   (e) $(2,5) \notin [4,1]$
3. Enumerate each of the three equivalence sets [1,5], [3,5], and [4,5].
4. The statements below test correctness in use of notation. Determine whether each of them is true, false, or without meaning. Justify your answers.

   (a) $(2,3) \in [3,1]$ (b) $(3,2) \notin [1,3]$
   (c) $[2,1] \in (2,1)$ (d) $[3,4] = [5,3]$
   (e) $[2,5] \neq [4,1]$ (f) $(1,1) = [2,2]$

6.4 ADDITION IN THE SECOND SYSTEM OF PAIRS

The fact that pairs are equivalent in sets of five requires that any definition proposed for addition of pairs must be tested by the criterion that additions of equivalent pairs shall give equivalent answers. Suppose that by a suggested definition a specific pair from $[A]$ added to a specific pair from $[B]$ gives a sum that is a member of $[C]$. Then it must be determined whether the addition of any pair whatever from $[A]$ and any pair whatever from $[B]$ always gives a pair that is a member of $[C]$. If it does not, then the proposed definition is clearly unsatisfactory and must be discarded. If it does and if, further, this same result holds for all equivalence sets in the system, then the definition is acceptable. In mathematical language, addition is said to be well defined.

Let us test our accustomed method for addition of pairs.

DEFINITION For pairs $A = (a_1,a_2)$ and $B = (b_1,b_2)$, $A + B \equiv (a_1 + b_1, a_2 + b_2)$; that is, $(a_1,a_2) + (b_1,b_2) = (a_1 + b_1,\ a_2 + b_2)$.

Let us choose addends from [3,5] and [1,5], respectively. Three such choices yield

$$\begin{aligned}(1,3) + (2,1) &= (3,4)\\(3,5) + (4,3) &= (2,3)\\(4,1) + (1,5) &= (5,1)\end{aligned}$$

Every one of the sums belongs in [4,5]. We are, therefore, encouraged to believe that whenever we add an element of [3,5] and an element of [1,5] we will get a sum in [4,5]. Furthermore, we hope that a similar situation will exist for additions using elements from any of the equivalence sets. How shall we determine whether this is so? We could, of course, make all 625 additions of two pairs each and study the resulting sums. If, however, we are adept with algebraic notation, a theoretical discussion will be both quicker and easier. We wish to know whether the following statement is true or false.

STATEMENT 1 For pairs A, B, C, and D, if A and C belong to the same equivalence set and if B and D belong to the same equivalence set, then the sums $A + B$ and $C + D$ are members of the same equivalence set. In symbols: if $[A] = [C]$ and $[B] = [D]$, then $[A + B] = [C + D]$.

Theorem 6.3-1 suggests a method for rewriting Statement 1 in terms of elements. For example, if (a_1,a_2) and (c_1,c_2) belong to the same equivalence set, then $a_1 + c_2 = a_2 + c_1$. Similarly, if $(a_1 + b_1,\ a_2 + b_2)$ and $(c_1 + d_1,\ c_2 + d_2)$ belong to the same equivalence set, then $(a_1 + b_1) + (c_2 + d_2) = (a_2 + b_2) + (c_1 + d_1)$.

STATEMENT 2 For pairs $A = (a_1,a_2)$, $B = (b_1,b_2)$, $C = (c_1,c_2)$, and $D = (d_1,d_2)$, and sums $A + B = (a_1 + b_1,\ a_2 + b_2)$ and $C + D = (c_1 + d_1,\ c_2 + d_2)$, if $a_1 + c_2 = a_2 + c_1$ and $b_1 + d_2 = b_2 + d_1$, then $(a_1 + b_1) + (c_2 + d_2) = (a_2 + b_2) + (c_1 + d_1)$.

Because addition of the two equalities in the "if" clause produces the desired equality in the "then" clause, we know that Statement 2 is true. Hence, Statement 1 is true and addition by this method is well defined.

THEOREM 6.4-1 *Law of Addition for Equality.* For pairs A, B, C, and D, if $[A] = [C]$ and $[B] = [D]$, then $[A + B] = [C + D]$.

We must now ask whether our method for addition of pairs is in accord with the association described above between a pair and the difference of its entries. We know that the definition

$$(a_1,a_2) + (b_1,b_2) = (a_1 + b_1,\ a_2 + b_2)$$

will be consistent with the described association only if the following equality holds:

$$(a_1 - a_2) + (b_1 - b_2) = (a_1 + b_1) - (a_2 + b_2)$$

In Exercise 3.4 you showed that the equality holds in the five-element system. Hence, the definition of addition of pairs preserves the association.

EXERCISE 6.4

1. Determine which of the following statements are true and which are false.

(a) $(3,5) + (2,2) \sim (5,2)$ (b) $(1,3) + (4,2) \sim (3,3)$
(c) $(2,4) + (3,5) \sim (1,5)$ (d) $(2,3) + (1,1) \sim (4,3)$
(e) $(3,3) + (1,4) \sim (2,5)$ (f) $(2,1) + (3,4) \sim (1,3)$

2. Determine which of the statements below are true and which are false.

(a) $(2,1) + (2,4) \sim (4,2) + (1,4)$ (b) $(3,2) + (1,5) \sim (2,3) + (5,1)$
(c) $(3,1) + (5,3) \sim (4,5) + (4,3)$ (d) $(1,4) + (2,3) \sim (2,5) + (3,4)$

3. Illustrate Theorem 6.4-1 using the pairs below.

(a) $A = (1,1)$, $B = (2,4)$, $C = (3,3)$, $D = (4,1)$
(b) $A = (3,1)$, $B = (5,2)$, $C = (4,2)$, $D = (2,4)$

4. Given the statement $[4,2] + [2,4] = [1,1]$:

(a) Interpret the statement in words.
(b) Determine whether the statement is true or false.

5. Determine which of the following statements are true and which are false.

(a) $[3,4] + [1,1] = [4,3]$ (b) $[2,2] + [3,1] = [3,1]$
(c) $[5,4] + [3,4] = [3,5]$ (d) $[1,3] + [4,4] = [5,2]$
(e) $[2,4] + [4,2] = [3,3]$ (f) $[4,1] + [4,1] = [2,1]$

6.5 A SYSTEM OF EQUIVALENCE SETS OF PAIRS OF THE SECOND SYSTEM

Results in the preceding section suggest that the second system of 25 pairs may be most easily handled as a system of five equivalence sets. Let us adapt the definitions of the preceding sections and then study addition by means of a table.

DEFINITION The sets of the system of equivalence sets will customarily be designated as [1,5], [2,5], [3,5], [4,5], [5,5].

THEOREM 6.5-1 For equivalence sets $[a,5]$ and $[b,5]$, $[a,5] = [b,5]$ if and only if $a = b$; $[a,5] + [b,5] = [a + b, 5]$.

EXERCISE 6.5

1. Copy and complete Table 6.2 for addition of equivalence sets.

Table 6.2 Addition Table for Equivalence Sets

+	[1,5]	[2,5]	[3,5]	[4,5]	[5,5]
[1,5]					
[2,5]					
[3,5]					
[4,5]					
[5,5]					

2. Make a new table by applying the following substitution of symbols to the table constructed in Exercise 1.

Equivalence set	[1,5]	[2,5]	[3,5]	[4,5]	[5,5]
Element	1	2	3	4	5

Note that this substitution of symbols is a very natural one. It makes formal acknowledgment of the association between general pair $(a,5)$ and the difference of its entries, namely, $a - 5 = a$.

6.6 AN ISOMORPHISM

The table constructed in Exercise 6.5.2 is identical with the addition table for the five-element system. It follows that, except for notation, the table in Exercise 6.5.1 is also identical with the element addition table. The additive properties of the equivalence-set system must, therefore, be the same as those of the element system. Hence, the set system forms a commutative group under addition. If we make use of the correspondence described in Exercise 6.5.2, we can make specific assertions.

(1) The set [5,5] is the unique identity set for addition.

(2) Each set $[A]$ has a unique additive inverse $[A]^*$. In particular, $[a,5]^* = [a^*,5]$.

(3) Every equation $[X] + [B] = [A]$ has the unique solution $[X] = [A] + [B]^*$.

(4) For sets $[A]$, $[B]$, and $[C]$, if $[A] + [B] = [A] + [C]$, then $[B] = [C]$.

As corollaries to statement (2) above we have the expected properties of the additive inverse, namely,

(2a) For sets $[A]$ and $[B]$, $[A]^{**} = [A]$; $([A] + [B])^* = [A]^* + [B]^*$; and if $[A] = [B]^*$, then $[A]^* = [B]$.

The element system and the set system are of exactly the same form. They differ in notation and in the manner of their construction, but they do not differ in their properties under addition. The systems are well characterized by the adjective *isomorphic*, which means "the same in form." A mathematician would describe our situation by saying that under the correspondence between element a and equivalence set $[a,5]$, the additive group of elements is isomorphic with the additive group of equivalence sets.

EXERCISE 6.6

1. Find equivalence set $[X]$ in each of the following equations.

(a) $[X] + [1,5] = [5,5]$ (b) $[X] + [2,5] = [3,5] + [2,5]$
(c) $[X]^* + [4,5] = [4,5] + [1,5]$ (d) $[X] + [5,5] = [2,5]$
(e) $[X] + [3,5] = [1,5]$ (f) $[X]^* + [4,5] = [5,5]$

2. Determine whether each of the statements below is true or false.

(a) If $[A] = [4,5]^*$, then $[4,5] = [A]^*$.
(b) If $[A] + [B] = [5,5]$, then $[A] = [B]^*$.
(c) If $[A] + [5,5] = [B]$, then $[A] = [B]$.
(d) $[2,5] + [3,5]^* = [2,5]^* + [3,5]$.

6.7 MULTIPLICATION IN THE SECOND SYSTEM OF PAIRS

Multiplication in the second system of pairs must be defined in such a way that if a pair from $[A]$ and a pair from $[B]$ have their product pair in $[C]$, then any pair whatever from $[A]$ and any pair whatever from $[B]$ must have their product pair in $[C]$. We test our accustomed definition: for pairs $A = (a_1,a_2)$ and $B = (b_1,b_2)$, $AB = (a_1b_1,a_2b_2)$. Consider the products

$$(1,3)(2,1) = (2,3)$$
$$(3,5)(4,3) = (2,5)$$
$$(4,1)(5,4) = (5,4)$$

In each case the first factor is a member of $[3,5]$ and the second of $[1,5]$. The products, however, lie in $[4,5]$, $[2,5]$, and $[1,5]$. Because these specific products do not conform to the criterion, the rule under which they are formed is not satisfactory and must be discarded. What shall we use in its place? Let us try the formula that results when we interpret the pairs as differences of their entries. The product $(a_1,a_2)(b_1,b_2)$ corresponds with $(a_1 - a_2)(b_1 - b_2)$. The formula in Exercise 3.4.7 shows that this last product is equal to the difference $(a_1b_1 + a_2b_2) - (a_1b_2 + a_2b_1)$. Translation into the language of pairs yields $(a_1b_1 + a_2b_2,\ a_1b_2 + a_2b_1)$.

DEFINITION For pairs $A = (a_1,a_2)$ and $B = (b_1,b_2)$ the **product**

$$AB = (a_1,a_2)(b_1,b_2) = (a_1b_1 + a_2b_2,\ a_1b_2 + a_2b_1)$$

Because of the mathematical analysis by which we determined the formula, we know that multiplication is well defined.

THEOREM 6.7-1 *Law of Multiplication for Equality.* For pairs A, B, C, and D, if $[A] = [C]$ and $[B] = [D]$, then $[AB] = [CD]$.

For the pairs considered above, all products must lie in the same equivalence set. Calculations confirm the fact that all lie in set [3,5].

EXERCISE 6.7

1. Determine which of the following statements are true and which are false.

(a) $(1,4)(3,5) \sim (3,2)$ (b) $(3,1)(2,4) \sim (1,5)$
(c) $(2,3)(3,2) \sim (1,4)$ (d) $(4,1)(2,1) \sim (3,1)$
(e) $(5,2)(2,3) \sim (3,1)$ (f) $(1,3)(4,4) \sim (4,4)$

2. Determine which of the following statements are true and which are false.

(a) $(2,1)(2,4) \sim (1,1)(4,4)$ (b) $(3,2)(1,5) \sim (2,3)(5,1)$
(c) $(1,3)(5,3) \sim (4,1)(4,5)$ (d) $(1,4)(2,3) \sim (2,5)(3,4)$

3. Illustrate Theorem 6.7-1 using the pairs of Exercise 6.4.3.

4. Given the statement $[1,4][2,2] = [1,1]$:
(a) Interpret the statement in words.
(b) Determine whether the statement is true or false.

5. Determine which of the following statements are true and which are false.

(a) $[4,2][3,5] = [3,2]$ (b) $[5,1][4,3] = [2,5]$
(c) $[3,5][2,1] = [3,1]$ (d) $[1,3][3,4] = [4,2]$
(e) $[2,4][2,2] = [3,3]$ (f) $[2,5][1,3] = [5,4]$

6.8 MULTIPLICATION OF EQUIVALENCE SETS OF PAIRS OF THE SECOND SYSTEM

When multiplication for pairs is interpreted for the system of equivalence sets, we get the following theorem.

THEOREM 6.8-1 For sets $[a,5]$ and $[b,5]$, the product $[a,5] \cdot [b,5] = [ab,5]$.

Let us study multiplication of sets by working problems.

EXERCISE 6.8

1. Construct a multiplication table for equivalence sets.

2. Make a new table as in Exercise 6.5.2.

3. Compare the table of preceding Exercise 2 with the multiplication table for the five-element system prepared in Exercise 2.2. (If all work has been correctly done, the two tables will be identical.)

4. Explain why the properties of the set system under multiplication must be the same as those of the element system.

5. What special properties has the identity set for addition?

6. Which set is the identity set for multiplication?

7. Discuss the multiplicative inverses of the sets.

8. What group properties has the set system under multiplication?

9. In each of the following equalities, find set $[X]$.

(a) $[3,5][X] = [1,5]$ (b) $[2,5][X] = [2,5][3,5]$
(c) $[X]'[3,5] = [3,5][4,5]$ (d) $[1,5][X] = [2,5]$
(e) $[X][4,5]' = [2,5]$ (f) $[2,5][X] = [5,5]$

10. Determine whether each of the following statements is true or false.
(a) $[2,5][3,5] = [3,5][2,5]$.
(b) If $[A] = [3,5]'$, then $[A]' = [3,5]$.
(c) If $[A][B]' = [1,5]$, then $[A] = [B]$.
(d) If $[X][5,5] = [3,5][5,5]$, then solution $[X] = [3,5]$ is unique.
(e) If $[A][1,5] = [B]$, then $[A]' = [B]'$.
(f) $[2,5][4,5]' = [2,5]'[4,5]$.

6.9 STRUCTURE OF THE SYSTEM OF EQUIVALENCE SETS

The addition and multiplication tables for the system of equivalence sets can be transformed into the corresponding tables for the five-element system, and vice versa. It follows that whenever properties are derived for the five-element system by use of the tables, then analogous properties hold for the system of sets. The sets, therefore, are a commutative group under addition and also, with the exception of the additive identity, under multiplication. Recall that distributivity for the elements was verified by using the entries in the tables. Hence, for the sets also, multiplication is distributive with respect to addition.

It is now evident that after long analysis, involving the introduction of much notation, we have the simple result that the second system of pairs is not really new but is essentially the five-element system in different notation. The pairs are separable into exactly five distinct equivalence sets which faithfully reproduce the properties of the parent system of elements.

We may react to this outcome either with disappointment that all our work seems to have produced nothing new or with pleasure at the power of an analysis that can transform one of two apparently disparate systems into the other. Whether we are pleased or disappointed, we may still find satisfaction in a useful consequence when, in the next two sections, an analogous system of pairs of natural numbers is set up. We shall then find that when a correspondence is constructed between the equivalence sets and the numbers, many sets remain unassociated. In this case the new system does not faithfully copy the parent system but has new elements and properties of its own.

6.10 PAIRS OF NATURAL NUMBERS (OPTIONAL)

Under addition the natural numbers (see Section 1.1) have closure, commutativity, associativity, and a law of cancellation. There is, however, no additive identity; hence, they cannot form a group under addition. The operation of multiplication has all the properties that hold for addition as well as the existence of an identity, namely, the natural number 1. Only this identity possesses a multiplicative inverse. Multiplication is distributive with respect to addition.

Ordered pairs of natural numbers can be formed and the usual notation employed to designate them; thus, (3,17) and (41,22) are specific pairs, and $N = (n_1,n_2)$ is a general pair, where n_1 and n_2 are natural numbers. Let us define equivalence as for the second system of pairs.

DEFINITION For pairs of natural numbers $M = (m_1,m_2)$ and $N = (n_1,n_2)$, $M \sim N$ if and only if $m_1 + n_2 = m_2 + n_1$.

Note, first, that all pairs of the form (n,n) are equivalent. For example, $(n,n) \sim (1,1)$ because $n + 1 = n + 1$. We may write $(1,1) \sim (2,2) \sim (3,3) \sim \cdots \sim (n,n) \sim \cdots$, where the three dots preceding the pair (n,n) indicate that pairs may have been omitted.

Let us next consider all pairs of form $(n, 1+n)$. Any two such pairs $(n, 1+n)$ and $(m, 1+m)$ are equivalent because $n + (1 + m) = (1 + n) + m$. We can determine these pairs by letting $n = 1, 2, 3, 4, \ldots$, in turn. We get $(1,2) \sim (2,3) \sim (3,4) \sim \cdots \sim (n, n+1) \sim \cdots$. Similarly, all pairs of the form $(1 + n, n)$ are equivalent; that is, $(2,1) \sim (3,2) \sim (4,3) \sim \cdots \sim (1+n, n) \sim \cdots$.

Continuing in this manner, we discover that pairs of natural numbers can be separated into equivalence sets as listed in Table 6.3. The sets contain every possible pair of natural numbers.

Table 6.3 Equivalence Sets of Pairs of Natural Numbers

. . .

$\{(1, m+1),\quad (2, m+2),\quad (3, m+3),\quad \ldots,\quad (n, m+n),\quad \ldots\} = [1, 1+m]$

. . .

$\{(1,4),\quad (2,5),\quad (3,6),\quad \ldots,\quad (n, 3+n),\quad \ldots\} = [1,4]$

$\{(1,3),\quad (2,4),\quad (3,5),\quad \ldots,\quad (n, 2+n),\quad \ldots\} = [1,3]$

$\{(1,2),\quad (2,3),\quad (3,4),\quad \ldots,\quad (n, 1+n),\quad \ldots\} = [1,2]$

$\{(1,1),\quad (2,2),\quad (3,3),\quad \ldots,\quad (n,n),\quad \ldots\} = [1,1]$

$\{(2,1),\quad (3,2),\quad (4,3),\quad \ldots,\quad (n+1, n),\quad \ldots\} = [2,1]$

$\{(3,1),\quad (4,2),\quad (5,3),\quad \ldots,\quad (n+2, n),\quad \ldots\} = [3,1]$

$\{(4,1),\quad (5,2),\quad (6,3),\quad \ldots,\quad (n+3, n),\quad \ldots\} = [4,1]$

. . .

$\{(m+1, 1),\quad (m+2, 2),\quad (m+3, 3),\quad \ldots,\quad (m+n, n),\quad \ldots\} = [1+m, 1]$

. . .

6.11 THE SYSTEM OF EQUIVALENCE SETS OF PAIRS OF NATURAL NUMBERS (OPTIONAL)

Addition and multiplication for pairs of natural numbers are defined by formulas similar to those employed for the second system of pairs of elements, namely,

$$\begin{aligned} (m_1,m_2) + (n_1,n_2) &= (m_1 + n_1,\ m_2 + n_2) \\ (m_1,m_2)(n_1,n_2) &= (m_1n_1 + m_2n_2,\ m_1n_2 + m_2n_1) \end{aligned}$$

If in the arguments of Sections 6.4 and 6.7 we interpret the letters a, b, c, and d to be natural numbers, then we can prove that when pairs from given equivalence sets are added or multiplied, their sums or products lie in unique sets. Pairs of natural numbers may, therefore, be studied as a system of equivalence sets subject to the following laws:

$$\begin{aligned} [m_1,m_2] &= [n_1,n_2] \qquad \text{if and only if} \qquad m_1 + n_2 = m_2 + n_1 \\ [m_1,m_2] + [n_1,n_2] &= [m_1 + n_1,\ m_2 + n_2] \\ [m_1,m_2][n_1,n_2] &= [m_1n_1 + m_2n_2,\ m_1n_2 + m_2n_1] \end{aligned}$$

Let us denote the sets as in the list of Table 6.3, namely,

$$\ldots,\ [1, 1+m],\ \ldots,\ [1,2],\ [1,1],\ [2,1],\ \ldots,\ [1+m, 1],\ \ldots$$

In the system so defined multiplication is distributive with respect to addition and both operations have the properties of closure, commutativity, and associativity. The system is a group under addition since [1,1] is its identity and all other sets are inverses in couples, namely, $[1 + m, 1]$ with $[1, 1 + m]$ because $[1 + m, 1] + [1, 1 + m] = [1,1]$. Set [2,1] is the multiplicative identity because

$$[1, 1 + m][2,1] = [3 + m, 3 + 2m] = [1, 1 + m]$$

and

$$[1 + m, 1][2,1] = [3 + 2m, 3 + m] = [1 + m, 1]$$

Investigation shows that $[M][N] = [2,1]$ for only two sets $[M] = [N] = [2,1]$ and $[M] = [N] = [1,2]$. Hence, each of [2,1] and [1,2] is its own multiplicative inverse and no other set has such an inverse. The system fails to be a group under multiplication. The multiplicative property of the additive identity holds, however, and there is a cancellation law for multiplication. That is, for sets $[M]$ and $[N]$, the product $[M][N] = [1,1]$ if and only if $[M] = [1,1]$ or $[N] = [1,1]$; and for set $[K] \neq [1,1]$, if $[K][M] = [K][N]$, then $[M] = [N]$. This system of sets is a richer one than its parent system. In this respect it is more interesting than the system of sets of pairs of elements of the five-element system with which the chapter began. That system was as limited as its parent system.

Let us define a correspondence by the two-line formula below.

Equivalence set	[2,1]	[3,1]	[4,1]	. . .	$[1 + m, 1]$	. . .
Natural number	1	2	3		m	

Each set is associated with a unique number, and each number with a unique set. Does this correspondence establish a useful association between sets and numbers? We cannot construct addition and multiplication tables as we did in the case of elements. Yet, we need the same information about these sets as we had for element sets. For example, we learned from the addition table that if two sets are added and if the elements to which they correspond are added, then the sum set corresponds with the sum element. We must use an algebraic argument to determine whether the sets with which we are presently concerned have an analogous property. Let us call the corresponding sets and numbers *images* of each other. We wish to know whether the sum and the product of the images of two natural numbers are equal to the images of their sum and product, respectively. We may write the correspondence in terms of general sets and numbers.

Set	$[1 + m, 1]$	$[1 + n, 1]$	$[1 + m + n, 1]$	$[1 + mn, 1]$
Number	m	n	$m + n$	mn

We first test for addition. We have

$$\begin{aligned}(\text{image of } m) + (\text{image of } n) &= [1 + m,\ 1] + [1 + n,\ 1]\\ &= [2 + m + n,\ 2] = [1 + m + n,\ 1]\\ &= (\text{image of } m + n)\end{aligned}$$

We test for multiplication in the same way. When we form the product of the images of m and n we find

$$\begin{aligned}[1 + m,\ 1][1 + n,\ 1] &= [(1 + m)(1 + n) + 1,\ (1 + m) + (1 + n)]\\ &= [(1 + m + n) + mn + 1,\ (1 + m + n) + 1]\\ &= [1 + mn,\ 1]\end{aligned}$$

Since this last bracket is the image of mn, we have

$$(\text{image of } m) \cdot (\text{image of } n) = (\text{image of } mn)$$

It follows that the correspondence produces an isomorphism between the numbers and the sets under the operations of addition and multiplication.

Note, however, that the formula does not include all the sets. The isomorphism associates the natural numbers with the sets of form $[m,1]$ for $m \neq 1$. It does not assign a natural number to any of the sets of form $[1,m]$. In a case like this, mathematicians say that the natural numbers are *embedded* in the system of sets; that is, the natural numbers are isomorphic to a part of the set system.

How shall we characterize the whole system of equivalence sets of pairs of natural numbers? We note that it can be separated into three subcollections: one containing the sets isomorphic to the natural numbers, one containing the additive inverses of these sets, and one containing only the identity for addition. We recall another system which contains the natural numbers, an additive inverse for each, and an additive identity. It is the system of integers discussed in Section 1.13. We can now say that a set of form $[1 + n,\ 1]$ corresponds to the positive integer n and its additive inverse $[1,\ 1 + n]$ to the integer $-n$. The identity sets $[1,1]$ and $[2,1]$ are associated with the identities 0 and 1 of the integer system. Furthermore, the sets $[2,1]$ and $[1,2]$, which are the only ones that have multiplicative inverses, correspond with the integers 1 and -1, which are also unique in this property.

It now becomes evident that in the formation of the pairs and their equivalence sets we have begun with the natural numbers and produced the integers. Constructions of new number systems are among the most important mathematical uses of pairs.

EXERCISE 6.11

1. Prove that addition of equivalence sets is

(a) Commutative (b) Associative

2. Prove that multiplication of equivalence sets is

(a) Commutative (b) Associative

3. Prove that multiplication is distributive with respect to addition in the system of sets.

4. Verify the following equalities.

(a) $[1 + m, 1] + [1,1] = [1 + m, 1]$ (b) $[1, 1 + m] + [1,1] = [1, 1 + m]$
(c) $[n_1,n_2] + [1,1] = [n_1,n_2]$ (d) $[1 + m, 1] + [1, 1 + m] = [1,1]$
(e) $[n_1,n_2] + [n_2,n_1] = [1,1]$

5. Explain why Exercise 4(c) establishes [1,1] as the additive identity of the system of sets.

6. Explain why Exercise 4(e) shows that sets $[n_1,n_2]$ and $[n_2,n_1]$ are additive inverses of each other.

7. Verify the following equalities.

(a) $[n_1,n_2][2,1] = [n_1,n_2]$ (b) $[1,2][1,2] = [2,1]$ (c) $[2,1][2,1] = [2,1]$

8. Explain why Exercise 7(a) establishes [2,1] as the multiplicative identity of the system of sets.

9. Explain why Exercises 7(b) and (c) show that each of [1,2] and [2,1] is its own multiplicative inverse.

6.12 PAIRS OF PAIRS

We have studied two of the many possible systems of pairs of elements. Our discussion has opened new mathematical insights for us. Would the combination of pairs with pairs be as fruitful? By what notation should we indicate a pair of pairs? If we choose to write one pair above another, we get a square array

$$\begin{pmatrix} P \\ Q \end{pmatrix} = \begin{pmatrix} p_1 & p_2 \\ q_1 & q_2 \end{pmatrix}$$

If we wish, we can set up rules of equality, addition, and multiplication that make an interesting system of these new mathematical entities. Such arrays are discussed in the next chapter, where they are treated as quadruples of elements rather than as pairs of pairs.

7

A SYSTEM OF MATRICES

In this chapter we shall study a system each of whose members is a quadruple of elements arranged in a square pattern called a matrix. We shall see that our accustomed definitions and methods are generally effective for these complicated entities. We shall gain an acquaintance with a field of mathematics that is fruitful in application, and, finally, we shall make profitable use of matrix methods.

7.1 DEFINITION AND NOTATION

A matrix is an array of items called *elements*, placed in rows and columns. In particular, a second-order square matrix is an array of four elements in two rows and two columns. It is called square because it has as many rows as columns, and it is called second-order because it has two of each.

It is customary to denote an element of a matrix by means of a lowercase letter with two subscripts. The first subscript gives the number of the row in which the element lies, and the second gives the number of its column. Thus, a_{11} (read "*a* sub one one" or "*a* one one") is the element in the first row and the first column; a_{12} is located in the first row and second column; and a_{21} in the second row and first column. Finally, a_{22} lies in the second row and second column. The whole array is set off with parentheses and may be designated by the capital letter that accords with the lowercase letter used for its entries. Thus,

$$A = \begin{pmatrix} a_{11} & a_{12} \\ a_{21} & a_{22} \end{pmatrix}$$

Occasionally it is more convenient to use separate letters within the matrix. When no ambiguity could arise, we may employ a symbol such as

$$\begin{pmatrix} x & y \\ z & w \end{pmatrix}$$

Each matrix has two diagonals, one running from upper left to lower right and called the *principal diagonal,* the other running from lower left to upper right and called the *secondary diagonal.* In matrix A above, a_{11} and a_{22} are principal diagonal elements, while a_{21} and a_{12} are secondary diagonal elements.

Let us choose the entries for our matrices from the five-element system. Hence, the system which we are about to study is composed of second-order square matrices whose elements are members of the five-element system.

7.2 EQUALITY OF MATRICES

As the first step in our discussion, let us discriminate between equality and inequality of matrices. For equality let us use the simple definition that has served us so well.

DEFINITION Two **matrices are equal** if and only if corresponding elements are equal. In symbols, for matrices

$$A = \begin{pmatrix} a_{11} & a_{12} \\ a_{21} & a_{22} \end{pmatrix} \quad \text{and} \quad B = \begin{pmatrix} b_{11} & b_{12} \\ b_{21} & b_{22} \end{pmatrix}$$

$A = B$ if and only if $a_{11} = b_{11}$, $a_{12} = b_{12}$, $a_{21} = b_{21}$, and $a_{22} = b_{22}$.

Let us now determine how many matrices our system contains. With five choices of element for each of the four entries, there are 625 distinct matrices. Because there are so many, we will not attempt to list them or to make addition or multiplication tables. We will find properties by using characteristics of the parent five-element system.

EXERCISE 7.2

1. Find elements a_{11}, a_{12}, a_{21}, a_{22}:

(a) When matrix A of the definition is set equal to $\begin{pmatrix} 3 & 4 \\ 1 & 5 \end{pmatrix}$.

(b) When $\begin{pmatrix} a_{11} & a_{12} \\ 2 & 4 \end{pmatrix} = \begin{pmatrix} 5 & 3 \\ a_{21} & a_{22} \end{pmatrix}$.

2. Find all sets of equal matrices in the following list.

$$A = \begin{pmatrix} 2 & 3 \\ 4 & 1 \end{pmatrix} \qquad B = \begin{pmatrix} 3 & 2 \\ 1 & 4 \end{pmatrix} \qquad C = \begin{pmatrix} 1 & 3 \\ 2 & 1 \end{pmatrix}$$

$$D = \begin{pmatrix} 4 & 1 \\ 2 & 3 \end{pmatrix} \qquad E = \begin{pmatrix} 3 & 1 \\ 1 & 2 \end{pmatrix} \qquad K = \begin{pmatrix} 1 & 4 \\ 3 & 2 \end{pmatrix}$$

$$L = \begin{pmatrix} 1^* & 4^* \\ 3^* & 2^* \end{pmatrix} \qquad M = \begin{pmatrix} 1' & 2' \\ 3' & 1' \end{pmatrix} \qquad N = \begin{pmatrix} 1^* & 4' \\ 2' & 3 \end{pmatrix}$$

3. Determine in which, if any, matrices of Exercise 2:
   (a) The principal diagonal has its elements equal to each other.
   (b) The secondary diagonal has its elements equal to each other.

4. Determine which sets of matrices in Exercise 2 have:
   (a) Corresponding principal-diagonal elements equal.
   (b) Corresponding secondary-diagonal elements equal.

7.3 ADDITION OF MATRICES

The first operation to be discussed is, as usual, addition. The simplest way to define it is through addition of corresponding entries.

DEFINITION The **sum of two matrices** is a matrix whose elements are the sums of corresponding elements of the given matrices. In symbols:

$$\begin{pmatrix} a_{11} & a_{12} \\ a_{21} & a_{22} \end{pmatrix} + \begin{pmatrix} b_{11} & b_{12} \\ b_{21} & b_{22} \end{pmatrix} = \begin{pmatrix} a_{11} + b_{11} & a_{12} + b_{12} \\ a_{21} + b_{21} & a_{22} + b_{22} \end{pmatrix}$$

This definition permits immediate practical application. Here is an example (even though the use of the five-element system makes the numbers in it absurd). Let a wholesale house supply apples and oranges to food stores A and B. Let the order for the first week be two bags of apples and three of oranges for store A and one bag of apples and four of oranges for store B; then the order can be displayed in the following matrix form:

$$\begin{array}{l} \\ \text{Apples} \\ \text{Oranges} \end{array} \begin{array}{cc} \text{Store } A & \text{Store } B \\ \left(\begin{array}{c} 2 \\ 3 \end{array} \right. & \left. \begin{array}{c} 1 \\ 4 \end{array} \right) \end{array}$$

Let the order for the second week be displayed as follows:

$$\begin{array}{l} \\ \text{Apples} \\ \text{Oranges} \end{array} \begin{array}{cc} \text{Store } A & \text{Store } B \\ \left(\begin{array}{c} 4 \\ 2 \end{array} \right. & \left. \begin{array}{c} 3 \\ 2 \end{array} \right) \end{array}$$

The cumulative order for the first two weeks is given by the sum

$$\begin{pmatrix} 2 & 1 \\ 3 & 4 \end{pmatrix} + \begin{pmatrix} 4 & 3 \\ 2 & 2 \end{pmatrix} = \begin{pmatrix} 1 & 4 \\ 5 & 1 \end{pmatrix}$$

Hence, during the fortnight the wholesaler must supply one bag of apples and five of oranges to store A and four bags of apples and one of oranges to store B. Note how much simpler it is to display this information in the right-hand member of the equation above than it is to write it out.

In seeking properties of addition of matrices, the most efficient procedure is to apply laws of addition of elements to the individual entries of the matrices. Consider first the question of closure. By definition, the sum of two matrices is a matrix. Is it unique? The answer is in the affirmative if each of the four entries in the sum matrix is a unique element. Consider the entry $a_{11} + b_{11}$. Because a_{11} and b_{11} are both members of the five-element system which is closed under addition, the sum $a_{11} + b_{11}$ is a unique element; hence, there is one and only one correct entry for the upper left-hand corner of the matrix. A similar argument holds for each of the other entries. The system is closed under addition.

EXERCISE 7.3

1. Find the following sums of matrices.

(a) $\begin{pmatrix} 1 & 2 \\ 3 & 4 \end{pmatrix} + \begin{pmatrix} 2 & 3 \\ 5 & 1 \end{pmatrix}$ (b) $\begin{pmatrix} 5 & 4 \\ 1 & 3 \end{pmatrix} + \begin{pmatrix} 2 & 5 \\ 4 & 4 \end{pmatrix}$

In each of Exercises 2–10 find all elements x, y, z, and w for which the statement is true.

2. $\begin{pmatrix} 1 & 2 \\ 4 & 3 \end{pmatrix} + \begin{pmatrix} 3 & y \\ 2 & 5 \end{pmatrix} = \begin{pmatrix} 3 & y \\ 2 & 5 \end{pmatrix} + \begin{pmatrix} 1 & 2 \\ 4 & 3 \end{pmatrix}$

3. $\begin{pmatrix} 2 & 3 \\ 1 & 5 \end{pmatrix} + \begin{pmatrix} x & 4 \\ 3 & 2 \end{pmatrix} = \begin{pmatrix} x & 4 \\ 3 & 2 \end{pmatrix} + \begin{pmatrix} 2 & 3 \\ 1 & 5 \end{pmatrix}$

4. $\begin{pmatrix} 3 & 2 \\ 5 & 1 \end{pmatrix} + \begin{pmatrix} x & y \\ z & w \end{pmatrix} = \begin{pmatrix} 3 & 4 \\ 2 & 5 \end{pmatrix}$

5. $\begin{pmatrix} 1 & 4 \\ 2 & 3 \end{pmatrix} + \begin{pmatrix} x & 1 \\ 5 & w \end{pmatrix} = \begin{pmatrix} 2 & y \\ z & 3 \end{pmatrix}$

6. $\begin{pmatrix} 5 & 1 \\ 2 & 3 \end{pmatrix} + \begin{pmatrix} x & y \\ z & w \end{pmatrix} = \begin{pmatrix} 5 & 1 \\ 2 & 3 \end{pmatrix}$

7. $\begin{pmatrix} x & y \\ z & w \end{pmatrix} + \begin{pmatrix} 5 & 5 \\ 5 & 5 \end{pmatrix} \neq \begin{pmatrix} x & y \\ z & w \end{pmatrix}$

8. $\begin{pmatrix} 3 & 2 \\ 4 & 5 \end{pmatrix} + \begin{pmatrix} 3^* & 2^* \\ 4^* & 5^* \end{pmatrix} = \begin{pmatrix} x & y \\ z & w \end{pmatrix}$

9. $\begin{pmatrix} 2 & 3 \\ 4 & 1 \end{pmatrix} + \begin{pmatrix} x & y \\ z & w \end{pmatrix} = \begin{pmatrix} 5 & 5 \\ 5 & 5 \end{pmatrix}$

10. $\begin{pmatrix} 2 & 3 \\ 4 & 1 \end{pmatrix} + \begin{pmatrix} x^* & y^* \\ z^* & w^* \end{pmatrix} = \begin{pmatrix} 5 & 5 \\ 5 & 5 \end{pmatrix}$

In each of Exercises 11–13 find the sum of the matrices in the brackets and then use this sum as an addend to determine whether the equality is correct.

11. $\left[\begin{pmatrix} 1 & 2 \\ 3 & 4 \end{pmatrix} + \begin{pmatrix} 3 & 5 \\ 2 & 1 \end{pmatrix}\right] + \begin{pmatrix} 4 & 4 \\ 3 & 2 \end{pmatrix} = \begin{pmatrix} 1 & 2 \\ 3 & 4 \end{pmatrix} + \left[\begin{pmatrix} 3 & 5 \\ 2 & 1 \end{pmatrix} + \begin{pmatrix} 4 & 4 \\ 3 & 2 \end{pmatrix}\right]$

12. $\begin{pmatrix} 2 & 3 \\ 1 & 5 \end{pmatrix} + \left[\begin{pmatrix} 3 & 3 \\ 2 & 4 \end{pmatrix} + \begin{pmatrix} 5 & 4 \\ 3 & 1 \end{pmatrix}\right] = \left[\begin{pmatrix} 1 & 2 \\ 1 & 5 \end{pmatrix} + \begin{pmatrix} 3 & 3 \\ 2 & 4 \end{pmatrix}\right] + \begin{pmatrix} 5 & 4 \\ 3 & 1 \end{pmatrix}$

13. $\left[\begin{pmatrix} 3 & 4 \\ 3 & 2 \end{pmatrix} + \begin{pmatrix} 1 & 5 \\ 2 & 2 \end{pmatrix}\right] + \begin{pmatrix} 2 & 1 \\ 5 & 4 \end{pmatrix} = \left[\begin{pmatrix} 2 & 1 \\ 5 & 4 \end{pmatrix} + \begin{pmatrix} 1 & 5 \\ 2 & 2 \end{pmatrix}\right] + \begin{pmatrix} 3 & 4 \\ 3 & 2 \end{pmatrix}$

14. What property of addition of matrices is suggested by each of the following sets of exercises chosen from Exercises 2–12 above?

(a) 2, 3 (b) 6, 7 (c) 8, 9, 10 (d) 11, 12

7.4 THE ADDITIVE GROUP OF MATRICES

According to the preceding exercise, addition of matrices may be commutative and associative. Let us investigate the equality $A + B = B + A$. Displaying the matrices, we have

$$\begin{aligned} A + B &= \begin{pmatrix} a_{11} & a_{12} \\ a_{21} & a_{22} \end{pmatrix} + \begin{pmatrix} b_{11} & b_{12} \\ b_{21} & b_{22} \end{pmatrix} = \begin{pmatrix} a_{11} + b_{11} & a_{12} + b_{12} \\ a_{21} + b_{21} & a_{22} + b_{22} \end{pmatrix} \\ B + A &= \begin{pmatrix} b_{11} & b_{12} \\ b_{21} & b_{22} \end{pmatrix} + \begin{pmatrix} a_{11} & a_{12} \\ a_{21} & a_{22} \end{pmatrix} = \begin{pmatrix} b_{11} + a_{11} & b_{12} + a_{12} \\ b_{21} + a_{21} & b_{22} + a_{22} \end{pmatrix} \end{aligned} \tag{1}$$

By the definition of equality, the two sum matrices are equal if and only if

$$\begin{aligned} a_{11} + b_{11} &= b_{11} + a_{11} \qquad & a_{12} + b_{12} &= b_{12} + a_{12} \\ a_{21} + b_{21} &= b_{21} + a_{21} \qquad & a_{22} + b_{22} &= b_{22} + a_{22} \end{aligned} \tag{2}$$

Commutativity of addition in the five-element system establishes relations (2). These in turn show that corresponding entries in the sum matrices of formulas (1) are equal. Hence $A + B = B + A$.

To prove associativity of addition we must demonstrate that for matrices A, B, and C,

$$(A + B) + C = A + (B + C) \tag{3}$$

Let us reckon the sums indicated in the formula.

$$(A + B) + C = \begin{pmatrix} (a_{11} + b_{11}) + c_{11} & (a_{12} + b_{12}) + c_{12} \\ (a_{21} + b_{21}) + c_{21} & (a_{22} + b_{22}) + c_{22} \end{pmatrix}$$

$$A + (B + C) = \begin{pmatrix} a_{11} + (b_{11} + c_{11}) & a_{12} + (b_{12} + c_{12}) \\ a_{21} + (b_{21} + c_{21}) & a_{22} + (b_{22} + c_{22}) \end{pmatrix}$$

Corresponding entries in the final sum matrices are equal because addition is associative in the five-element system. Hence, the sum matrices are equal and we have the desired equality (3).

These arguments are reminiscent of those made for commutativity and associativity of addition in the first system of pairs. The proofs developed here differ from those in Sections 5.4 and 5.5 only in their consideration of four components instead of two. It is, therefore, not surprising that the similarity between pairs and matrices extends to the composition of identity and inverse. The matrix having 5 for every entry is an identity for addition. Let us call it the *five-matrix* and give it the appropriate designation F. Furthermore, each matrix has an additive inverse whose entries are the additive inverses of the corresponding entries in the given matrix. All group requirements have been met, and the following properties are assured for the matrix system.

THEOREM 7.4-1 The system of matrices forms a commutative group under addition.

THEOREM 7.4-2 The five-matrix

$$F = \begin{pmatrix} 5 & 5 \\ 5 & 5 \end{pmatrix}$$

is the unique identity for addition.

THEOREM 7.4-3 Every matrix A has a unique additive inverse designated by A^*.

$$\text{If} \quad A = \begin{pmatrix} a_{11} & a_{12} \\ a_{21} & a_{22} \end{pmatrix} \quad \text{then} \quad A^* = \begin{pmatrix} a_{11}^* & a_{12}^* \\ a_{21}^* & a_{22}^* \end{pmatrix}$$

Corollary 7.4-3 For matrices A and B, $A^{**} = A$; $(A + B)^* = A^* + B^*$; and if $A = B^*$, then $A^* = B$.

THEOREM 7.4-4 For matrices A and B, equation $B + X = A$ has the unique solution $X = B^* + A$.

THEOREM 7.4-5 *Law of Cancellation for Addition.* For matrices A, B, and C, if $A + B = A + C$, then $B = C$.

EXERCISE 7.4

1. Find the inverse under addition for each matrix below.

(a) $\begin{pmatrix} 2 & 3 \\ 5 & 4 \end{pmatrix}$ (b) $\begin{pmatrix} 3^* & 5 \\ 4^* & 1 \end{pmatrix}$ (c) $\begin{pmatrix} 4^* & 1^* \\ 2^* & 5^* \end{pmatrix}$ (d) $\begin{pmatrix} 2 & 5 \\ 4 & 3 \end{pmatrix}^*$

2. Solve each of the following equations for matrix X.

(a) $X + \begin{pmatrix} 2 & 3 \\ 5 & 4 \end{pmatrix} = F$

(b) $X + \begin{pmatrix} 1 & 4 \\ 3 & 2 \end{pmatrix}^* = \begin{pmatrix} 5 & 2 \\ 1 & 3 \end{pmatrix}$

(c) $X + \begin{pmatrix} 2 & 5 \\ 5 & 2 \end{pmatrix} = \begin{pmatrix} 1 & 5 \\ 5 & 1 \end{pmatrix}$

(d) $X^* + \begin{pmatrix} 3 & 3^* \\ 4 & 1 \end{pmatrix} = \begin{pmatrix} 2^* & 4 \\ 3 & 1^* \end{pmatrix} + \begin{pmatrix} 3 & 3^* \\ 4 & 1 \end{pmatrix}$

(e) $X + \begin{pmatrix} 1 & 2 \\ 3 & 4 \end{pmatrix}^* = \begin{pmatrix} 1^* & 2^* \\ 3^* & 4^* \end{pmatrix} + \begin{pmatrix} 4 & 5 \\ 2 & 1 \end{pmatrix}$

3. Show that F, the five-matrix, is its own additive inverse and is unique in this property.

4. Consider the sum of matrices A, B, and C. Explain why adding from left to right gives the same result as adding from right to left; that is, explain why $(A + B) + C = (C + B) + A$.

7.5 SCALAR MATRICES

Under addition the properties of the matrix system resemble those of the first system of pairs and of the parent five-element system. The dissimilarities between the systems lie chiefly in the total count of their members and the way in which their elements are written. One further distinction is of special interest. The additive group of elements has exactly one subgroup: that consisting of the single element 5. The additive groups of pairs and of matrices have other subgroups besides the single member F. Those of the matrix system are the most varied and interesting; we shall study some of them in this and the following sections.

A well-known subgroup of the matrix group has its members restricted by the following rules: $a_{11} = a_{22}$, $a_{12} = a_{21} = 5$. A typical matrix can be written

$$E = \begin{pmatrix} e & 5 \\ 5 & e \end{pmatrix} \tag{4}$$

where e is any element. These matrices are called *scalar matrices*. The entries in the principal diagonal are equal to each other and the entries in the secondary diagonal are the additive identity of the element system. Two matrices of special importance are F, where $e = 5$, and a matrix called I, where $e = 1$. How many scalar matrices are there? Only one element symbol appears in the type form, and since it may be any of the elements, there are precisely five matrices in this set. We should be able to study them by means of an addition table.

Table 7.1 Addition Table for Scalar Matrices

$+$	$\begin{pmatrix} 1 & 5 \\ 5 & 1 \end{pmatrix}$	$\begin{pmatrix} 2 & 5 \\ 5 & 2 \end{pmatrix}$	$\begin{pmatrix} 3 & 5 \\ 5 & 3 \end{pmatrix}$	$\begin{pmatrix} 4 & 5 \\ 5 & 4 \end{pmatrix}$	$\begin{pmatrix} 5 & 5 \\ 5 & 5 \end{pmatrix}$
$\begin{pmatrix} 1 & 5 \\ 5 & 1 \end{pmatrix}$	$\begin{pmatrix} 2 & 5 \\ 5 & 2 \end{pmatrix}$	$\begin{pmatrix} 3 & 5 \\ 5 & 3 \end{pmatrix}$	$\begin{pmatrix} 4 & 5 \\ 5 & 4 \end{pmatrix}$	$\begin{pmatrix} 5 & 5 \\ 5 & 5 \end{pmatrix}$	$\begin{pmatrix} 1 & 5 \\ 5 & 1 \end{pmatrix}$
$\begin{pmatrix} 2 & 5 \\ 5 & 2 \end{pmatrix}$	$\begin{pmatrix} 3 & 5 \\ 5 & 3 \end{pmatrix}$	$\begin{pmatrix} 4 & 5 \\ 5 & 4 \end{pmatrix}$	$\begin{pmatrix} 5 & 5 \\ 5 & 5 \end{pmatrix}$	$\begin{pmatrix} 1 & 5 \\ 5 & 1 \end{pmatrix}$	$\begin{pmatrix} 2 & 5 \\ 5 & 2 \end{pmatrix}$
$\begin{pmatrix} 3 & 5 \\ 5 & 3 \end{pmatrix}$	$\begin{pmatrix} 4 & 5 \\ 5 & 4 \end{pmatrix}$	$\begin{pmatrix} 5 & 5 \\ 5 & 5 \end{pmatrix}$	$\begin{pmatrix} 1 & 5 \\ 5 & 1 \end{pmatrix}$	$\begin{pmatrix} 2 & 5 \\ 5 & 2 \end{pmatrix}$	$\begin{pmatrix} 3 & 5 \\ 5 & 3 \end{pmatrix}$
$\begin{pmatrix} 4 & 5 \\ 5 & 4 \end{pmatrix}$	$\begin{pmatrix} 5 & 5 \\ 5 & 5 \end{pmatrix}$	$\begin{pmatrix} 1 & 5 \\ 5 & 1 \end{pmatrix}$	$\begin{pmatrix} 2 & 5 \\ 5 & 2 \end{pmatrix}$	$\begin{pmatrix} 3 & 5 \\ 5 & 3 \end{pmatrix}$	$\begin{pmatrix} 4 & 5 \\ 5 & 4 \end{pmatrix}$
$\begin{pmatrix} 5 & 5 \\ 5 & 5 \end{pmatrix}$	$\begin{pmatrix} 1 & 5 \\ 5 & 1 \end{pmatrix}$	$\begin{pmatrix} 2 & 5 \\ 5 & 2 \end{pmatrix}$	$\begin{pmatrix} 3 & 5 \\ 5 & 3 \end{pmatrix}$	$\begin{pmatrix} 4 & 5 \\ 5 & 4 \end{pmatrix}$	$\begin{pmatrix} 5 & 5 \\ 5 & 5 \end{pmatrix}$

You can easily verify the entries displayed in Table 7.1. Observation of the table suggests that a correspondence be set up with the five elements, as follows:

$$\begin{array}{lcccccc} \text{Matrix} \begin{pmatrix} e & 5 \\ 5 & e \end{pmatrix} & \begin{pmatrix} 1 & 5 \\ 5 & 1 \end{pmatrix} & \begin{pmatrix} 2 & 5 \\ 5 & 2 \end{pmatrix} & \begin{pmatrix} 3 & 5 \\ 5 & 3 \end{pmatrix} & \begin{pmatrix} 4 & 5 \\ 5 & 4 \end{pmatrix} & \begin{pmatrix} 5 & 5 \\ 5 & 5 \end{pmatrix} \\ \text{Element } e & 1 & 2 & 3 & 4 & 5 \end{array} \tag{5}$$

When the correspondence is made, the resulting table in terms of elements is identical with the addition table for the five-element system. It follows that, except for notation, Table 7.1 is also identical with the elements table. All characteristics of elements that arose from the addition table must, therefore, hold for the scalar matrices. Among these properties is that of being an additive group. Hence, scalar matrices are a subgroup of the group of all matrices.

It is clear that the systems of elements and of scalar matrices are the same in form. They are called *isomorphic* (see Section 6.6). In mathematical terms, two groups are isomorphic under a correspondence such as that described in formula (5), when each member of one has a correspondent or image in the other that exactly imitates its behavior with respect to the group operation. For example, a correspondence associating elements a and b with matrices A and B associates the element produced by adding a and b with the matrix produced by adding A and B, and vice versa. Two isomorphic groups possess identical properties with respect to their operations.

7.6 FURTHER SUBGROUPS OF THE ADDITIVE GROUP OF MATRICES (OPTIONAL)

Let us next consider a set of matrices formed according to the less restrictive rule: $a_{11} = a_{22}$, $a_{12} = a_{21}$. A typical matrix can be written

$$A = \begin{pmatrix} a_1 & a_2 \\ a_2 & a_1 \end{pmatrix} \tag{6}$$

where a_1 and a_2 are any elements. Note that F and I are both of this form: F when $a_1 = a_2 = 5$, and I when $a_1 = 1$ and $a_2 = 5$. Since the two symbols in the general matrix A can each be any of the five elements, there are 25 matrices in this set. An addition table would not be practicable. Let us make an algebraic investigation, beginning with the property of closure. We add two arbitrary matrices and find

$$A + B = \begin{pmatrix} a_1 & a_2 \\ a_2 & a_1 \end{pmatrix} + \begin{pmatrix} b_1 & b_2 \\ b_2 & b_1 \end{pmatrix} = \begin{pmatrix} a_1 + b_1 & a_2 + b_2 \\ a_2 + b_2 & a_1 + b_1 \end{pmatrix} \tag{7}$$

In the sum matrix, the entries in the principal diagonal are equal to each other, as are those in the secondary diagonal. Hence, they satisfy the restrictions laid down in form (6), and we have closure of addition.

Observation reveals a further important characteristic of this addition. If we should associate matrix A with pair (a_1,a_2) of the first pair system and matrix B with pair (b_1,b_2), then from formula (7) we could associate matrix $A + B$ with pair $(a_1 + b_1,\ a_2 + b_2)$. Each of the 25 matrices can be associated with one of the 25 pairs by the following addition-preserving correspondence:

$$\text{Matrix} \begin{pmatrix} a_1 & a_2 \\ a_2 & a_1 \end{pmatrix} \qquad \text{Pair } (a_1,a_2)$$

Such a correspondence is an isomorphism. Hence the matrices possess the same properties under addition as the first system of pairs. They form a commutative group, a subgroup of the matrix group.

In the preceding examples, the establishment of an isomorphism determined the characteristics of the new system. In the next example, the two criteria required by Theorem 2.11-1 will be investigated.

Consider matrices having three arbitrary element symbols:

$$A = \begin{pmatrix} 5 & a_1 \\ a_2 & a_3 \end{pmatrix} \tag{8}$$

There are 125 matrices in the set. If we add two general ones we get

$$A + B = \begin{pmatrix} 5 & a_1 \\ a_2 & a_3 \end{pmatrix} + \begin{pmatrix} 5 & b_1 \\ b_2 & b_3 \end{pmatrix} = \begin{pmatrix} 5 & a_1 + b_1 \\ a_2 + b_2 & a_3 + b_3 \end{pmatrix}$$

Clearly the sum matrix is of the proper form, and the property of closure holds for addition. Matrix

$$A^* = \begin{pmatrix} 5 & a_1^* \\ a_2^* & a_3^* \end{pmatrix}$$

is of form (8), so that every matrix A of the set has its additive inverse A^* also a member of the set. It follows by Theorem 2.11-1 that these 125 matrices form a commutative group, a subgroup of the matrix group.

EXERCISE 7.6

1. Let the typical matrix of a set have the form

$$\begin{pmatrix} a & 5 \\ 5 & a^* \end{pmatrix}$$

How many matrices are in the set? Construct an addition table and determine whether the set is a subgroup of the matrix system.

2. Consider matrices of the form

$$\begin{pmatrix} a_1 & 5 \\ 5 & a_2 \end{pmatrix}$$

Show that under addition they are isomorphic to the first system of pairs and hence are a subgroup of the matrix system.

★3. Use Theorem 2.11-1 to show that under addition the matrices of the form

$$\begin{pmatrix} a & b \\ c & a^* \end{pmatrix}$$

are a subgroup of the matrix system.

★**4.** Determine whether the sets of matrices whose type forms are given below are subgroups of the additive group of matrices.

$$\text{(a)}\quad \begin{pmatrix} a & 2a \\ 3a & 4a \end{pmatrix} \qquad \text{(b)}\quad \begin{pmatrix} a+1 & a \\ a & a+1 \end{pmatrix}$$

$$\text{(c)}\quad \begin{pmatrix} a_1 & a_1^* \\ a_2^* & a_2 \end{pmatrix} \qquad \text{(d)}\quad \begin{pmatrix} a_1 & 1 \\ 1 & a_2 \end{pmatrix}$$

7.7 MULTIPLICATION OF MATRICES

Our experience with the system of pairs warns us not to define multiplication of matrices by the simple expedient of multiplying corresponding entries; if we do, then every matrix different from F that contains at least one entry of 5 will be a divisor of F. Instead, let us use a definition which more nearly resembles that for the second system of pairs. Purely mathematical justifications for the method abound in any advanced text on the subject. Although it does not eliminate the occurrence of divisors of F, it gives matrices a wide usefulness.

DEFINITION The **product of matrices** A and B is defined to be the matrix formed according to the following rule:

$$\begin{aligned} AB &= \begin{pmatrix} a_{11} & a_{12} \\ a_{21} & a_{22} \end{pmatrix} \begin{pmatrix} b_{11} & b_{12} \\ b_{21} & b_{22} \end{pmatrix} \\ &= \begin{pmatrix} a_{11}b_{11} + a_{12}b_{21} & a_{11}b_{12} + a_{12}b_{22} \\ a_{21}b_{11} + a_{22}b_{21} & a_{21}b_{12} + a_{22}b_{22} \end{pmatrix} \end{aligned}$$

The element in the *first row* and *first column* of the product matrix is the sum of the products of the elements in the *first row* of matrix A with the corresponding elements in the *first column* of matrix B. The element in the *first row* and *second column* of the product matrix is the sum of the products of the elements in the *first row* of A with the corresponding elements in the *second column* of B. Similarly, the element in the *second row* and *first column* of the product matrix is the sum of the products of the elements in the *second row* of A with the corresponding elements in the *first column* of B. And, finally, the element in the *second row* and *second column* of the product matrix is the sum of the products of the elements in the *second row* of A with the corresponding elements in the *second column* of B.

EXERCISE 7.7

1. Show that the following products are correct.

$$\text{(a)}\quad \begin{pmatrix} a & b \\ c & d \end{pmatrix} \begin{pmatrix} e & f \\ g & h \end{pmatrix} = \begin{pmatrix} ae + bg & af + bh \\ ce + dg & cf + dh \end{pmatrix}$$

(b) $\begin{pmatrix} 1 & 2 \\ 3 & 4 \end{pmatrix}\begin{pmatrix} 5 & 4 \\ 3 & 2 \end{pmatrix} = \begin{pmatrix} 1 & 3 \\ 2 & 5 \end{pmatrix}$

(c) $\begin{pmatrix} 5 & 4 \\ 3 & 2 \end{pmatrix}\begin{pmatrix} 1 & 2 \\ 3 & 4 \end{pmatrix} = \begin{pmatrix} 2 & 1 \\ 4 & 4 \end{pmatrix}$

(d) $\begin{pmatrix} 1 & 3 \\ 4 & 2 \end{pmatrix}\begin{pmatrix} 1 & 1 \\ 3 & 3 \end{pmatrix} = \begin{pmatrix} 5 & 5 \\ 5 & 5 \end{pmatrix}$

(e) $\begin{pmatrix} 1 & 3 \\ 4 & 2 \end{pmatrix}\begin{pmatrix} 4 & 3 \\ 2 & 4 \end{pmatrix} = \begin{pmatrix} 5 & 5 \\ 5 & 5 \end{pmatrix}$

In each of Exercises 2–7 determine whether the statement is true or false.

2. $\begin{pmatrix} 2 & 1 \\ 4 & 2 \end{pmatrix}\begin{pmatrix} 1 & 1 \\ 1 & 1 \end{pmatrix} = \begin{pmatrix} 2 & 1 \\ 4 & 2 \end{pmatrix}\begin{pmatrix} 2 & 4 \\ 4 & 5 \end{pmatrix}$

3. $\begin{pmatrix} 3 & 4 \\ 2 & 1 \end{pmatrix}\begin{pmatrix} 1 & 5 \\ 5 & 1 \end{pmatrix} = \begin{pmatrix} 1 & 5 \\ 5 & 1 \end{pmatrix}\begin{pmatrix} 3 & 4 \\ 2 & 1 \end{pmatrix}$

4. $\begin{pmatrix} 1 & 3 \\ 3 & 4 \end{pmatrix}\begin{pmatrix} 1 & 3 \\ 4 & 2 \end{pmatrix} \neq \begin{pmatrix} 2 & 4 \\ 2 & 3 \end{pmatrix}\begin{pmatrix} 1 & 3 \\ 4 & 2 \end{pmatrix}$

5. $\begin{pmatrix} 4 & 1 \\ 2 & 3 \end{pmatrix}\begin{pmatrix} 3 & 5 \\ 2 & 1 \end{pmatrix} = \begin{pmatrix} 3 & 5 \\ 2 & 1 \end{pmatrix}\begin{pmatrix} 4 & 1 \\ 2 & 3 \end{pmatrix}$

6. $\begin{pmatrix} 2 & 3 \\ 4 & 2 \end{pmatrix}\begin{pmatrix} 1 & 2 \\ 3 & 4 \end{pmatrix} \neq \begin{pmatrix} 2 & 3 \\ 4 & 2 \end{pmatrix}\begin{pmatrix} 1 & 2 \\ 3 & 4 \end{pmatrix}$

7. $\begin{pmatrix} 3 & 5 \\ 4 & 2 \end{pmatrix}\begin{pmatrix} 2 & 5 \\ 1 & 3 \end{pmatrix} = \begin{pmatrix} 2 & 5 \\ 1 & 3 \end{pmatrix}\begin{pmatrix} 3 & 5 \\ 4 & 2 \end{pmatrix}$

In Exercises 8–11, first find the matrices in the brackets, then verify the equalities.

8. $\left[\begin{pmatrix} 1 & 2 \\ 2 & 3 \end{pmatrix}\begin{pmatrix} 5 & 3 \\ 2 & 4 \end{pmatrix}\right]\begin{pmatrix} 2 & 1 \\ 3 & 5 \end{pmatrix} = \begin{pmatrix} 1 & 2 \\ 2 & 3 \end{pmatrix}\left[\begin{pmatrix} 5 & 3 \\ 2 & 4 \end{pmatrix}\begin{pmatrix} 2 & 1 \\ 3 & 5 \end{pmatrix}\right]$

9. $\left[\begin{pmatrix} 3 & 4 \\ 2 & 1 \end{pmatrix}\begin{pmatrix} 2 & 5 \\ 1 & 3 \end{pmatrix}\right]\begin{pmatrix} 1 & 3 \\ 4 & 5 \end{pmatrix} = \begin{pmatrix} 3 & 4 \\ 2 & 1 \end{pmatrix}\left[\begin{pmatrix} 2 & 5 \\ 1 & 3 \end{pmatrix}\begin{pmatrix} 1 & 3 \\ 4 & 5 \end{pmatrix}\right]$

10. $\left[\begin{pmatrix} 3 & 2 \\ 1 & 4 \end{pmatrix} + \begin{pmatrix} 1 & 3 \\ 5 & 2 \end{pmatrix}\right]\begin{pmatrix} 2 & 4 \\ 3 & 1 \end{pmatrix} = \left[\begin{pmatrix} 3 & 2 \\ 1 & 4 \end{pmatrix}\begin{pmatrix} 2 & 4 \\ 3 & 1 \end{pmatrix}\right] + \left[\begin{pmatrix} 1 & 3 \\ 5 & 2 \end{pmatrix}\begin{pmatrix} 2 & 4 \\ 3 & 1 \end{pmatrix}\right]$

11. $\begin{pmatrix} 2 & 4 \\ 3 & 1 \end{pmatrix}\left[\begin{pmatrix} 3 & 2 \\ 1 & 4 \end{pmatrix} + \begin{pmatrix} 1 & 3 \\ 5 & 2 \end{pmatrix}\right] = \left[\begin{pmatrix} 2 & 4 \\ 3 & 1 \end{pmatrix}\begin{pmatrix} 3 & 2 \\ 1 & 4 \end{pmatrix}\right] + \left[\begin{pmatrix} 2 & 4 \\ 3 & 1 \end{pmatrix}\begin{pmatrix} 1 & 3 \\ 5 & 2 \end{pmatrix}\right]$

12. What property of multiplication of matrices is suggested or refuted by each of the following sets of exercises chosen from Exercises 1–11 above?

(a) 1(b), 1(c), 5, 7
(b) 8, 9
(c) 10, 11
(d) The product matrices in Exercise 3.

7.8 AN EXAMPLE

The operation of multiplication is more complicated in definition than in practice. An example illustrates its usefulness. Because the entries in the matrices are drawn from the five-element system, the numbers in the illustration are absurd, but the type of problem and the method of solution are typical of some encountered in practical applications.

In a world in which the five-element system is the usual and practical number system, a contractor is building two different housing developments, one at Urban and one at Polis. At Urban he puts up three houses of type A and two houses of type B. At Polis he plans two houses of type A and four of type B. Now, an A house requires four units of steel and three units of wood, whereas a B house requires one unit of steel and two of wood. The matrix whose rows are the numbers of houses of types A and B in the two developments is

$$\begin{array}{l} \\ \text{Urban} \\ \text{Polis} \end{array} \begin{array}{c} \;\;A \quad B \\ \begin{pmatrix} 3 & 2 \\ 2 & 4 \end{pmatrix} \end{array}$$

Let us form a second matrix whose first column is the amount of steel required for the two types of houses and whose second column is the amount of wood required:

$$\begin{array}{l} \\ A \\ B \end{array} \begin{array}{c} \text{Steel} \quad \text{Wood} \\ \begin{pmatrix} 4 & 3 \\ 1 & 2 \end{pmatrix} \end{array}$$

Then the product of these two matrices, in the proper order, records the amount of steel and wood that must be delivered to the two housing developments.

$$\begin{pmatrix} 3 & 2 \\ 2 & 4 \end{pmatrix}\begin{pmatrix} 4 & 3 \\ 1 & 2 \end{pmatrix} = \begin{array}{c} \text{Steel} \quad \text{Wood} \\ \begin{pmatrix} 4 & 3 \\ 2 & 4 \end{pmatrix} \end{array} \begin{array}{l} \\ \text{Urban} \\ \text{Polis} \end{array}$$

From the rows of the product matrix we see that four units of steel and three units of wood must be on hand at Urban, and two units of steel and four of wood at Polis. For those who have been distressed by the absurdity of the numbers in the answers, let us temporarily abandon the five-element system and use the natural number system. Now we have

$$\begin{pmatrix} 3 & 2 \\ 2 & 4 \end{pmatrix}\begin{pmatrix} 4 & 3 \\ 1 & 2 \end{pmatrix} = \begin{pmatrix} 14 & 13 \\ 12 & 14 \end{pmatrix}$$

It follows that 14 units of steel and 13 of wood must be on hand at the first development and 12 of steel and 14 of wood at the second. Note that the entries in the product matrix yield the desired answers because of the rule used for multiplication. Furthermore, the product matrix presents the results in a simple and clear form that is easy to interpret. These advantages remain when discussing any number of housing developments and any number of materials. Computers can be programmed to reckon the product matrix; hence, the desired results can be made quickly and easily available.

EXERCISE 7.8

1. Two different plans are being considered for an index of the cost of citrus fruits. According to plan A, the total cost of four bags of oranges and of two bags of grapefruit will be used. According to plan B, the total cost of three bags of each will be used. The cities of Urban and Polis are in a country in which the five-element system is the usual and practical number system. In Urban the price of a bag of oranges is \$4, while that of a bag of grapefruit is \$3. In Polis the prices are \$2 and \$4, respectively. Using matrices find the cost of citrus fruits in the two cities as measured by the two indices. (Suggestion: let the first row of the first matrix represent in order the number of bags of oranges and of grapefruit used in index A. Let the second row represent those in index B. Let the first column of the second matrix contain, in corresponding order, the prices of the two commodities in Urban. Let the second column contain prices in Polis. In order to interpret the product matrix, recall the rule for multiplication.)

2. Rework Exercise 1 using the natural numbers instead of the five-element system.

3. The Fibrox Company with plants in Urban and Polis (see Exercise 1 above) manufactures two dress materials named Artex and Bestex. Orders for October call for two bolts of Artex and three of Bestex to be made in the Urban plant, and for four bolts of each material to be made in the Polis plant. A bolt of Artex requires 1 pound of wool and 4 of cotton, while a bolt of Bestex requires 3 pounds of wool and 2 of cotton. Using matrices determine how many pounds of wool and of cotton must be available at each of the plants to meet the October orders.

4. Rework Exercise 3 using the natural numbers instead of the five-element system.

5. (a) In the city of Urban (see Exercise 1 above) the business firms of Martus and Tradus mount similar advertising campaigns using fancy paper hats and shopping bags. Martus plans to distribute four hats and three bags while Tradus plans for two hats and four bags. Each hat requires 3 yards of paper and 2 yards of cord. Each bag requires 4 yards of paper and 3 yards of cord. Use matrices to determine how many yards of paper and of cord are required by each business firm.

(b) Suppose that the necessary materials can be purchased from either of two supply houses, Qualo and Supro. Qualo charges 4 cents a yard for paper and 1 cent a yard for cord. Supro charges 3 cents and 2 cents, respectively, for these materials. Using this new information and the matrix of part (a) determine for each of the business firms the cost of materials when purchased at each of the supply houses.

[Suggestion: if the answer to part (a) is a matrix whose first row records the number of yards of paper and cord required for Martus and whose second row records the amounts for Tradus, then use a second matrix whose first column represents in order the cost of paper and cord per yard when supplied by Qualo and whose second column records the costs at Supro. In order to achieve the correct answer matrix and to interpret it, recall the rule for multiplication.]

★6. (a) Using the information in Exercise 5 determine the cost of one hat and of one bag when made of materials supplied by Qualo and Supro.

(b) Using the matrix of part (a) determine the cost of the advertising campaign for each of the business firms when materials are purchased at each of the supply houses. Your answers should be the same as those for Exercise 5(b).

7.9 PROPERTIES OF MULTIPLICATION OF MATRICES

A satisfactory definition of multiplication must provide closure. The symbolic definition in Section 7.7 does not ensure that the product matrix is unique. Consider the entry in the first row and first column. By closure of multiplication and addition of elements each of the product $a_{11}b_{11}$ and $a_{12}b_{21}$ is a unique element and so is their sum $a_{11}b_{11} + a_{12}b_{21}$. Consequently, this entry in the product matrix is a unique element. Similar reasoning shows that each of the other entries is a unique element of the five-element system. The product matrix is, therefore, a unique matrix of the matrix system.

THEOREM 7.9-1 The system of matrices is closed under multiplication.

Now let us look at commutativity. In the multiplications of Exercise 7.7 there were some cases in which the property holds and some cases in which it does not. For the system as a whole we lack the unanimity of result that would let us declare the operation commutative.

Will we fare any better when we consider associativity? Our experience in examples suggests that we may. A study of the two products $(AB)C$ and $A(BC)$ for general matrices establishes the property, although the complicated formula for the product of matrices introduces considerable notational difficulty into the calculations. Only students who enjoy algebraic manipulation should attempt Exercise 7.9.10 below, which calls for verification of the associative law. These students will find pleasure and satisfaction in the neatness with which the notation implements the calculations.

Let us now extend the usual definition for the multiplication of three factors to the matrix system and state the associative law.

DEFINITION For matrices A, B, and C, the product ABC is defined to be $(AB)C$.

THEOREM 7.9-2 In the system of matrices, multiplication is associative.

Exercises 7.7.10 and 7.7.11 suggest that multiplication of matrices may be both left and right distributive with respect to addition. We can prove that this conjecture is true. In doing so, the same types of notational difficulties are encountered as in the proof of associativity. For this reason, we again suggest that only students who enjoy algebraic manipulation attempt to answer Exercise 7.9.11 below, which calls for verification of the right-distributive law.

THEOREM 7.9-3 In the system of matrices, multiplication is distributive with respect to addition.

EXERCISE 7.9

1. Find the following products of matrices.

(a) $\begin{pmatrix} 1 & 2 \\ 2 & 3 \end{pmatrix}\begin{pmatrix} 3 & 4 \\ 4 & 5 \end{pmatrix}\begin{pmatrix} 1 & 3 \\ 2 & 4 \end{pmatrix}$ (b) $\begin{pmatrix} 2 & 1 \\ 3 & 4 \end{pmatrix}\begin{pmatrix} 1 & 5 \\ 5 & 1 \end{pmatrix}\begin{pmatrix} 2 & 3 \\ 5 & 2 \end{pmatrix}$

2. Let A be any matrix. Let E be a scalar matrix (see definition in Section 7.5).

(a) Show that the statement $AE = EA$ is true.

(b) Write a theorem on commutativity of multiplication that is established by the equality of (a).

(c) Describe the effect on any matrix A of multiplication on right or left by scalar matrix E.

3. (a) Write the scalar matrix in which $e = 1$. Denote this matrix by the symbol I.

(b) For a general matrix A, find AI; find IA.

(c) What do the results in (a) and (b) suggest concerning matrix I?

4. (a) Write the scalar matrix in which $e = 2$; in which $e = 3$.

(b) Multiply the two matrices of (a).

(c) What does the product found in (b) suggest about the matrices of (a)?

5. (a) Let $e \neq 5$. Multiply scalar matrix E by the matrix

$$\begin{pmatrix} e' & 5 \\ 5 & e' \end{pmatrix}$$

(b) What does the product found in (a) suggest concerning the two matrices that were multiplied?

6. When entry e is specialized to be 5, the general scalar matrix E becomes the matrix F. Find AF; find FA.

7. Show that for each pair of factors below, the product is F.

(a) $\begin{pmatrix} 2 & 1 \\ 3 & 4 \end{pmatrix}\begin{pmatrix} 1 & 3 \\ 3 & 4 \end{pmatrix}$ (b) $\begin{pmatrix} 1 & 3 \\ 3 & 4 \end{pmatrix}\begin{pmatrix} 2 & 1 \\ 1 & 3 \end{pmatrix}$

8. What do the answers to Exercises 6 and 7 suggest about the multiplicative properties of F?

9. The entries in our matrices are members of the five-element system in which multiplication is commutative. Explain why this property does not carry over to multiplication of matrices.

★★10. Prove that multiplication of matrices is associative. Use the customary notation, employing lowercase letters and subscripts. The proof will demonstrate the power of good notation in simplifying algebraic manipulation and in checking accidental errors.

★★11. (a) Using symbols, write a definition of the statement: multiplication of matrices is right distributive with respect to addition.

(b) Prove that the statement in part (a) is correct. Use notation as suggested in Exercise 10.

(c) Can we assert without further proof that multiplication of matrices is left distributive with respect to addition? Explain.

★★12. Because of the distributive law, the following relations can be expected to hold for additive inverses (see Section 2.14): $A^*B^* = AB$ and $(AB)^* = A^*B$ for matrices A and B. Show that this conjecture is correct.

7.10 SCALAR MATRICES UNDER MULTIPLICATION

The five scalar matrices are isomorphic under addition to the five elements by correspondence (5) in Section 7.5. It will now be shown that the correspondence defines an isomorphism also under multiplication. Following the method of our earlier study, we construct a multiplication table for the matrices. The entries in Table 7.2 can be easily verified. Under the correspondence, the matrix table transforms into the multiplication table for elements, and the isomorphism is established. Hence, the four matrices having $e \neq 5$ are a commutative group under multiplication. The usual con-

Table 7.2 Multiplication Table for Scalar Matrices

$\times$	$\begin{pmatrix}1&5\\5&1\end{pmatrix}$	$\begin{pmatrix}2&5\\5&2\end{pmatrix}$	$\begin{pmatrix}3&5\\5&3\end{pmatrix}$	$\begin{pmatrix}4&5\\5&4\end{pmatrix}$	$\begin{pmatrix}5&5\\5&5\end{pmatrix}$
$\begin{pmatrix}1&5\\5&1\end{pmatrix}$	$\begin{pmatrix}1&5\\5&1\end{pmatrix}$	$\begin{pmatrix}2&5\\5&2\end{pmatrix}$	$\begin{pmatrix}3&5\\5&3\end{pmatrix}$	$\begin{pmatrix}4&5\\5&4\end{pmatrix}$	$\begin{pmatrix}5&5\\5&5\end{pmatrix}$
$\begin{pmatrix}2&5\\5&2\end{pmatrix}$	$\begin{pmatrix}2&5\\5&2\end{pmatrix}$	$\begin{pmatrix}4&5\\5&4\end{pmatrix}$	$\begin{pmatrix}1&5\\5&1\end{pmatrix}$	$\begin{pmatrix}3&5\\5&3\end{pmatrix}$	$\begin{pmatrix}5&5\\5&5\end{pmatrix}$
$\begin{pmatrix}3&5\\5&3\end{pmatrix}$	$\begin{pmatrix}3&5\\5&3\end{pmatrix}$	$\begin{pmatrix}1&5\\5&1\end{pmatrix}$	$\begin{pmatrix}4&5\\5&4\end{pmatrix}$	$\begin{pmatrix}2&5\\5&2\end{pmatrix}$	$\begin{pmatrix}5&5\\5&5\end{pmatrix}$
$\begin{pmatrix}4&5\\5&4\end{pmatrix}$	$\begin{pmatrix}4&5\\5&4\end{pmatrix}$	$\begin{pmatrix}3&5\\5&3\end{pmatrix}$	$\begin{pmatrix}2&5\\5&2\end{pmatrix}$	$\begin{pmatrix}1&5\\5&1\end{pmatrix}$	$\begin{pmatrix}5&5\\5&5\end{pmatrix}$
$\begin{pmatrix}5&5\\5&5\end{pmatrix}$	$\begin{pmatrix}5&5\\5&5\end{pmatrix}$	$\begin{pmatrix}5&5\\5&5\end{pmatrix}$	$\begin{pmatrix}5&5\\5&5\end{pmatrix}$	$\begin{pmatrix}5&5\\5&5\end{pmatrix}$	$\begin{pmatrix}5&5\\5&5\end{pmatrix}$

sequences now hold. When these are combined with the information gained from Exercises 7.9.2–7.9.5, we get the theorems below.

THEOREM 7.10-1 Matrix

$$I = \begin{pmatrix} 1 & 5 \\ 5 & 1 \end{pmatrix}$$

is the unique multiplicative identity of the matrix system.

THEOREM 7.10-2 Under multiplication, the scalar matrix

$$E = \begin{pmatrix} e & 5 \\ 5 & e \end{pmatrix}$$

has the following properties:

(a) E is commutative with every matrix.

(b) For $A = \begin{pmatrix} a_{11} & a_{12} \\ a_{21} & a_{22} \end{pmatrix}$, $EA = AE = \begin{pmatrix} ea_{11} & ea_{12} \\ ea_{21} & ea_{22} \end{pmatrix}$.

(c) When $e \neq 5$, $E' = \begin{pmatrix} e' & 5 \\ 5 & e' \end{pmatrix}$.

With regard to the additive identity F we found the usual relations $AF = FA = F$ for every matrix A. It is not true, however, that when a product of matrices is F then at least one of the matrices is F. In Exercise 7.9.7 above, some pairs of matrices give the product F even though neither is F. Such matrices are called divisors of the additive identity (see the definition in Section 4.6).

THEOREM 7.10-3 For every matrix A, $AF = FA = F$. There are divisors of the additive identity F in the matrix system.

We will find that, as in the six-element system and the first system of pairs, one member of the matrix system will be the additive identity, some members will be divisors of that identity, and each of the remaining members will have a multiplicative inverse.

EXERCISE 7.10

1. Show that the following matrix equalities hold.

(a) $\begin{pmatrix} 1 & 2 \\ 3 & 1 \end{pmatrix}\begin{pmatrix} 4 & 3 \\ 1 & 2 \end{pmatrix} = \begin{pmatrix} 1 & 2 \\ 3 & 1 \end{pmatrix}$ (b) $\begin{pmatrix} 2 & 3 \\ 1 & 4 \end{pmatrix}\begin{pmatrix} 2 & 3 \\ 1 & 4 \end{pmatrix} = \begin{pmatrix} 2 & 3 \\ 1 & 4 \end{pmatrix}$

2. Do the equalities of Exercise 1 contradict the theorem which asserts that matrix I is the unique multiplicative identity? Justify your answer.

3. Show that each product below is equal to I.

(a) $\begin{pmatrix} 1 & 1 \\ 3 & 4 \end{pmatrix}\begin{pmatrix} 4 & 4 \\ 2 & 1 \end{pmatrix}$ (b) $\begin{pmatrix} 2 & 3 \\ 4 & 5 \end{pmatrix}\begin{pmatrix} 5 & 4 \\ 2 & 4 \end{pmatrix}$

(c) $\begin{pmatrix} 4 & 4 \\ 2 & 1 \end{pmatrix}\begin{pmatrix} 1 & 1 \\ 3 & 4 \end{pmatrix}$ (d) $\begin{pmatrix} 5 & 4 \\ 2 & 4 \end{pmatrix}\begin{pmatrix} 2 & 3 \\ 4 & 5 \end{pmatrix}$

4. List two pairs of matrices that are multiplicative inverses of each other.

5. Show that each product below is equal to F.

(a) $\begin{pmatrix} 1 & 3 \\ 4 & 2 \end{pmatrix}\begin{pmatrix} 2 & 2 \\ 1 & 1 \end{pmatrix}$ (b) $\begin{pmatrix} 4 & 3 \\ 3 & 1 \end{pmatrix}\begin{pmatrix} 1 & 2 \\ 2 & 4 \end{pmatrix}$

(c) $\begin{pmatrix} 1 & 2 \\ 2 & 4 \end{pmatrix}\begin{pmatrix} 3 & 1 \\ 1 & 2 \end{pmatrix}$ (d) $\begin{pmatrix} 1 & 1 \\ 4 & 4 \end{pmatrix}\begin{pmatrix} 1 & 3 \\ 4 & 2 \end{pmatrix}$

6. List some pairs of matrices that are divisors of F.

7. Find x, y, z, and w if

$$\begin{pmatrix} 1 & 5 \\ 5 & 2 \end{pmatrix}\begin{pmatrix} x & y \\ z & w \end{pmatrix} = \begin{pmatrix} 2 & 3 \\ 4 & 5 \end{pmatrix}$$

[Suggestion: multiply the matrices in the left member and get

$$\begin{pmatrix} x & y \\ 2z & 2w \end{pmatrix} = \begin{pmatrix} 2 & 3 \\ 4 & 5 \end{pmatrix}$$

Now use the definition of equality of matrices.]

8. In each equation below complete the second matrix by finding x, y, z, or w.

(a) $\begin{pmatrix} 1 & 3 \\ 2 & 1 \end{pmatrix}\begin{pmatrix} x & 5 \\ 5 & w \end{pmatrix} = \begin{pmatrix} 3 & 2 \\ 1 & 4 \end{pmatrix}$ (b) $\begin{pmatrix} 2 & 3 \\ 4 & 1 \end{pmatrix}\begin{pmatrix} 1 & y \\ z & 2 \end{pmatrix} = F$

(c) $\begin{pmatrix} 1 & 2 \\ 3 & 4 \end{pmatrix}\begin{pmatrix} x & 1 \\ 4 & w \end{pmatrix} = I$ (d) $\begin{pmatrix} 2 & 3 \\ 2 & 4 \end{pmatrix}\begin{pmatrix} 2 & y \\ z & 1 \end{pmatrix} = I$

9. The answer to Exercise 8(c) provides a matrix N such that $MN = I$, where

$$M = \begin{pmatrix} 1 & 2 \\ 3 & 4 \end{pmatrix}$$

Is it true that $NM = I$? If so, then N is a multiplicative inverse for M.

10. Using the answer to Exercise 8(d), determine whether the matrix

$$\begin{pmatrix} 2 & 3 \\ 2 & 4 \end{pmatrix}$$

has a multiplicative inverse.

7.11 DETERMINANTS OF MATRICES

Associated with matrix A is an element called *the determinant of* A and denoted by the symbol $|A|$. The determinant is the sum of products of the entries in A as described below, and can be written in various forms by using the relation $(ab)^* = a^*b = ab^*$ of Theorem 2.14-1. One of the advantages of our rule for multiplying matrices is the important property that the determinant of a product of matrices is the product of their determinants. Because of the notational difficulties in the verification of this statement, however, only students who enjoy algebraic manipulation should attempt Exercise 7.11.10.

DEFINITION For matrix

$$A = \begin{pmatrix} a_{11} & a_{12} \\ a_{21} & a_{22} \end{pmatrix}$$

the element $a_{11}a_{22} + (a_{21}a_{12})^*$ is called the **determinant** of A and is denoted by $|A|$.

THEOREM 7.11-1 For matrix A of the definition,

$$|A| = a_{11}a_{22} + a_{21}^*a_{12} = a_{11}a_{22} + a_{21}a_{12}^*$$

THEOREM 7.11-2 For matrices A and B, $|AB| = |A| \cdot |B|$.

Whenever matrix A has determinant 5, its product with any matrix B also has determinant 5 because $|AB| = |A| \cdot |B| = 5|B| = 5$. Furthermore, a product of matrices has determinant 5 only when at least one of the factors has such a determinant. Finally, the determinant of a product of matrices is different from 5 if and only if each matrix has a determinant different from 5. When a matrix has determinant 5 it is called *singular;* otherwise it is *nonsingular*. We give a formal definition, then summarize this discussion in Theorem 7.11-3.

DEFINITION Matrix A is **singular** if and only if $|A| = 5$. It is **nonsingular** if and only if $|A| \neq 5$.

THEOREM 7.11-3 A product of matrices is singular if and only if at least one of the factors is singular. A product of matrices is nonsingular if and only if all factors are nonsingular.

EXERCISE 7.11

1. Show that the matrices below all have determinant 3.

(a) $\begin{pmatrix} 1 & 2 \\ 3 & 4 \end{pmatrix}$ (b) $\begin{pmatrix} 1 & 1 \\ 2 & 5 \end{pmatrix}$ (c) $\begin{pmatrix} 2 & 3 \\ 1 & 3 \end{pmatrix}$ (d) $\begin{pmatrix} 3 & 5 \\ 4 & 1 \end{pmatrix}$

2. Find all elements x, y, z, and w for which each of the following statements is true.

(a) $A = \begin{pmatrix} 1 & 2 \\ 3 & w \end{pmatrix}$ and $|A| = 2$ (b) $B = \begin{pmatrix} 4 & y \\ 1 & 2 \end{pmatrix}$ and $|B| = 3$

(c) $C = \begin{pmatrix} x & 2 \\ 3 & 1 \end{pmatrix}$ and $|C|' = 2$ (d) $D = \begin{pmatrix} 4 & 1 \\ z & 2 \end{pmatrix}$ and $|D| = 5$

(e) $M = \begin{pmatrix} x & 4 \\ 2 & w \end{pmatrix}$, $x = w$, and $|M| = 1$

3. For matrices $A = \begin{pmatrix} 1 & 5 \\ 3 & 3 \end{pmatrix}$ and $B = \begin{pmatrix} 2 & 3 \\ 5 & 2 \end{pmatrix}$

(a) Find $|A|$; find $|B|$.
(b) Find $A + B$; find $|A + B|$.
(c) Determine whether $|A + B| = |A| + |B|$.
(d) Find AB; find $|AB|$.
(e) Show that $|AB| = |A| \cdot |B|$.

4. Repeat Exercise 3 for $A = \begin{pmatrix} 2 & 3 \\ 4 & 2 \end{pmatrix}$ and $B = \begin{pmatrix} 2 & 2 \\ 1 & 2 \end{pmatrix}$.

5. (a) Find $|I|$.
(b) Show that when $AA' = I$, then $|A| \cdot |A'| = 1$.
(c) What can be said of the relation between elements $|A|$ and $|A'|$?
(d) Is your conclusion in (c) consistent with the values of the determinants of the matrices in Exercise 7.10.3?
(e) Can a singular matrix have a multiplicative inverse? Justify your answer.

6. (a) Show that $|E| = e^2$ for scalar matrix E.
(b) Explain why all scalar matrices except F are nonsingular.

7. (a) Find $|F|$.
(b) Show that when $AB = F$, then $|A| \cdot |B| = 5$.
(c) Show that the relation in (b) is consistent with the values of the determinants of the matrices in Exercise 7.10.5.
(d) Can a pair of nonsingular matrices be divisors of F? Justify your answer.

8. Explain why the set of nonsingular matrices is closed under multiplication.

★9. Prove Theorem 7.11-1.

★★10. (a) Write matrices A and B in general terms. Write matrix AB.
(b) Find $|A|$, $|B|$, $|AB|$.
(c) Show that $|A| \cdot |B| = |AB|$.

7.12 THE ADJOINT MATRIX

In this section we shall find formulas for matrices that are divisors of the additive identity or else multiplicative inverses of each other. Let us begin by associating with every matrix a second matrix called its *adjoint.* To write the adjoint, we interchange the elements in the principal diagonal and replace those in the secondary diagonal by their additive inverses.

DEFINITION For matrix

$$A = \begin{pmatrix} a_{11} & a_{12} \\ a_{21} & a_{22} \end{pmatrix}$$

the **adjoint of** A, denoted by $\tilde{A}$ (read "A adjoint" or "A tilde"), is the matrix

$$\begin{pmatrix} a_{22} & a_{12}^* \\ a_{21}^* & a_{11} \end{pmatrix}$$

The following exercise will give you practice in finding and using adjoints.

EXERCISE 7.12-A

1. Show that the pairs of matrices indicated in the following parts of Exercise 7.10 are adjoints of each other.

(a) Exercise 3(a), 3(c). (b) Exercise 5(a), 5(b).
(c) Exercise 8(b).

2. For each pair of matrices in Exercise 1 find the product in the reverse order. What do your answers suggest concerning the product of a matrix with its adjoint?

3. For each matrix A below find the adjoint $\tilde{A}$, then find the products $A\tilde{A}$ and $\tilde{A}A$.

(a) $\begin{pmatrix} 2 & 3 \\ 4 & 5 \end{pmatrix}$ (b) $\begin{pmatrix} 1 & 3 \\ 2 & 1 \end{pmatrix}$ (c) $\begin{pmatrix} 1 & 3 \\ 1 & 2 \end{pmatrix}$

4. A matrix A for which $A = \tilde{A}$ is called self-adjoint. Show that F and I are self-adjoint.

In the exercises above you found the product of a matrix and its adjoint to be a scalar matrix. To show that this result is universal, we multiply general matrix A with its adjoint:

$$\begin{aligned} A\tilde{A} &= \begin{pmatrix} a_{11} & a_{12} \\ a_{21} & a_{22} \end{pmatrix} \begin{pmatrix} a_{22} & a_{12}^* \\ a_{21}^* & a_{11} \end{pmatrix} \\ &= \begin{pmatrix} a_{11}a_{22} + a_{12}a_{21}^* & a_{11}a_{12}^* + a_{12}a_{11} \\ a_{21}a_{22} + a_{22}a_{21}^* & a_{21}a_{12}^* + a_{22}a_{11} \end{pmatrix} \end{aligned}$$

Each entry in the principal diagonal of the product matrix is the determinant of A, and each entry in the secondary diagonal is 5. It follows that $A\tilde{A}$ is the scalar matrix

$$\begin{pmatrix} |A| & 5 \\ 5 & |A| \end{pmatrix}$$

Multiplication of A on the left by $\tilde{A}$ results in this same product.

When $A \neq F$ but $|A| = 5$, then clearly $A\tilde{A} = F$, and A and $\tilde{A}$ are divisors of F. When $|A| \neq 5$, then $|A|$ has a multiplicative inverse $|A|'$ and the triple product

$$A\tilde{A}\begin{pmatrix} |A|' & 5 \\ 5 & |A|' \end{pmatrix}$$

is I; hence, the multiplicative inverse of A is

$$\tilde{A}\begin{pmatrix} |A|' & 5 \\ 5 & |A|' \end{pmatrix}$$

Note that because multiplication by adjoint and scalar matrices is commutative, this inverse is effective on the left as well as on the right. Let us summarize our results.

THEOREM 7.12-1 For matrix A the following equality holds:

$$A\tilde{A} = \tilde{A}A = \begin{pmatrix} |A| & 5 \\ 5 & |A| \end{pmatrix}$$

THEOREM 7.12-2 For every matrix A one of the following properties holds:

(a) $|A| = 5$, and $A = F$.

or (b) $|A| = 5$, but $A \neq F$; then A and $\tilde{A}$ are a pair of divisors of F.

or (c) $|A| \neq 5$; then

$$A' = \tilde{A}\begin{pmatrix} |A|' & 5 \\ 5 & |A|' \end{pmatrix}$$

Corollary 7.12-2 For nonsingular matrix

$$A = \begin{pmatrix} a_{11} & a_{12} \\ a_{21} & a_{22} \end{pmatrix}$$

the formula for A' can be written

$$\begin{pmatrix} a_{22}|A|' & a_{12}^*|A|' \\ a_{21}^*|A|' & a_{11}|A|' \end{pmatrix}$$

EXERCISE 7.12-B

5. (a) Find the multiplicative inverse of the matrix

$$\begin{pmatrix} 2 & 3 \\ 4 & 5 \end{pmatrix}$$

(b) Compare your answer with that in Exercise 7.10.3(b).

6. For each matrix below, find a multiplicative inverse or find a mate that is a divisor of the additive identity.

$$\text{(a)}\ \begin{pmatrix} 4 & 1 \\ 4 & 2 \end{pmatrix} \qquad \text{(b)}\ \begin{pmatrix} 2 & 3 \\ 1 & 2 \end{pmatrix} \qquad \text{(c)}\ \begin{pmatrix} 3 & 4 \\ 4 & 2 \end{pmatrix}$$

7. Given matrix $M = \begin{pmatrix} 1 & 2 \\ 2 & 2 \end{pmatrix}$:

(a) Find $|M|$.
(b) Find M'.
(c) Explain the difference in notation between $|M'|$ and $|M|'$.
(d) Find $|M'|$ and $|M|'$.
(e) Why does $|M'| = |M|'$?

8. Explain why each entry in the secondary diagonal of the product $A\tilde{A}$ is 5.

★9. Find the product $\tilde{A}A$. Hence, show that $\tilde{A}A = A\tilde{A}$.

★10. For a nonsingular matrix A:
(a) Verify the formula for A' in Corollary 7.12-2.
(b) Show that $A'A = I$.

★11. Explain how Theorem 7.12-2 confirms the prediction made at the end of the discussion in Section 7.10.

★12. Prove that the set of nonsingular matrices is a group under multiplication.

7.13 SOLUTION OF MATRIX EQUATIONS

The set of nonsingular matrices is a multiplicative group. As in the case of the first system of pairs, some group theorems can be extended for all matrices. In particular, we have the theorems below.

THEOREM 7.13-1 For matrices A and B with A nonsingular, each equation $AX = B$ and $XA = B$ has a unique solution for matrix X given, respectively, by the formula $X = A'B$ and $X = BA'$.

THEOREM 7.13-2 *Law of Cancellation for Multiplication.* For matrices A, B, and C, with A nonsingular, if $AB = AC$ or $BA = CA$, then $B = C$.

The presence of divisors of the additive identity should make unpredictable the number of solutions of a matrix equation of the form

$$(X + A)(X + B) = F$$

There are, of course, the two evident solutions $X = A^*$ and $X = B^*$, since each of these reduces one of the factors to F. There may, however, be further solutions that transform the factors into an associated pair of divisors of F whose product is F. This can be illustrated with an example. The matrix equation

$$\left[X + \begin{pmatrix} 1 & 2 \\ 3 & 1 \end{pmatrix}\right]\left[X + \begin{pmatrix} 5 & 2 \\ 4 & 2 \end{pmatrix}\right] = F$$

has the two obvious solutions

$$X = \begin{pmatrix} 1 & 2 \\ 3 & 1 \end{pmatrix}^* = \begin{pmatrix} 4 & 3 \\ 2 & 4 \end{pmatrix} \quad \text{and} \quad X = \begin{pmatrix} 5 & 2 \\ 4 & 2 \end{pmatrix}^* = \begin{pmatrix} 5 & 3 \\ 1 & 3 \end{pmatrix}$$

because they give, respectively, the true equalities

$$F \cdot \begin{pmatrix} 4 & 5 \\ 1 & 1 \end{pmatrix} = F \quad \text{and} \quad \begin{pmatrix} 1 & 5 \\ 4 & 4 \end{pmatrix} \cdot F = F$$

Here are three further solutions:

$$X = \begin{pmatrix} 1 & 2 \\ 3 & 1 \end{pmatrix} \quad \text{which yields} \quad \begin{pmatrix} 2 & 4 \\ 1 & 2 \end{pmatrix}\begin{pmatrix} 1 & 4 \\ 2 & 3 \end{pmatrix} = F$$

$$X = \begin{pmatrix} 3 & 1 \\ 2 & 4 \end{pmatrix} \quad \text{which yields} \quad \begin{pmatrix} 4 & 3 \\ 5 & 5 \end{pmatrix}\begin{pmatrix} 3 & 3 \\ 1 & 1 \end{pmatrix} = F$$

$$X = \begin{pmatrix} 5 & 3 \\ 5 & 4 \end{pmatrix} \quad \text{which yields} \quad \begin{pmatrix} 1 & 5 \\ 3 & 5 \end{pmatrix}\begin{pmatrix} 5 & 5 \\ 4 & 1 \end{pmatrix} = F$$

EXERCISE 7.13

1. Given matrices

$$A = \begin{pmatrix} 4 & 2 \\ 5 & 1 \end{pmatrix} \quad \text{and} \quad B = \begin{pmatrix} 1 & 2 \\ 3 & 1 \end{pmatrix}$$

(a) Show that matrix A is nonsingular.
(b) Is matrix B singular or nonsingular?
(c) Find A'.
(d) Solve each of the equations below for matrix X.

$$\text{(i)} \quad AX = B \qquad \text{(ii)} \quad XA = B$$

(e) Determine whether each solution in (d) is singular or nonsingular.

2. Repeat Exercise 1 for matrices

$$A = \begin{pmatrix} 2 & 3 \\ 1 & 5 \end{pmatrix} \quad \text{and} \quad B = \begin{pmatrix} 4 & 2 \\ 3 & 1 \end{pmatrix}$$

When possible, solve each of Exercises 3–8 for matrix X. When a solution exists determine whether it is unique. (Note: Theorems 7.13-1 and 7.13-2 hold for situations in which matrix A is nonsingular. An equation in which A is singular must be studied and dealt with as it appears.)

3. $\begin{pmatrix} 2 & 3 \\ 1 & 2 \end{pmatrix} X = \begin{pmatrix} 2 & 3 \\ 1 & 2 \end{pmatrix}\begin{pmatrix} 5 & 2 \\ 4 & 4 \end{pmatrix}$ **4.** $\begin{pmatrix} 1 & 2 \\ 4 & 3 \end{pmatrix} X = \begin{pmatrix} 1 & 2 \\ 4 & 3 \end{pmatrix}\begin{pmatrix} 3 & 1 \\ 4 & 3 \end{pmatrix}$

5. $\begin{pmatrix} 1 & 1 \\ 2 & 3 \end{pmatrix} X = \begin{pmatrix} 1 & 2 \\ 4 & 3 \end{pmatrix}\begin{pmatrix} 1 & 1 \\ 2 & 3 \end{pmatrix}$ **6.** $\begin{pmatrix} 4 & 5 \\ 5 & 5 \end{pmatrix} X = I$

7. $\begin{pmatrix} 4 & 5 \\ 5 & 5 \end{pmatrix} X = F$ **8.** $\begin{pmatrix} 4 & 2 \\ 2 & 1 \end{pmatrix} X = \begin{pmatrix} 3 & 4 \\ 4 & 2 \end{pmatrix}$

9. Given the equation

$$\left[X + \begin{pmatrix} 4 & 2 \\ 3 & 4 \end{pmatrix}\right]\left[X + \begin{pmatrix} 4 & 5 \\ 2 & 3 \end{pmatrix}\right] = F$$

(a) Find the two obvious solutions.
(b) Show that each of the following matrices is also a solution for X.

(i) $\begin{pmatrix} 2 & 2 \\ 4 & 4 \end{pmatrix}$ (ii) $\begin{pmatrix} 3 & 2 \\ 2 & 1 \end{pmatrix}$

10. Given the equation

$$\left[X + \begin{pmatrix} 2 & 4 \\ 1 & 3 \end{pmatrix}\right]\left[X + \begin{pmatrix} 3 & 5 \\ 2 & 4 \end{pmatrix}\right] = F$$

(a) Find the two obvious solutions.
(b) Show that each of the following matrices is also a solution for X.

(i) $\begin{pmatrix} 1 & 3 \\ 2 & 4 \end{pmatrix}$ (ii) $\begin{pmatrix} 5 & 3 \\ 5 & 3 \end{pmatrix}$

(iii) $\begin{pmatrix} 4 & 2 \\ 1 & 4 \end{pmatrix}$ (iv) $\begin{pmatrix} 3 & 1 \\ 2 & 5 \end{pmatrix}$

(v) $\begin{pmatrix} 4 & 5 \\ 5 & 1 \end{pmatrix}$ (vi) $\begin{pmatrix} 2 & 2 \\ 3 & 3 \end{pmatrix}$

7.14 THE SOLUTION OF SIMULTANEOUS LINEAR EQUATIONS

The system of matrices with its extensive multiplicative properties can be used in remarkably efficient formulas that give common solutions for systems of equations, of which the pair below is a typical example.

$$2x + 3y = 1 \qquad (9)$$
$$4x + 2y = 3 \qquad (10)$$

Such equations are called *simultaneous linear equations;* simultaneous, because, if a common solution exists, the equations are true together, and linear because under certain circumstances they can be shown to represent straight lines on a graph. There are many methods for solving these equations; here formulas employing matrices and their determinants are used.

Let us begin by forming the matrix A of the coefficients of x and y. For the equations above we have

$$A = \begin{matrix} \text{Equation (9)} \\ \text{Equation (10)} \end{matrix} \overset{\begin{matrix} x & y \end{matrix}}{\begin{pmatrix} 2 & 3 \\ 4 & 2 \end{pmatrix}}$$

We next write a new matrix by replacing each coefficient of x by the constant term of its equation. For our example, we replace elements 2 and 4 in the first column of matrix A with corresponding constant terms 1 and 3. We designate the new matrix by X, so that

$$X = \begin{pmatrix} 1 & 3 \\ 3 & 2 \end{pmatrix}$$

In a similar fashion, we form a new matrix, called Y, by replacing the coefficients of y with the corresponding constant terms. We have

$$Y = \begin{pmatrix} 2 & 1 \\ 4 & 3 \end{pmatrix}$$

We now reckon the determinants of the three matrices and use them in the formulas

$$x = |X| \cdot |A|' \qquad \text{and} \qquad y = |Y| \cdot |A|'$$

We have $x = 3 \cdot 2' = 4$ and $y = 2 \cdot 2' = 1$. Substituting these values of x and y in the original equations, we find that $2 \cdot 4 + 3 \cdot 1 = 3 + 3 = 1$ and $4 \cdot 4 + 2 \cdot 1 = 1 + 2 = 3$. Hence, the formulas produce correct solutions. It can be shown that they are the only solutions.

The rule that we have used is known in mathematical literature as *Cramer's rule.* Its general formulation is the next theorem.

THEOREM 7.14-1 *Cramer's Rule.* Given simultaneous linear equations

$$\begin{aligned} a_{11}x + a_{12}y &= b_1 \\ a_{21}x + a_{22}y &= b_2 \end{aligned} \qquad (11)$$

Let A be the matrix of coefficients of x and y, and let X and Y be matrices formed from A by writing the constant terms in place of the coefficients of

x and y, respectively, so that

$$A = \begin{pmatrix} a_{11} & a_{12} \\ a_{21} & a_{22} \end{pmatrix} \qquad X = \begin{pmatrix} b_1 & a_{12} \\ b_2 & a_{22} \end{pmatrix} \qquad Y = \begin{pmatrix} a_{11} & b_1 \\ a_{21} & b_2 \end{pmatrix}$$

Then, if matrix A is nonsingular so that $|A| \neq 5$, the unique solutions of Equations (11) are given by the formulas $x = |X| \cdot |A|'$ and $y = |Y| \cdot |A|'$.

If matrix A had been singular, its determinant would have had no multiplicative inverse, and we could not have applied Cramer's rule. For example, the equations

$$\begin{aligned} 2x + 3y &= 1 \\ 4x + y &= 2 \end{aligned}$$

for which $|A| = 2 \cdot 1 + (3 \cdot 4)^* = 2 + 2^* = 5$, are equivalent in the sense that multiplication of the first by element 2 produces the second; hence, we might as well have been given only the first one. It has five solutions found by letting x equal 1, 2, 3, 4, 5, in turn. Corresponding values of y are 3, 4, 5, 1, and 2. There are five pairs of solutions: 1 and 3; 2 and 4; 3 and 5; 4 and 1; and 5 and 2. That is, there are as many solutions as there are elements in our system.

As a second example, the equations

$$\begin{aligned} 2x + 3y &= 1 \\ 4x + y &= 3 \end{aligned}$$

again have $|A| = 5$. When we multiply both members of the first equation by 2 we get the contradictory result

$$\begin{aligned} 4x + y &= 2 \\ 4x + y &= 3 \end{aligned}$$

There are no values of x and y that will make the sum $4x + y$ equal to 2 and to 3 at the same time. In this case the equations have no solution.

EXERCISE 7.14

1. Solve by Cramer's rule:

(a) $\begin{aligned} 2x + 3y &= 4 \\ 4x + 4y &= 5 \end{aligned}$ (b) $\begin{aligned} 3x + 4y &= 3 \\ 2x + 3y &= 3 \end{aligned}$

(c) $\begin{aligned} 2x + y &= 3 \\ 3x + 2y &= 2 \end{aligned}$ (d) $\begin{aligned} 3x + 5y &= 2 \\ 2x + 4y &= 2 \end{aligned}$

2. Solve, when possible, each pair of simultaneous equations below. [Note: Cramer's rule does not apply (why?); hence, special analyses must be made.]

(a) $3x + 4y = 2$, $4x + 2y = 1$ (b) $3x + 2y = 1$, $2x + 3y = 3$

(c) $3x + 4y = 2$, $4x + 2y = 2$ (d) $2x + 3y = 1$, $x + 4y = 3$

3. (a) Show that there are five solutions for the equation $2x + 3y = 1$, and find the solutions.

(b) Show that there are five solutions for the equation $4x + 2y = 3$, and find the solutions.

(c) Show that exactly one solution satisfies both of the equations in (a) and (b).

4. (a) Show that in Exercise 3 the solution was found for the pair of equations with which this section opened, and thus note the pertinence of the adjective "simultaneous."

(b) Use the method of Exercise 3 to solve the equations of Exercise 1(a) and (b).

(c) Use the method of Exercise 3 to show that the equations of Exercise 2(a) have five solutions and those of Exercise 2(b) have no solutions.

5. Explain why the method of Exercise 3 would not be applicable if the five-element system were replaced by the real number system.

6. Cramer's rule holds in the real number system. In this system determinants of matrices are found by the same formula as in the five-element system. Show, therefore, that for the equations with which this section opened

(a) $|A| = -8$, $|X| = -7$, and $|Y| = 2$.

(b) Solutions are $x = \dfrac{-7}{-8} = \dfrac{7}{8}$ and $y = \dfrac{2}{-8} = -\dfrac{1}{4}$.

7. Use Cramer's rule to find solutions of the equations of Exercise 1 when the five-element system is replaced by the real number system.

7.15 A PRACTICAL PROBLEM

Here is one of the many actual situations that lead to the use of simultaneous linear equations: for each article produced in a factory, different stages of production take different amounts of time, and the machines used can be run continuously for different numbers of hours. Therefore, the daily output must be carefully reckoned to avoid "bottlenecks" or continuously growing accumulations of unfinished articles awaiting some of the slower processes. In order to present an example suited to our matrix system, we must limit ourselves to a two-stage, two-product situation. Furthermore, the use of the five-element system will make the numbers seem silly. Nevertheless the type of problem and the method of solution deserve study.

In a small bakery two part-time workers, Jean and Laura, are placed in charge of the complete process of making and packaging fancy lemon and chocolate cookies. A batch of chocolate cookies takes 3 hours to prepare and 1 hour to bake and package. A batch of lemon cookies takes 2 hours to prepare and 2 hours to bake and package. Hence, for x batches of chocolate cookies and y batches of lemon cookies, the time required for preparation is $3x + 2y$ hours, and the time spent baking and packaging is $1x + 2y$ hours. Jean, who will do all the preparation, can work 3 hours a day. Laura, who will do the baking and packaging, can work 2 hours a day. How many batches of cookies of each kind should they plan to make per day?

To start the solution, let x be the proper number of batches of chocolate cookies and y the proper number of batches of lemon cookies. Then, x and y must be determined so that the preparation time, which we found to be $3x + 2y$ hours, will equal the 3 hours that Jean can devote to it; that is, $3x + 2y = 3$. Furthermore, x and y must also meet the requirement that the baking and packaging time, namely $x + 2y$ hours, will equal the 2 hours that Laura can work; that is, $x + 2y = 2$. The x and y which satisfy both equations give the number of batches of each kind of cookie that will keep both participants busy during their working time. We must find the solution of the simultaneous linear equations

$$\begin{aligned} 3x + 2y &= 3 \\ x + 2y &= 2 \end{aligned}$$

Using Cramer's rule we write

$$A = \begin{pmatrix} 3 & 2 \\ 1 & 2 \end{pmatrix} \qquad X = \begin{pmatrix} 3 & 2 \\ 2 & 2 \end{pmatrix} \qquad Y = \begin{pmatrix} 3 & 3 \\ 1 & 2 \end{pmatrix}$$

whence $|A| = 4$, $|X| = 2$, and $|Y| = 3$. Because $|A|' = 4$, $x = 2 \cdot 4 = 3$ and $y = 3 \cdot 4 = 2$. Hence, the two workers should plan to bake three batches of chocolate cookies and two batches of lemon cookies per day.

For those readers who have been distressed by the absurdity of the numbers used in this problem, let us temporarily abandon the five-element system and substitute the ordinary number system in its place. The number of hours that Jean and Laura can work are modified from 3 and 2 to 8 and 4, respectively. Then the equations become

$$\begin{aligned} 3x + 2y &= 8 \\ x + 2y &= 4 \end{aligned}$$

Cramer's rule still applies, so we form

$$A = \begin{pmatrix} 3 & 2 \\ 1 & 2 \end{pmatrix} \qquad X = \begin{pmatrix} 8 & 2 \\ 4 & 2 \end{pmatrix} \qquad Y = \begin{pmatrix} 3 & 8 \\ 1 & 4 \end{pmatrix}$$

Now $|A| = 4$, $|X| = 8$, and $|Y| = 4$. Consequently, the number of batches of chocolate cookies that should be baked is $x = 8/4 = 2$, and the number of batches of lemon cookies is $y = 4/4 = 1$.

This problem required two equations in two unknowns for solution. One advantage of Cramer's rule is that it can be applied to any number n of simultaneous linear equations in n unknowns. Furthermore, electronic computers can be programmed to handle it. When the coefficients and constant terms are put into the computer, the machine finds the determinants and presents the answers. The burden of arithmetic can, however, become very heavy. As the number of rows and columns in a matrix increase, the number of terms in the determinant and the number of factors in each term also increase. For a slow computer making about 3000 multiplications per second, the solution of 10 equations in 10 unknowns could take more than a month. In actual practice, therefore, a considerable amount of theory, which is not studied here, must be used to find shorter methods of handling the arithmetic.

EXERCISE 7.15

1. In the city of Urban where the five-element system is the usual and practical number system, George and Jim go into the business of making toy tractors and fire engines. George takes 2 hours to stamp out the parts of a tractor and 3 hours to stamp out the parts of a fire engine. Jim takes 1 hour to assemble a tractor and 2 hours to assemble a fire engine. If each man works 4 hours a day, how many tractors and how many fire engines should be made? (Suggestion: let x represent the number of tractors and y the number of fire engines that should be made in a day. Then what does $2x + 3y$ represent? What does $x + 2y$ represent?)

2. In the city of Urban (see Exercise 1) foods named Nutri and Gurmi are sold. An ounce of Nutri contains 2 units of vitamin A and 3 of vitamin B. An ounce of Gurmi contains 4 units of vitamin A and 2 of vitamin B. Suppose that the daily requirement of vitamins is 2 units of A and 2 of B. How many ounces of Nutri and Gurmi should be eaten to provide the daily requirements of vitamins? (Suggestion: let x be the number of ounces of Nutri and y be the number of ounces of Gurmi to be eaten per day. Then what does $2x + 4y$ represent? What does $3x + 2y$ represent?)

3. The manager of a grocery store in the city of Urban (see Exercise 1) finds that his supply of peach and currant jelly must be replenished. He has \$2 to pay for the new stock and has room on his shelves for four jars of it. If peach jelly costs \$1 a jar and currant jelly costs \$2 a jar, how many jars of each should he buy?

7.16 THE SYSTEM OF MATRICES IN CRYPTOGRAPHY

In Section 3.10 we studied a use of the five-element system in cryptography. Now we will see how the matrix system may be used to disguise messages further.

With our correspondent we choose a nonsingular matrix M as the fundamental matrix for the work; for example,

$$M = \begin{pmatrix} 1 & 2 \\ 3 & 4 \end{pmatrix}$$

Given a message, encode it into elements by an agreed cipher and separate the resulting sequence into consecutive sets of four elements each. Let one of the sets be designated by $a_1a_2a_3a_4$. Assemble this quadruple into a matrix using a fixed pattern such as

$$A = \begin{pmatrix} a_1 & a_2 \\ a_3 & a_4 \end{pmatrix}$$

We multiply A on the right by M, and then use the four corresponding entries in the product matrix as the elements in the final message. When a friendly decoder receives the transmitted series of elements, he separates it into consecutive sets of four, assembles each set into the proper matrix form, and multiplies on the right by M'. The entries in the product matrices now form for him the initial encoded message.

Let us illustrate our method with the first message used in Section 3.10. We encoded *Nan has. a sash* into the elements 3132 1451 4142, using the cipher

$$\begin{matrix} \text{a} & \text{h} & \text{n} & \text{s} & . \\ 1 & 2 & 3 & 4 & 5 \end{matrix} \tag{12}$$

We assemble the first quadruple of elements, 3132, into the matrix

$$\begin{pmatrix} 3 & 1 \\ 3 & 2 \end{pmatrix}$$

and multiply this matrix on the right by M:

$$\begin{pmatrix} 3 & 1 \\ 3 & 2 \end{pmatrix}\begin{pmatrix} 1 & 2 \\ 3 & 4 \end{pmatrix} = \begin{pmatrix} 1 & 5 \\ 4 & 4 \end{pmatrix}$$

The first four elements in the message will, therefore, be 1544. We treat the second and third quadruples of elements in the same fashion:

$$\begin{pmatrix} 1 & 4 \\ 5 & 1 \end{pmatrix}\begin{pmatrix} 1 & 2 \\ 3 & 4 \end{pmatrix} = \begin{pmatrix} 3 & 3 \\ 3 & 4 \end{pmatrix}$$

$$\begin{pmatrix} 4 & 1 \\ 4 & 2 \end{pmatrix}\begin{pmatrix} 1 & 2 \\ 3 & 4 \end{pmatrix} = \begin{pmatrix} 2 & 2 \\ 5 & 1 \end{pmatrix}$$

The final message reads 1544 3334 2251 or, in letters, a.ss nnns hh.a.

The decoder assembles the quadruples of elements into matrices and multiplies on the right by M', namely,

$$\begin{pmatrix} 3 & 1 \\ 4 & 2 \end{pmatrix}$$

to get

$$\begin{pmatrix} 1 & 5 \\ 4 & 4 \end{pmatrix}\begin{pmatrix} 3 & 1 \\ 4 & 2 \end{pmatrix} = \begin{pmatrix} 3 & 1 \\ 3 & 2 \end{pmatrix}$$

$$\begin{pmatrix} 3 & 3 \\ 3 & 4 \end{pmatrix}\begin{pmatrix} 3 & 1 \\ 4 & 2 \end{pmatrix} = \begin{pmatrix} 1 & 4 \\ 5 & 1 \end{pmatrix}$$

$$\begin{pmatrix} 2 & 2 \\ 5 & 1 \end{pmatrix}\begin{pmatrix} 3 & 1 \\ 4 & 2 \end{pmatrix} = \begin{pmatrix} 4 & 1 \\ 4 & 2 \end{pmatrix}$$

He obtains the elements of the message, 3132 1451 4142, or, in letters, nanh as.a sash.

EXERCISE 7.16

1. In this exercise use cipher (12) and the fundamental matrix M of the illustrative example.

(a) Show that the message *Hans has Ann's sash.* encodes into .n.h .nhh hnas nnhs

(b) Decode the message hh.a ah.n sn.s nsas

2. Use cipher (12) and the fundamental matrix

$$\begin{pmatrix} 1 & 5 \\ 1 & 2 \end{pmatrix}$$

to encode or decode the following messages.

(a) *Shah Hassan. has a. sash*

(b) .has a.nh s.as nshs .n.h hnsn .nh.

★3. You are the "enemy." You intercept the message below. You have good reason to believe that cipher (12) was used and that the last word in the message is *sash*. Can you decipher the following message?

hsnh .sas .snh a..s aahn

7.17 RETROSPECT AND PROSPECT

In our study of matrices we have seen again that the properties of a new system depend on the characteristics of the parent system and on the definitions established for the relations between and the operations on the new numbers. We have found that the system of matrices exhibits many of the usual properties of an algebraic system. It differs from some in our previous study in that it fails to exhibit commutativity under multiplication and possesses divisors of its additive identity. It is a more sophisticated system than the earlier ones, and it has, correspondingly, more sophisticated and interesting applications. No new methods of analysis were introduced, but those with which we were already familiar were used in more complicated situations.

Our work in the first seven chapters has shown that the arithmetic and algebra of real numbers taught in elementary and secondary school comprise only one of the many possible arithmetics and algebras. We can no longer consider the facts of that arithmetic and algebra to be immutable truths, because the properties of the systems that we have developed are equally true in the sense that they develop logically from their foundations. To be sure, the real number system fits best with the transactions and calculations of our daily lives and can be used to describe with close accuracy many of the events of the physical world of which we are a part; its arithmetic and algebra are highly useful whereas, in contrast, those of the five-element system appear to be only enticing games. Yet as mathematical constructs all of these systems have equal validity.

In this point of view we are distinctly modern. Even in the nineteenth century it was generally assumed that the arithmetic and algebra that had been developed with so much labor over so many centuries were the true and only ones. Since that time the emphasis on abstractness and generalization in mathematics has not only led to new developments in old systems, but has given rise to new systems whose investigation, as we have seen, requires the play of mathematical imagination.

Our work has given us considerable experience with the method of mathematical argument. In the next two chapters we shall study the validity of that method.

II

LOGIC AND SETS

In Part I we developed algebraic systems whose dissimilarities from each other began with dissimilarities in fundamental postulates and definitions. We assumed that if the foundations of a given system were accepted and if the methods of reasoning were sound, then the resulting theorems were true for that system. In Part II we shall study the reasoning process itself and discuss the significance of assertions such as "Statement q follows from statement p." In doing so, we shall use symbolism in order not to be sidetracked by differing connotations of words and varying interpretations of sentences.

As our investigation proceeds in Chapter 8, we shall develop an algebra of logic that both resembles and differs from the algebras of Part I. In Chapter 9 we shall consider collections of things, some or all of which have common characteristics. We shall discover that these collections and the laws they obey also form an algebra, a companion algebra to the algebra of logic.

8

THE ALGEBRA OF LOGIC

In our study of mathematical systems we have argued from known properties to new characteristics that we hoped would be true. Sentences of the form *if* . . . , *then* . . . occur frequently in our exposition. They attest to the fact that a good mathematical argument consists of a chain of true statements. Proofs and derivations are concerned with only relevant material. They present all the necessary information but suppress comments that may be interesting and important in other connections but of no immediate use in the current discussion. Such economy and elegance is no accident in mathematical writing. It occurs because the basic tool of the subject is a two-valued logic which holds that a statement is either true or false.

This logic is the one by which we profess to conduct much of our reasoning through a whole range of considerations, from a mathematical theorem to a casual comment on the weather. It embodies the laws of validity of verbal thinking. The first symbolic formulation of logic was made during the early years of the past century by the British mathematician George Boole. He observed that the nature of language imposes regulations on its use in argument. When we examine these rules, we shall find that they can be expressed in symbols and that so expressed they form a mathematical system. Let us now begin the study of the algebra of logic invented by George Boole.

8.1 STATEMENTS AND CONNECTIVES

All reasoning uses sentences together with certain words and phrases that connect or modify those sentences. For mathematical arguments five of these connectives or modifiers are sufficient, namely:

and	Example: The five-element system is closed under addition *and* the five-element system is closed under multiplication.
or	This connective is used in the ordinary legal sense of and/or. Example: Element a equals 5 *or* element b equals 5.
if . . . , then . . .	Example: *If* $a = 5$, *then* $ab = 5$.
if and only if	Example: An element has a multiplicative inverse *if and only if* the element is different from 5.
not	Example: Subtraction is *not* commutative.

Although the last of these words is a modifier rather than a connective, the name *logical connective* is used to denote each of the five words or phrases. To each is given a name and a symbol.

Connective	Symbol	Name
p **and** q	$p \wedge q$	conjunction
p **or** q	$p \vee q$	alternation
if p**, then** q	$p \rightarrow q$	conditional
p **if and only if** q	$p \leftrightarrow q$	biconditional
not p	$\sim p$	denial

The sentences above can now be written in words and symbols as follows:

The five-element system is closed under addition $\wedge$ the five-element system is closed under multiplication.

Element a equals 5 $\vee$ element b equals 5.

$a = 5 \rightarrow ab = 5.$

An element has a multiplicative inverse $\leftrightarrow$ the element is different from 5.

$\sim$ (subtraction is commutative).

Note that the symbol for *not* is written in front of the word statement that it modifies, although the customary phraseology would be different.

Thus, in the last sentence "not 'subtraction is commutative' " is written rather than "subtraction is not commutative."

So far, only declarative sentences have been used, and this custom will be continued. In our work sentences are defined to be declarative sentences. Acceptable examples are

(a) The five-element system is closed under addition.
(b) In the five-element system subtraction is commutative.
(c) In the five-element system $x + 3 = 2$.
(d) It is associative.

Note that sentence (a) is true, sentence (b) is false, sentence (c) is true of element 4 and false of all other elements, sentence (d) is true of some operations and false of others. When a sentence such as (a) or (b) is true or false as it stands, it is said to have a truth value: *true* for a true sentence and *false* for a false one. Such a sentence is called a *statement*. Clearly, a sentence such as (c) or (d) can be converted into a statement by the substitution of the name of an element for x in (c) or of an operation for "It" in (d). If we replace x by 3, we get the false statement "In the five-element system $3 + 3 = 4$." If we replace "It" by "Addition in the five-element system," we get the true statement "Addition in the five-element system is associative." As they stand, and until substitutions are made that give them truth values, sentences (c) and (d) are called *open sentences*.

The logical connectives can be applied, of course, to nonmathematical sentences. Consider the two statements "The rabbit hurries by" and "Alice follows the rabbit." These can be combined by the connectives to give comments of disparate meanings.

(1) The rabbit hurries by *and* Alice follows the rabbit.
(2) The rabbit does *not* hurry by.
(3) *If* the rabbit hurries by, *then* Alice follows the rabbit.
(4) The rabbit hurries by *or* Alice follows the rabbit.
(5) The rabbit hurries by *if and only if* Alice follows the rabbit.

In order to generalize our discussion let us introduce letters to stand for statements. For example, if

p represents the statement "The rabbit hurries by"
q represents the statement "Alice follows the rabbit" (i)

then statements (1)–(5) can be written in the following concise form.

(1) $p \wedge q$
(2) $\sim p$
(3) $p \rightarrow q$
(4) $p \vee q$
(5) $p \leftrightarrow q$

Such expressions in terms of statement letters and symbols for the logical connectives are called *statement patterns*. A given pattern has many word translations that differ according to the statements assigned to the letters. Each translation is called an *instance* of the pattern.

Let us now consider more elaborate patterns:

(6) $(p \vee r) \rightarrow q$
(7) $\sim(p \wedge r)$
(8) $\sim p \wedge r$

Note that when the connective *not* modifies a single letter, it is customary to write the symbol and letter without parentheses. Thus, pattern (8) is written $\sim p \wedge r$ instead of $(\sim p) \wedge r$. When this connective modifies a whole expression, as in pattern (7), it will be translated by the phrase *it is not true that.*

Let us expand dictionary (i) and let r stand for the statement "The rabbit takes a watch out of his waistcoat pocket." Reasonable translations are listed below.

(6) *If* the rabbit hurries by *or* takes a watch out of his waistcoat pocket, *then* Alice follows him.
(7) *It is not true that* the rabbit hurries by *and* takes a watch out of his waistcoat pocket.
(8) The rabbit *does not* hurry by, *and* he takes a watch out of his waistcoat pocket.

All the numbered statements above are called *composite statements.* Each composite statement has one *principal connective* that tells its essential nature. The principal connective operates on briefer composite statements called *immediate components.* Each immediate component is, in turn, composed of one or more simple statements joined by the requisite number of logical connectives. These simple statements are called *ultimate components.*

For example, statement (6) is a conditional having principal connective *if . . . , then . . . .* The immediate components are the alternation "The rabbit hurries by or takes a watch out of his waistcoat pocket" and the simple statement "Alice follows the rabbit." The ultimate components are the three statements "The rabbit hurries by," "The rabbit takes a watch out of his waistcoat pocket," and "Alice follows the rabbit." The same type of analysis can be made of a statement pattern. For example, pattern (7) is a denial having principal connective $\sim$. There is only one immediate component, namely, the conjunction $p \wedge r$. The ultimate components are p and r. Let us summarize our definitions.

A **sentence** is a declarative sentence.

The **truth value** of a sentence is *true* for a sentence that is true as it stands and *false* for a sentence that is false as it stands.

A **statement** is a sentence that has a truth value.

A **composite statement** is a statement formed from one or more simple statements called **ultimate components** by use of the logical connectives.

In a composite statement there is one **principal connective** that joins or modifies briefer composite statements called **immediate components.**

A **statement letter** is a letter that is used in place of a statement.

A **statement pattern** is a symbolic expression composed of statement letters combined by use of symbols for the logical connectives.

An **instance of a statement pattern** is a word statement that results when the statement letters and connective symbols in the pattern are replaced by specific word statements and connectives.

EXERCISE 8.1

Use the following dictionary:

p:	the cat grins	q:	the mock turtle sobs
r:	the gryphon dances	s:	the dormouse sleeps

1. Translate into symbols:
(a) The dormouse does not sleep.
(b) The dormouse sleeps and the gryphon does not dance.
(c) If the cat grins, then the mock turtle does not sob.
(d) The mock turtle sobs if and only if the gryphon dances.

2. Translate into word statements:

(a) $\sim s$ (b) $r \wedge s$
(c) $p \vee \sim s$ (d) $s \leftrightarrow p$

3. For each statement in Exercise 1, determine its type (conjunction, conditional, etc.), its principal connective, its immediate components, and its ultimate components.

4. For each statement pattern in Exercise 2, determine its type, its principal connective, its immediate components, and its ultimate components.

8.2 PUNCTUATION

The first four logical connectives combine their immediate components into new composite statements. They can be considered binary operations on the set of statements. The fifth connective, which is really a modifier, acts on one immediate component to produce a statement having a different meaning. It is a unary operator.

Every statement or statement pattern must clearly indicate the immediate components of each connective. For patterns this is easily done by using parentheses. For example, the set of symbols $p \vee q \wedge r$ is not a statement pattern as it stands, but by introducing parentheses we can make two different patterns:

$$p \vee (q \wedge r) \qquad \text{and} \qquad (p \vee q) \wedge r \tag{1}$$

Each of these displays precisely the principal connective, namely, $\vee$ in the first and $\wedge$ in the second, and the immediate components.

In considering translations of patterns into word sentences, it will be helpful to establish some rules of punctuation. These rules will be observed consistently because we will use punctuation with statements as we use parentheses with patterns. The comma is the weakest mark of punctuation. A semicolon is stronger, and repeated semicolons are even stronger. A period, of course, ends the statement.

Let us apply these rules in writing instances of patterns (1) using the dictionary of Exercise 8.1. If the alternation is principal connective, we have "The cat grins, or the mock turtle sobs and the gryphon dances." With the conjunction as principal connective, the pattern translates into "The cat grins or the mock turtle sobs, and the gryphon dances." Clearly, the two instances have diverse meanings.

When writing a conditional, a comma is placed before the word *then*. Thus, for $q \rightarrow r$ we have "If the mock turtle sobs, then the gryphon dances." The pattern $q \rightarrow (r \leftrightarrow s)$ translates easily into "If the mock turtle sobs, then the gryphon dances if and only if the dormouse sleeps." How would we punctuate the translation of $(q \rightarrow r) \leftrightarrow s$? Because we use a comma for the conditional, we will need a semicolon to set off the principal connective *if and only if*. When this phrase requires punctuation, identical symbols are placed after each word *if*. The pattern above, therefore, translates into "If the mock turtle sobs, then the gryphon dances if; and only if; the dormouse sleeps."

The clause *it is not true that* is reserved to indicate that the denial modifies a composite statement rather than a single ultimate component. For example, the denial is the principal connective in the pattern $\sim(p \vee q)$. The translation is "It is not true that the cat grins or the mock turtle sobs."

EXERCISE 8.2

Using the dictionary below, translate into words the patterns in Exercises 1–6 or explain why a translation cannot be made.

p: the sea is wet $\quad$ q: the sands are dry $\quad$ r: the moon is bright

1. (a) $(p \vee q) \wedge r$
(b) $p \vee (q \wedge r)$

2. (a) $\sim p \rightarrow q$
(b) $\sim(p \rightarrow q)$

3. (a) $p \vee q \leftrightarrow r$
(b) $p \leftrightarrow q \vee r$

4. (a) $(p \wedge q) \leftrightarrow r$
(b) $p \wedge (q \leftrightarrow r)$

5. (a) $p \rightarrow \sim q \wedge r$
(b) $p \wedge \sim q \rightarrow r$

6. (a) $\sim p \vee \sim q$
(b) $\sim(p \vee q)$

7. Given the statement "if $2 + 3 = 2 + 3$, then $3 = 3$." In order to translate this into symbols a statement letter is required for each of the ultimate components, namely, "$2 + 3 = 2 + 3$," and "$3 = 3$." Let p represent $2 + 3 = 2 + 3$, and let q represent $3 = 3$. Then the statement above yields the pattern $p \rightarrow q$. Using the same dictionary, translate into symbols each of the following statements.
(a) $2 + 3 \neq 2 + 3$ or $3 = 3$.
(b) It is not true that $2 + 3 = 2 + 3$ and $3 = 3$.
(c) $2 + 3 \neq 2 + 3$ if and only if $3 \neq 3$.
(d) $2 + 3 = 2 + 3$ and $3 = 3$, or $3 \neq 3$.

In Exercises 8–12 composite statements are listed in pairs. For each exercise, set up a dictionary and translate into symbols or explain why a translation cannot be made. The five-element system is used.

8. (a) $2 + 3 = 5$ and $2 = 3^*$ or $5 \neq 5$.
(b) $2 + 3 \neq 5$ or $2 = 3^*$ and $5 = 5$.

9. (a) It is not true that $2 = 2^*$ or $2 = 3^*$.
(b) $2 \neq 2^*$ or $2 = 3^*$.

10. (a) If $2 \cdot 5 = 5$, then 2 is the multiplicative identity or 5 is the additive identity.
(b) If 2 is not the multiplicative identity and 5 is the additive identity, then $2 \cdot 5 \neq 5$.

11. (a) $3 = 2^*$ if and only if $3 + 2 = 5$.
(b) $3 = 2^*$ and $3 + 2 = 5$, or $3 \neq 2^*$ and $3 + 2 \neq 5$.

12. (a) $2^* = 3$ or $2' = 3$ if and only if $3 + 4 = 2$.
(b) $2' = 3$ if and only if $2^* = 3$ or $3 + 4 = 2$.

8.3 EXAMPLES OF LOGICAL SENTENCES

Now that we have had practice in discovering the patterns which underlie statements, let us analyze some of the definitions and theorems of Chapter 5 on the first system of pairs. To be sure, we will be dealing with sentences and sentence patterns, for we will be considering general pairs rather than specific pairs. The rules governing the composition of sentences are, however, the same as those governing the composition of statements.

Let us begin with the definition of equality of pairs in Section 5.2.

DEFINITION If $P = (p_1, p_2)$ and $Q = (q_1, q_2)$, then $P = Q$ if and only if $p_1 = q_1$ and $p_2 = q_2$.

There are five sentences in this definition. We assign letters as follows:

$$\begin{array}{llll} p\colon & P = (p_1,p_2) & q\colon & Q = (q_1,q_2) \\ r\colon & P = Q & s\colon & p_1 = q_1 \\ t\colon & p_2 = q_2 & & \end{array}$$

The definition is now seen to be an instance of the pattern

$$(p \wedge q) \rightarrow (r \leftrightarrow (s \wedge t)) \tag{2}$$

Note that the placement of the semicolon requires translation (2) rather than

$$((p \wedge q) \rightarrow r) \leftrightarrow (s \wedge t)$$

Theorem 5.4-1 asserts that if P and Q are pairs, then $P + Q = Q + P$. Here we have three sentences:

$$\begin{array}{llll} p\colon & P \text{ is a pair} & r\colon & P + Q = Q + P \\ q\colon & Q \text{ is a pair} & & \end{array}$$

The desired sentence pattern is, therefore, $(p \wedge q) \rightarrow r$.

EXERCISE 8.3

Find the sentence pattern of which each theorem below is an instance.

1. Theorem 5.6-5.
2. The last part of Corollary 5.11-2.
3. Theorem 5.11-4.

8.4 TRUTH VALUES OF STATEMENTS

We have defined a statement to be a declarative sentence that is true or false as it stands. The truth value of a composite statement depends on the truth value of its components. The rules governing truth values of the simplest composite statements, those employing only one logical connective, are postulates that, accepted without proof as the basis for argument, form the foundation for further work. Like many sets of postulates they are abstracted from practical experience. In summarizing them from extensive study of the usage of language in reasoning, George Boole set up a model of argumentative thinking. We shall study this model. It is a mathematical construct. It has some application to our thought processes, but it is a theoretical, not a practical, study. A change in the rules would create a different system of logic. In place of this two-valued logic of true and false, for example, we could set up a logic of three or more values. Many such dissimilar systems are under mathematical investigation.

In presenting the postulates for truth values, we shall first discuss examples illustrating customary decisions about the truth or falsity of a given type of composite statement, and then abstract a rule from these usual decisions. Our first examples are drawn from the five-element system.

The statement "5 has a multiplicative inverse" is false. Its denial, "5 has no multiplicative inverse," is true. On the other hand, the true statement "$2 = 2$" has a false denial "$2 \neq 2$." These contrasting situations illustrate the rule below.

RULE 1 A statement and its denial have opposite truth values.

Next consider four conjunctions.

(a) $2 = 2$ and $1 = 4^*$ (b) $2 = 2$ and $1 = 4$
(c) $2 = 2^*$ and $1 = 4^*$ (d) $2 = 2^*$ and $1 = 4$

The first conjunction has truth value true for both components; hence, it is true as a whole. The last one, however, has truth value false for both components and is false as a composite statement. In (b) and (c) one part of the statement is true but the other is not; therefore, the statement as a whole is not true. This discussion makes clear the reasonableness of the second rule.

RULE 2 A conjunction is true if and only if both of its immediate components are true.

With an alternation, conditions for truth value true are less restrictive than those in Rule 2. Consider the four statements below.

(e) $2 = 2$ or $1 = 4^*$ (f) $2 = 2$ or $1 = 4$
(g) $2 = 2^*$ or $1 = 4^*$ (h) $2 = 2^*$ or $1 = 4$

Statement (e) is clearly true; statements (f) and (g) are also true because in each case one of the choices offered us is true. Only statement (h), which is false in both components, is false as a whole. We thus have a third rule.

RULE 3 An alternation is false if and only if both of its immediate components are false.

In a conditional, the immediate component that follows the word "if" is called the *antecedent* and the other component is called the *consequent*. Let us deal first with conditionals having true antecedents.

(i) If the moon is a satellite of earth, then it revolves around earth.
(j) If the moon is a satellite of earth, then it revolves around Mars.

Clearly statement (i), having a true consequent, is true, and statement (j), having a false consequent, is false. For conditionals with false antecedents, however, the choice of truth value is not so obvious.

(k) If the moon is made of green cheese, then moon rocks are cheese crumbs.
(l) If the moon is made of green cheese, then it reflects the light of the sun.

Statements (k) and (l) are equally reasonable or nonsensical. The falsity of the antecedent makes the truth or falsity of the consequent immaterial It is customary to assign truth value true to both statements according to the rule below.

RULE 4 A conditional is false if and only if its antecedent is true and its consequent is false.

Let us replace the conditional with the biconditional in statements (i) to (l).

(m) The moon is a satellite of earth if and only if it revolves around earth.
(n) The moon is a satellite of earth if and only if it revolves around Mars.
(o) The moon is made of green cheese if and only if moon rocks are cheese crumbs.
(p) The moon is made of green cheese if and only if it reflects the light of the sun.

An intuitive assessment of *true* for statements (m) and (o) and of *false* for (n) and (p) arises from the observation that the components in the true statements have like truth values while those in the false statements have opposite truth values.

RULE 5 A biconditional is true if and only if its immediate components have identical truth values.

Let us apply these rules and find truth values of composite statements containing more than one connective. Consider the illustrative examples of Section 8.1.

(6) $(p \vee r) \rightarrow q$
(7) $\sim(p \wedge r)$
(8) $\sim p \wedge r$

Arbitrarily assign truth values to the ultimate components in dictionary (i) of that section.

Symbol	Statement	Truth Value
p	The rabbit hurries by.	false
q	Alice follows the rabbit.	false
r	The rabbit takes a watch out of his waistcoat pocket.	true

Let us deal first with composite statement (6) "If the rabbit hurries by or takes a watch out of his waistcoat pocket, then Alice follows him." The statement is a conditional. Its antecedent is the alternation of a false and a true component. The alternation is true; that is, the antecedent of the conditional is true. The consequent, however, is false. By Rule 4, the statement is false.

Composite statement (7) says "It is not true that the rabbit hurries by and takes a watch out of his waistcoat pocket." This denial has one immediate component, the conjunction of a false and a true statement. The conjunction is false, whence the denial is true. Statement (7) is true.

Composite statement (8) is expressible as "The rabbit does not hurry by and he takes a watch out of his waistcoat pocket." The first component of this conjunction is the denial of a false statement and has truth value true. Since two true statements are conjoined, the conjunction is true. It follows that statement (8) is true.

EXERCISE 8.4

1. Under what circumstances does a denial have truth value true? Truth value false?

2. When is a conjunction false?

3. When is an alternation true?

4. When is a conditional true?

5. When is a biconditional false?

6. The following conditional is stated for the five-element system.

$$\text{If } 2 + 2 = 4, \text{ then } 3 = 2^* \text{ or } 3 = 1^*$$

(a) What is the antecedent of the conditional?
(b) What is the consequent of the conditional?
(c) What is the truth value of the conditional?
(d) By use of punctuation, change the conditional to an alternation.
(e) What is the truth value of the alternation of part (d)?

8.5 TRUTH TABLES

The determination of truth values of composite statement patterns is facilitated by the use of *truth tables.* These display in an orderly fashion the truth values of the components and of the patterns themselves. To make a truth table, first write the pattern. Beneath the ultimate components, write all possible arrangements of truth values for these components, using the letter **T** to represent *true* and **F** to represent *false.* Next, write the resulting truth values of the immediate components under their principal connectives. Finally, write the resulting truth values of the composite pattern under its principal connective. An illustration will make the procedure clear.

EXAMPLE Construct a truth table for the statement pattern $(p \vee q) \rightarrow (p \wedge q)$.

Procedure Consider all possible assignments of truth values to the statement letters p and q, the ultimate components of the pattern. For each, there is a choice of T or F. Hence, there are four combinations of truth values for p and q. These combinations are listed in a specific order in Table 8.1. The truth values are written in order TFTF under p, the first statement letter to appear in the formula, and in order TTFF under the second statement letter q. This arbitrary order is employed so that truth tables constructed by different people for the same pattern can be easily compared.

Table 8.1

$(p$	$\vee$	$q)$	$\rightarrow$	$(p$	$\wedge$	$q)$
T		**T**				
F		**T**				
T		**F**				
F		**F**				

Returning to the statement pattern, consider the truth values of the immediate components. These follow from the different assignments of truth values to the ultimate components. In the first line of the truth table, p and q are both true; hence, the alternation $p \vee q$ is true and so is the conjunction $p \wedge q$. Write **T** in the first line under the symbol $\vee$ in the first component and under $\wedge$ in the second component. In the second line of the table we have p false and q true. The alternation of these components is true, but their conjunction is false. Hence, enter **T** under $\vee$ and **F** under $\wedge$. Continuing in this way, we get the entries in the remaining lines of Table 8.2.

Table 8.2

$(p$	$\vee$	$q)$	$\rightarrow$	$(p$	$\wedge$	$q)$
T	**T**	T			**T**	
F	**T**	T			**F**	
T	**T**	F			**F**	
F	**F**	F			**F**	

Finally, use the rule on the truth value of a conditional to write, beneath the principal connective, the resulting truth value of the composite statement pattern. Truth Table 8.3 is now complete. It shows that some

Table 8.3

$(p$	$\vee$	$q)$	$\rightarrow$	$(p$	$\wedge$	$q)$
T	T	T	**T**		T	
F	T	T	**F**		F	
T	T	F	**F**		F	
F	F	F	**T**		F	

instances of this pattern are true and some are false. Here are some illustrations from the five-element system. For the dictionary

$$p\colon\ 2 = 2 \qquad q\colon\ 1 = 4^* \tag{ii}$$

p and q are both true. Hence, the first line of the truth table shows that the statement "If 2 = 2 or 1 = 4*, then 2 = 2 and 1 = 4*" is true. On the other hand, the second line shows that the instance using the dictionary below is false.

$$p\colon\ 1 = 2 \qquad q\colon\ 2 = 3^* \tag{iii}$$

This instance is "If 1 = 2 or 2 = 3*, then 1 = 2 and 2 = 3*."

The pattern $p \vee \sim p$ has only one ultimate component; hence, its truth table has only two lines. Note that in the first line of Table 8.4 where p is

Table 8.4

p	$\vee$	$\sim p$
T	T	F
F	T	T

true, the denial $\sim p$ is false and the letter F is entered under the symbol $\sim$ of $\sim p$. Similarly, in the second line where p is false, the letter T is entered under the symbol $\sim$ of the component $\sim p$. This pattern always has truth value true; hence, all of its instances are true. Many other statement patterns also have truth value true for every possible combination of truth values for the ultimate components. Such patterns are said to be *logically valid.*

DEFINITION A **logically valid** statement pattern is a pattern that has truth value true under all assignments of truth values for the ultimate components.

These patterns, also called *tautologies,* are theorems of the algebra of logic. They are always true because of the five postulates on truth value that we have adopted. Some of them have names that describe the laws which they embody. For example, the alternation $p \vee \sim p$ is called the law of the excluded middle. It asserts that a statement is true or its denial is true; there is no "middle ground." It indicates that we are using a two-valued logic.

EXERCISE 8.5

1. Given the statement pattern $(p \wedge q) \rightarrow p$:

(a) Construct a truth table for this pattern. After the table has been completed, check it against Table 8.5.

(b) Why is the pattern logically valid? What can be said about all instances of it?

(c) Give an instance that satisfies the third line of the table.

Table 8.5

$(p$	$\wedge$	$q)$	$\rightarrow$	p
T	T	T	T	T
F	F	T	T	F
T	F	F	T	T
F	F	F	T	F

2. Construct a truth table for the statement pattern $(p \vee q) \leftrightarrow (\sim p \wedge q)$. In making the table note that there will be two steps in finding the truth value of the second immediate component. First, the denial $\sim p$ must be evaluated, and then the conjunction $(\sim p \wedge q)$. This analysis is shown in Table 8.6.

(a) Complete the truth table.

(b) Write an instance of the pattern using dictionary (ii) in this section. Read the truth value of the instance from the table.

(c) Follow directions for part (b) using dictionary (iii).

Table 8.6

$(p$	$\vee$	$q)$	$\leftrightarrow$	$(\sim p$	$\wedge$	$q)$
T		T		F		F
F		T		T		T
T		F		F		F
F		F		T		F

3. Show that the statement pattern $((p \rightarrow q) \wedge \sim q) \rightarrow \sim p$ is logically valid. In the instances below, drawn from the five-element system, we use the word *when* in place of *if* in order to make the translation smooth. For each statement tell which line of the truth table determines its value.

(a) If when $1 = 2$, then $3 = 4$; and $3 \neq 4$;; then $1 \neq 2$.
(b) If when $2 = 2$, then $3 = 4$; and $3 \neq 4$;; then $2 \neq 2$.

4. In the patterns tested so far, all statement letters appeared in the first immediate component. This need not happen. In the symbolic statement $p \rightarrow (p \vee q)$, the second letter appears as the last one in the pattern.

(a) Construct a truth table for this pattern. Note that a partial answer is presented in Table 8.7.

(b) Does the truth table justify any assertion about all instances of the pattern?

(c) Give two instances of the pattern. For each instance tell which line of the table gives its truth value.

Table 8.7

p	$\rightarrow$	$(p$	$\vee$	$q)$
T				T
F				T
T				F
F				F

5. Consider the statement pattern

$$((p \rightarrow q) \wedge (p \rightarrow r)) \leftrightarrow (p \rightarrow (q \wedge r))$$

This pattern contains three statement letters. For each of them there is a choice of T or F. Hence, there are $2 \cdot 2 \cdot 2 = 8$ triples of truth values for the ultimate components. To construct the truth table write four repetitions of the pair TF for p (the first letter to appear) and two repetitions of the quadruple TTFF for the second letter q; under the third letter r, write the sequence TTTTFFFF. These basic entries are shown in Table 8.8.

(a) Copy and complete the truth table for this statement pattern.
(b) Write an instance of each of the third, fifth, and sixth lines of the table.

(c) What can be said of the truth values of all instances of this pattern?
(d) Is the pattern logically valid?

Table 8.8

((p	$\rightarrow$	q)	$\wedge$	(p	$\rightarrow$	r))	$\leftrightarrow$	(p	$\rightarrow$	(q	$\wedge$	r))
T		T				T						
F		T				T						
T		F				T						
F		F				T						
T		T				F						
F		T				F						
T		F				F						
F		F				F						

6. Show that the following statement pattern is logically valid.

$$((p \rightarrow r) \wedge (q \rightarrow r)) \leftrightarrow ((p \vee q) \rightarrow r)$$

7. Show that the pattern $p \wedge \sim p$ always has truth value false.

8. Every conditional is associated with three other conditionals: the converse, the inverse, and the contrapositive, defined as follows. If the conditional is $p \rightarrow q$, then its *converse* is $q \rightarrow p$, its *inverse* is $\sim p \rightarrow \sim q$, and its *contrapositive* is $\sim q \rightarrow \sim p$.

(a) Complete the truth table presented in Table 8.9.

(b) Are any two of the conditionals true together and false together? If so, which pair or pairs?

(c) Do any two of these conditionals always have opposite truth values? If so, which pair or pairs?

Table 8.9

Truth value		Conditional	Converse	Inverse	Contrapositive
p	q	$p \rightarrow q$	$q \rightarrow p$	$\sim p \rightarrow \sim q$	$\sim q \rightarrow \sim p$
T	T				
F	T				
T	F				
F	F				

8.6 RELATIONS BETWEEN PATTERNS: LOGICAL IMPLICATION

Let us introduce the useful device of employing capital letters such as P and Q to represent statement patterns. Thus, P might be the pattern $p \vee \sim p$, and Q might be the pattern $p \rightarrow q$. In this case, P is always true, but Q is not. Or P might be the pattern $p \wedge q$ while Q might be $p \vee q$.

In this case, neither P nor Q is always true; but whenever P is true, then Q is also. Or, for a third choice, P might be $\sim(p \leftrightarrow q)$ and Q might be the pattern consisting of the single component r. In this case, the truth values of P and Q are completely independent of each other.

Let us now survey possible relations between statement patterns and study one of them. Clearly, patterns are either identical with or different from each other. If they are identical, the adjective *equivalent* properly applies to them. If they differ, they may still bear some relation to each other in the sense that their truth values exhibit some orderly association. For example, the patterns $p \vee q$ and $q \vee p$ are true together and false together. They deserve to be called equivalent. On the other hand, patterns p and $\sim p$ have exactly opposite truth values; any instance of one completely denies the corresponding instance of the other. They are said to negate each other. These relations arise from the forms of the patterns involved. They result from the truth-value rules that are fundamental postulates of our system of logic. For that reason, we say that $p \vee q$ and $q \vee p$ are *logical equivalents* of each other and that p and $\sim p$ are *logical negations* of each other.

The relations above are of the "all or none" type. A less stringent relation is the very useful logical implication. For pattern P to imply pattern Q, it is required only that whenever P is true, Q shall also be true. For example, pattern P: $p \wedge q$ logically implies pattern Q: $p \vee q$ because, as we have seen above, whenever the conjunction is true, so also is the alternation. Of course, Q cannot logically imply P because in at least one case when the alternation is true (that is, when p: T, q: F), the conjunction is false.

DEFINITION Pattern P **logically implies** pattern Q if and only if every assignment of truth values to the ultimate components that makes P true also makes Q true.

The pattern P of the definition is called the *premise*, or, if several statements are conjoined to form P, then each is called a premise. Similarly, Q is called the *conclusion*. For example, the two premises $p \vee q$ and $\sim q$ logically imply the conclusion p; because Table 8.10 shows that the conjunction of the premises is true only in line 3, and in this line the conclusion is true.

Table 8.10

p	q	$(p$	$\vee$	$q)$	$\wedge$	$\sim q$	p
T	T				F		T
F	T				F		F
T	F				T		T
F	F				F		F

When patterns P and Q of the definition are translated by the same dictionary, it is customary to say that the instance of P logically implies the instance of Q. Hence, the implication just discussed shows that the premises

$$2 = 3^* \quad \text{or} \quad 3 = 3^* \qquad \text{and} \qquad 3 \neq 3^*$$

imply the conclusion

$$2 = 3^*$$

We have the following definition.

DEFINITION A first statement **logically implies** a second if and only if the statements are corresponding instances, respectively, of two patterns of which the first logically implies the second. Corresponding instances are obtained by use of the same dictionary.

The chief mathematical use of logical implication is in drawing valid conclusions. As an example, consider the conjunction of two premises one of which is the theorem "if $a = 5$, then a has no multiplicative inverse" and the second of which is the assertion "$a = 5$." We are led inescapably to the conclusion "a has no multiplicative inverse" because pattern $p \wedge (p \rightarrow \sim q)$ logically implies pattern $\sim q$. In Exercise 8.6 you will be asked to establish this implication and the law of syllogism which says that the conjunction of the conditionals $p \rightarrow q$ and $q \rightarrow r$ logically implies $p \rightarrow r$. Let us now show how this law is used in the construction of a familiar proof. Recall the chain of arguments presented in Section 5.4 to prove that addition of pairs is commutative. We set up a dictionary and examine the structure of this proof:

a: $P = (p_1, p_2)$ is a pair

b: $Q = (q_1, q_2)$ is a pair

c: p_1, q_1, p_2, q_2 are elements of the five-element system

d: $p_1 + q_1 = q_1 + p_1$

e: $p_2 + q_2 = q_2 + p_2$

f: $(p_1 + q_1, p_2 + q_2) = (q_1 + p_1, q_2 + p_2)$

g: $P + Q = Q + P$

Theorem 5.4-1 can now be stated as $(a \wedge b) \rightarrow g$. The steps in the proof are translated into symbols as follows:

Arguments	*Reasons*
(1) $(a \wedge b) \rightarrow c$	(1) Definition of pair.
(2) $c \rightarrow (d \wedge e)$	(2) Addition is commutative in the five-element system.

Arguments	*Reasons*
(3) $(a \wedge b) \rightarrow (d \wedge e)$	(3) Steps 1 and 2 and law of syllogism.
(4) $(d \wedge e) \rightarrow f$	(4) Definition of equality of pairs.
(5) $(a \wedge b) \rightarrow f$	(5) Steps 3 and 4 and law of syllogism.
(6) $f \rightarrow g$	(6) Definition of addition of pairs.
(7) $(a \wedge b) \rightarrow g$	(7) Steps 5 and 6 and law of syllogism.

EXERCISE 8.6

1. Show that the premise $p \wedge (p \rightarrow \sim q)$ logically implies the conclusion $\sim q$.

2. Show that the premise $(p \rightarrow q) \wedge (q \rightarrow r)$ logically implies the conclusion $p \rightarrow r$.

3. Show that the conjunction of the premises $p \rightarrow q$ and $\sim q$ logically implies $\sim p$.

In each of Exercises 4–9 determine whether the implication from the premise(s) in part (a) to the conclusion(s) in part (b) is valid. Justify your answers.

4. (a) $2 = 3^*$ or $2 = 4^*$. $2 \neq 4^*$.
(b) $2 = 3^*$.

5. (a) $3' = 2$. $2 = 3^*$.
(b) $3' = 3^*$.

6. (a) If $4 = 2^*$, then $4 + 2 = 5$. $4 + 2 \neq 5$.
(b) $4 \neq 2^*$.

7. (a) If $1^* = 1'$, then $4^* = 4'$.
(b) If $1^* \neq 1'$, then $4^* \neq 4'$.

8. (a) If $2 = 3$, then $2 + 1 = 3 + 1$. $2 \neq 3$.
(b) $2 + 1 \neq 3 + 1$.

9. (a) If $2 + 3 = 5$ or $2^* = 3$, then $3^* = 2$.
(b) If $2 + 3 = 5$, then $3^* = 2$. If $2^* = 3$, then $3^* = 2$.

★10. Prove that logical implication is transitive; that is, for patterns P, Q, and R, if P logically implies Q and Q logically implies R, then P logically implies R.

8.7 LOGICAL EQUIVALENCE

Let us turn now to patterns which, like $p \vee q$ and $q \vee p$, have identical truth values for corresponding assignments of truth values to their ultimate components. Their instances using a common dictionary have like truth values. We describe such patterns and their instances by the phrase "logically equivalent." It is not necessary for the patterns to be compounded from the same ultimate components; for example, the conjunction $(p \vee \sim p) \wedge q$ and the simple pattern q are equivalent. Hence, the statement "$1 = 1$ or $1 \neq 1$, and $2 = 3^*$" has exactly the same truth value as the simple statement "$2 = 3^*$." In ordinary usage such extraneous statements as $p \vee \sim p$ are employed to give emphasis. An example is the saying, attributed to the river, "Man may come or man may go, but I go on forever."

DEFINITION Two statement patterns are **logically equivalent** if and only if they have identical truth values under every assignment of truth values to the ultimate components.

DEFINITION Two statements are **logically equivalent** if and only if they are corresponding instances of logically equivalent statement patterns.

One of the chief uses of logical equivalence is the substitution of one pattern for another. For example, the conditional $p \rightarrow q$ and its contrapositive $\sim q \rightarrow \sim p$ are logically equivalent (see Exercise 8.5.8); hence, mathematicians frequently use the device of establishing the contrapositive of a theorem whenever that is easier to prove than the theorem itself. Consider the following statement in the five-element system:

STATEMENT If $b \neq c$, then $a + b \neq a + c$.

One way to determine the truth or falsity of the statement is, of course, to investigate every pair of unlike elements. This method would require many calculations. A better way is to consider the contrapositive of the statement. This is easy to write: if $a + b = a + c$, then $b = c$. This will be recognized as the true law of cancellation for addition. Because the *proposed* statement is equivalent to its contrapositive, it also is true.

EXERCISE 8.7

1. Show that the pattern $(p \vee \sim p) \wedge q$ is logically equivalent to q.

2. Show that the biconditional $p \leftrightarrow q$ is logically equivalent to the conjunction of the conditionals $p \rightarrow q$ and $q \rightarrow p$.

3. Show that the conjunction $p \wedge (q \vee r)$ is logically equivalent to the alternation $(p \wedge q) \vee (p \wedge r)$.

In each of Exercises 4–7 translate into symbols the statements in parts (a) and (b). Determine whether the resulting statement patterns are logically equivalent.

4. (a) It is not true that $1 = 4$ or $1' = 4$.
(b) $1 \neq 4$ and $1' \neq 4$.

5. (a) If $2 + 3 = 5$, then $2^* = 3$.
(b) If $2 + 3 \neq 5$, then $2^* \neq 3$.

6. (a) $2 + 3^* = 4$ if and only if $2 - 3 = 4$.
(b) $2 + 3^* \neq 4$ if and only if $2 - 3 \neq 4$.

7. (a) If $2 = 3'$, then $2 \neq 4'$.
(b) $2 \neq 3'$ or $2 \neq 4'$.

★**8.** Show that when two patterns are logically equivalent, each logically implies the other.

★**9.** Show that logical equivalence is reflexive; that is, every pattern is logically equivalent to itself.

★**10.** Show that logical equivalence is symmetric; that is, if pattern P is logically equivalent to pattern Q, then Q is equivalent to P.

★**11.** Show that logical equivalence is transitive; that is, for patterns P, Q, and R, if P is equivalent to Q and Q is equivalent to R, then P is equivalent to R.

8.8 LOGICAL NEGATION

When two patterns such as $p \vee q$ and $\sim(p \vee q)$ always have opposite truth values, they are called *logical negations* of each other. Thus $p \rightarrow q$ and $\sim(p \rightarrow q)$ are logical negations. Patterns $p \rightarrow q$ and $p \rightarrow \sim q$ are not, however, because although they differ in truth value when p is true, they are alike in truth value when p is false.

DEFINITION Two statement patterns are **logical negations** of each other if and only if under every assignment of truth values to the ultimate components they have opposite truth values.

DEFINITION Two statements are **logical negations** of each other if and only if they are corresponding instances of patterns which are logical negations of each other.

In Exercise 8.8 the student is asked to show that the formula $p \wedge \sim q$ is a logical negation of the conditional $p \rightarrow q$. This has a useful practical application because of its affirmation that an instance of $p \wedge \sim q$ presents a counterexample to a proposed conditional $p \rightarrow q$. To illustrate, assume that in the five-element system we propose as a possible theorem the assertion "If $ab = ac$, then $b = c$." The general form of a logical negation of this statement is "$ab = ac$, but $b \neq c$." A specific instance of this negation is the true statement "$2 \cdot 5 = 3 \cdot 5$, but $2 \neq 3$." This counterexample shows that there is at least one instance in which the negation is true and the proposed statement is false. The statement, therefore, cannot qualify as a theorem.

EXERCISE 8.8

1. Show that the pattern $p \wedge \sim q$ is a logical negation of pattern $p \rightarrow q$.

2. Show that patterns $p \wedge q$ and $\sim p \vee \sim q$ are logical negations of each other, and give instances.

3. Show that patterns $p \vee q$ and $\sim p \wedge \sim q$ are logical negations of each other, and give instances.

In each of Exercises 4–6 the statements in parts (a) and (b) have opposite truth values. For each exercise, determine whether the statements in the pair are logical negations of each other.

4. (a) If $1^* = 4$, then $1' = 4$.
(b) If $1^* \neq 4$, then $1' \neq 4$.

5. (a) $1 + 2 = 3$ or $1 + 2 = 4$.
(b) $1 + 2 = 3$ and $1 + 2 = 4$.

6. (a) $5 \cdot 2 = 5 \cdot 3$ if and only if $2 \neq 3$.
(b) $5 \cdot 2 \neq 5 \cdot 3$ if and only if $2 \neq 3$.

8.9 THE LOGICAL SYSTEM AS AN ALGEBRAIC SYSTEM (OPTIONAL)

This section shows that the chapter is well named because we are, indeed, dealing with an algebraic system. The symbolic formulation of our work is considered in order to determine the elements of the system and the relations that may hold between them. The elements are the statement patterns. These range in complexity from a single statement letter to patterns containing as many ultimate components as we please. The relations are three in number: logical equivalence, logical implication, and logical negation. The operations of the system are the logical connectives. All have the property of closure because each connective combines the patterns on which it operates into a more complicated composite statement pattern.

To determine further properties of the connectives we must consider the truth values of the resulting composite patterns. Let us establish the property of commutativity for conjunction. We must show that for any two patterns P and Q, the conjunction $P \wedge Q$ is logically equivalent to $Q \wedge P$. Note first that each assignment of truth values to the ultimate components of P and Q produces a unique truth value for P and a unique truth value for Q. Hence, all possible such assignments result in exactly four distinct combinations of truth values for P and Q:

P:	T	F	T	F
Q:	T	T	F	F

It follows that the truth table constructed for $P \wedge Q$ is identical with the table for $p \wedge q$; similarly, the table for $Q \wedge P$ is the same as that for $q \wedge p$. The tables show that the conjunctions $P \wedge Q$ and $Q \wedge P$ are equivalent, and the operation of conjunction is commutative.

By analogous arguments it can be shown that when any composite statement pattern has its statement letters replaced by pattern letters, the truth tables of the original and new patterns remain identical. In Exercise 8.9 you are asked to prove that alternation is commutative, that conjunction and alternation are both associative, and that each is distributive with respect to the other. The operations possess a further property that we have not previously encountered. The conjunction $P \wedge P$ and the alternation $P \vee P$ are logically equivalent to P; that is, the conjunction or alternation of a pattern with itself yields the original pattern. The adjective *idempotent* is used to describe this situation, and the operations are said to be idempotent. The adjective signifies that the combination of an element with itself under an operation produces the element. Thus, in the five-element system, 5 is an idempotent element under addition because $5 + 5 = 5$. Both 5 and 1 are idempotent elements under multiplication because $5 \cdot 5 = 5$ and $1 \cdot 1 = 1$. Frequently the only idempotent element with respect to an operation is the identity. In the logical system, however, the property is more widely held.

When the conditional operates on two patterns it transforms them into the antecedent and consequent of a conditional statement. Except for closure it possesses none of the properties found for the first two operations. The biconditional, however, is commutative and associative although not idempotent or distributive with respect to any of the other operations. Finally, the denial acts upon a single pattern to produce a new pattern with exactly opposite truth value. It is a unary operation.

Having determined the properties of the logical connectives considered as operations, we concern ourselves with the existence of identity elements. If all we require of an identity for a connective is that it produce a pattern logically equivalent to the given pattern, then some identities do exist. For example, any logically valid pattern satisfies this weak definition for the conjunction and the biconditional, and any pattern with truth value always false is an identity for the alternation. The conditional, however, has no identity. When we seek inverses, we find that there are no patterns which fulfill our usual definition of inverse. It is clear that the properties of the logical system are dissimilar in many respects from those of the other systems we have studied.

EXERCISE 8.9

1. Establish the logical equivalence of the patterns in the following pairs and determine which properties of logical operations are thereby confirmed.

(a) $P \vee Q$ and $Q \vee P$
(b) $(P \wedge Q) \wedge R$ and $P \wedge (Q \wedge R)$
(c) $(P \vee Q) \vee R$ and $P \vee (Q \vee R)$
(d) $P \wedge (Q \vee R)$ and $(P \wedge Q) \vee (P \wedge R)$
(e) $P \vee (Q \wedge R)$ and $(P \vee Q) \wedge (P \vee R)$
(f) $P \wedge P$ and P
(g) $P \vee P$ and P
(h) $P \leftrightarrow Q$ and $Q \leftrightarrow P$
(i) $(P \leftrightarrow Q) \leftrightarrow R$ and $P \leftrightarrow (Q \leftrightarrow R)$

2. Explain why any logically valid pattern can be an identity for the conjunction.

3. Explain why any pattern always having truth value *false* can be an identity for the alternation.

4. Explain why the conditional has no identity.

5. Explain why any logically valid pattern can be an identity for the biconditional.

8.10 OPEN SENTENCES

In Section 8.1 a statement was defined to be a sentence that has a truth value as it stands. In the succeeding sections properties of statements and statement patterns were studied. When application was made to the material of earlier chapters, however, sentences rather than statements were often used in the illustrative examples. Theorems and definitions are expressed in general terms, and they cannot be translated adequately into logical symbols by use of statements alone. Consider the assertion that no element has a multiplicative inverse equal to 5. We could, to be sure, conjoin the five statements $1' \neq 5$, $2' \neq 5$, $3' \neq 5$, $4' \neq 5$, and $5' \neq 5$; but the impact of the phrase "no element has" would be lost in the long conjunction. Furthermore, the comparable statement for the system of integers would not yield to such a representation. We must adjoin to the logical system new constituents that will permit us to deal with assertions like the one above.

Let us begin our study with two sentences introduced in Section 8.1. The sentence

(c) In the five-element system $x + 3 = 2$

is true of the element 4 and false of all other elements. Similarly, the sentence

(d) It is associative

is true of some operations and false of others. Such sentences are called *open sentences.* Their truth values cannot be immediately determined, but are left open for further consideration. The letter x in sentence (c) and the word "It" in (d) are *place-holders* for names. If the place-holder x is replaced by the numeral *4*, sentence (c) is turned into the true statement

(c′) In the five-element system $4 + 3 = 2$.

If the place-holder "It" in sentence (d) is replaced by the words "Subtraction in the five-element system," we get the false statement

(d′) Subtraction in the five-element system is associative.

Instead of words such as "it," we shall customarily use letters of the alphabet such as x and y for place-holders and call them *variables.*

The totality of things whose names are under consideration for substitution in a given open sentence is called the universe of discourse or, simply, the *universe*. In sentence (c) the universe is clearly defined to be the five-element system. In sentence (d), however, we may choose it in any convenient way. We could let it consist of all operations in the five-element system, of the operation of addition in all systems in which addition is defined, or of the operations of addition and multiplication in the system of matrices. When the name of a thing chosen from the universe is substituted for the variable, the open sentence becomes a statement making an assertion about that thing. The resulting statement is called an *instance* of the open sentence. In the examples above, (c′) is a true instance of sentence (c) and (d′) is a false instance of sentence (d). Let us summarize our definitions.

A **place-holder** is a word or symbol that holds the place for the name of a thing. A place-holder is often called a *variable* and may be denoted by a letter of the alphabet.

An **open sentence** is a sentence containing a place-holder. It is true of certain things or false of certain things whose names the place-holder may represent.

The **universe of discourse** of an open sentence is the totality of things whose names may be substituted for the place-holder in the sentence.

An **instance** of an open sentence is the statement that results when the place-holder is replaced by the name of a thing from the universe.

Open sentences, like statements, may be associated by connectives. The symbols $x + 3 = 2 \wedge 4x = 1$ represent the sentence "$x + 3 = 2$ and $4x = 1$." For the universe of elements of the five-element system, the instance formed when x is replaced by the symbol 4, namely "$4 + 3 = 2$ and $4 \cdot 4 = 1$," is true. When x is replaced by 5, the resulting instance is false.

EXERCISE 8.10

1. Let the universe of discourse be the five-element system. Determine of which, if any, things in the universe each open sentence below is true.

(a) x is the identity for multiplication. (b) $y = 3^* \wedge y + 3 = 2$.
(c) $z = 3' \vee z + 3 = 2$. (d) $4e = 2 \rightarrow e + 2 = 1$.
(e) $\sim(a^* = a)$. (f) $x + 2 = 5 \leftrightarrow 2x = 3$.

2. Give a false instance of each sentence in Exercise 1.

8.11 OPEN-SENTENCE PATTERNS

Our work will proceed more smoothly if we have a good notation for open sentences. Let us use a capital letter, such as P or Q, as a symbol for a sentence and follow the capital letter by a lowercase letter to indicate the variable. Thus, Px could represent the open sentence $x + 3 = 2$, and Qx could stand for $4x = 5$. If x is replaced by another variable, the new open sentence says about the new variable what the original sentence said about the old one. For the two sentences above, Py asserts that $y + 3 = 2$, and Qy that $4y = 5$.

In earlier work we used statement letters and statement patterns independently of instances. In the same way, let us use open-sentence symbols and open-sentence patterns, for example, $Px \rightarrow Qx$. Sometimes statement letters and open-sentence symbols will be combined as in the pattern $p \wedge (Qx \vee Rx)$. In these examples the same variable is used in all open-sentence symbols in a given pattern. We shall continue this custom and thus deal with patterns in only one variable. The study of open sentences and patterns in two or more variables is interesting but beyond the intended scope of our work. Here are three definitions.

> An **open-sentence symbol** is a capital letter followed by a lowercase letter. The capital letter represents the open sentence and the lowercase letter represents the variable.
>
> An **open-sentence pattern** (in one variable) is a symbolic expression in which open-sentence symbols and, sometimes, statement letters are combined by use of symbols for the logical connectives. The same variable is used in all open-sentence symbols throughout the pattern.
>
> An **instance of an open-sentence pattern** is the word statement that results when all symbols for open sentences, statements, and logical connectives are replaced by their word equivalents.

EXERCISE 8.11

In this exercise, let the universe of discourse be the five-element system. Let open sentences be designated as follows:

Px: x is the additive identity
Qx: x has a multiplicative inverse
Rx: $x^* = x'$

Translate each expression below into word form. If the result is a statement, determine whether it is true or false. If the result is an open sentence, determine of which items from the universe, if any, it is true and of which, if any, it is false.

1. $P2$
2. $\sim P3 \wedge R3$
3. $Q4 \vee R4$
4. $R4 \rightarrow Q4$
5. Py
6. $Qy \wedge \sim Qy$
7. $Pz \rightarrow Qz$
8. $Qx \rightarrow Px$

8.12 LOGICAL RELATIONS BETWEEN OPEN-SENTENCE PATTERNS

Let us deal first with the problem of constructing a truth table for an open-sentence pattern. The discussion parallels that in Section 8.9 for statement-letter patterns. Consider the specific pattern $Px \wedge Qx$. Choose any item from the universe. The substitution of the name of this item for the variable x transforms Px into a statement having a truth value and, at the same time, transforms Qx into a statement having a truth value. Hence, for any given item, corresponding instances of Px and Qx have a fixed pair of truth values. Because we are using a two-valued logic, we have only four distinguishable sets of truth values for these instances, namely,

Px:	T	F	T	F
Qx:	T	T	F	F

It follows that the truth table for $Px \wedge Qx$ is identical with the table for $p \wedge q$. By analogous arguments it can be shown that when any composite open-sentence pattern has its open-sentence symbols replaced by statement letters, the truth tables of the original and new patterns remain identical. If either pattern is logically valid, then so also is the other (see definition in Section 8.5).

THEOREM 8.12-1 If in any composite open-sentence pattern the open-sentence symbols are replaced by statement letters, or if in any statement pattern the statement letters are replaced by open-sentence symbols, then the original and resulting patterns have identical truth tables.

Corollary 8.12-1 (a) The property of logical validity for open-sentence patterns corresponds to the same property for statement patterns.

(b) The relations of logical equivalence, logical negation, and logical implication for open-sentence patterns correspond, respectively, to the same relations for statement patterns.

EXERCISE 8.12

1. Explain why each of the open-sentence patterns below is logically valid.
   (a) $\sim(Px \wedge Qx) \leftrightarrow (\sim Px \vee \sim Qx)$
   (b) $((Px \vee q) \wedge \sim Px) \rightarrow q$
   (c) $((Px \rightarrow Qx) \wedge (Qx \rightarrow r)) \rightarrow (Px \rightarrow r)$

In each of Exercises 2 and 3 determine whether the implication from the premises in part (a) to the conclusion in part (b) is valid. Justify your answers.

2. (a) If $a = 5$, then $3a = 5$. $3a \neq 5$.
   (b) $a \neq 5$.

3. (a) If $2x = 3$, then $x = 4$. If $3x = 2$, then $x = 4$.
(b) If $2x = 3$, then $3x = 2$.

In Exercises 4 and 5 determine whether the open sentence in part (a) is logically equivalent to the open sentence in part (b). Justify your answers.

4. (a) If $a = 5$, then $2a = 5$.
(b) $2a = 5$ or $a \neq 5$.

5. (a) $e = 5$.
(b) e is the additive identity of the five-element system.

In Exercises 6 and 7 determine whether the open sentences in parts (a) and (b) are logical negations of each other. Justify your answers.

6. (a) P has a multiplicative inverse and is not a divisor of F.
(b) P is a divisor of F and does not have a multiplicative inverse.

7. (a) If $a = 4$, then $a' = 4$.
(b) If $a \neq 4$, then $a' \neq 4$.

8.13 QUANTIFIED SENTENCES

A sentence may contain a variable yet be a statement. Consider the following sentences for which the universe is the five-element system.

$x = x^*$.

There is an x such that $x = x^*$.

For every x, $x = x^*$.

There is an x such that $x \neq x^*$.

For every x, $x \neq x^*$.

The first is an open sentence, true of element 5 and false of element 3. The second is a statement because it asserts the existence of an element that is its own additive inverse. This statement is true, since such an element, namely 5, does exist. The third is also a statement because it asserts that every element is its own additive inverse. This statement is false since $3 \neq 3^*$. Note that one contrary instance (counterexample) is sufficient to demonstrate the falsity of a statement that begins with the words "for every." The fourth sentence is the true statement that some element is not its own additive inverse. The last one is the false statement that every element fails to be its own additive inverse.

If we let Px stand for the open sentence $x = x^*$, we can summarize the discussion above by giving symbol and word equivalents.

Px	x is its own additive inverse.
There is an x, (Px)	Some element is its own additive inverse.
For every x, (Px)	Every element is its own additive inverse.
There is an x, $(\sim Px)$	Some element is not its own additive inverse.
For every x, $(\sim Px)$	Every element fails to be its own additive inverse.

The last four statements are called quantified sentences because they concern the number or quantity of items from the universe that have the indicated characteristics. The sentence "There is an x, (Px)" asserts that there exists an item in the universe of which Px is true. The phrase "there is an x" is called the *existential* quantifier. The traditional symbol for it is $\exists x$, formed by reversing the capital letter E, the first letter of the word *exists*. The sentence "For every x, (Px)" asserts that Px is universally true of all items; hence, the phrase "for every x" is called the *universal* quantifier. Its traditional symbol is $\forall x$, formed by inverting the capital letter A, the first letter of the word *all*. The formal definitions are:

> The following phrases are **logical quantifiers:** "there is an x," denoted by $\exists x$ and called the **existential** quantifier; and "for every x," denoted by $\forall x$ and called the **universal** quantifier.
>
> A **quantified sentence** is a statement made by prefixing a logical quantifier to an open sentence, the variable being alike in each.
>
> A **quantified-sentence symbol** is made by prefixing a symbol for a logical quantifier to a symbol for an open sentence, the variable being alike in each.

Quantified-sentence patterns and their instances have definitions analogous to those for open-sentence patterns and instances.

Let us consider the translation of four new sentences into symbols. The universe is the set of linear equations in the five-element system, and open sentence Px is "x has a solution."

(a) Some linear equations have solutions.
(b) All linear equations have solutions.
(c) Some linear equations do not have solutions.
(d) Every linear equation fails to have a solution.

Sentence (a) is a true statement. We have solved many linear equations while studying the five-element system. If we rephrase the sentence to say "There is a linear equation that has a solution," we get the translation $\exists x(Px)$.

Sentence (b) is a false statement because the equation $5x = 3$ has no solution. We can rewrite it in the form "For every linear equation, the equation has a solution," or in more concise notation, $\forall x(Px)$.

Sentence (c) is the true statement that there is a linear equation that does not have a solution, for example, $5x = 3$ cited above. Using symbol Px, its expression is $\exists x(\sim Px)$.

Sentence (d) is a false statement. It is contradicted by the many linear equations that we have solved. The following translation is obvious: $\forall x(\sim Px)$.

EXERCISE 8.13

1. Let the universe be the five-element system. Choose an appropriate dictionary and translate each of the following sentences into symbols.

(a) Some elements have additive inverses.
(b) Every element has an additive inverse.
(c) Some elements do not have additive inverses.
(d) All elements fail to have additive inverses.

2. Let the universe be the five-element system. Let open sentence Px be "x has a multiplicative inverse." Translate into words the quantified-sentence symbols below.

(a) $\exists x(Px)$ (b) $\forall x(Px)$
(c) $\exists x(\sim Px)$ (d) $\forall x(\sim Px)$

3. (a) Show that in each of Exercises 1 and 2 above, statement (a) presents counterexamples to (d).

(b) Would it be reasonable to call statements (a) and (d) logical negations of each other? Justify your answer.

4. In each of Exercises 1 and 2 what relation exists between statements (b) and (c)?

8.14 ON THE TRANSLATION OF QUANTIFIED-SENTENCE PATTERNS

In the preceding section we discussed four typical forms of quantification of one open sentence, and we determined a set of instances for a universe consisting of all linear equations in the five-element system and the dictionary Px: x has a solution. Let us now enlarge the universe to include all equations in the five-element system, and adjoin to the dictionary the open sentence Qx: x is linear. The variable x no longer refers to a linear equation. Instead, it denotes any equation whatever. If we wish to restrict an observation to linear equations, we must do so by using open sentence Qx. We shall discuss the translation into symbols of each of the word statements (a)–(d).

(a) *Some linear equations have solutions.* The symbolic sentence must explain that there is in the universe an equation that is both linear and solvable; that is, there is an item of which open sentences Qx and Px are simultaneously true. We have $\exists x(Qx \wedge Px)$.

(b) *All linear equations have solutions.* In Section 8.13 sentence (b) was derived from sentence (a) by changing from the existential to the universal quantifier. Shall we use a corresponding substitution here? What does it mean to write $\forall x(Qx \wedge Px)$? This symbolic sentence has the literal translation that every equation in the universe is both solvable and linear. It makes the sweeping assertion that in the five-element system every equation is a solvable linear equation. This is quite different from the intended assertion; statement (b) sets up the modest requirement that when an equation is linear, then it is solvable. Expressed in a form that indicates the quantification more clearly, it states that for every equation, if the equation is linear, it is solvable. The symbolic statement $\forall x(Qx \to Px)$ is now obvious.

(c) *Some linear equations do not have solutions.* Interpretation of this declaration in terms of quantifiers yields the word sentence that in the universe there is an equation that is linear but not solvable. It produces the symbolic sentence $\exists x(Qx \wedge \sim Px)$.

(d) *Every linear equation fails to have a solution.* Like statement (b), this comment concerns only linear equations. It can be rewritten to say that for every equation, if the equation is linear, then it is not solvable. The desired symbolic expression is $\forall x(Qx \to \sim Px)$.

Finally, we note a wording that is unacceptable for any quantified sentence. It is the sequence "all . . . do not . . . ." Consider the statement "All linear equations do not have solutions." In ordinary usage this statement could mean that although some linear equations may have solutions, still not all of them do. In this sense it refers to pattern $\sim\forall x(Qx \to Px)$. The sentence can, however, be understood to mean that all linear equations fail to have solutions. In this sense it refers to the pattern $\forall x(Qx \to \sim Px)$. Because of these two possible interpretations, this ambiguous form of statement should never be used.

EXERCISE 8.14

Let the universe be the integers -5, -3, -1, 2, 4, 6. Note that in this universe all negative integers are odd, and all positive integers are even. Let open sentences be Px: x is even. Qx: x is positive. Here $\sim Px$ can be translated as "x is odd," and $\sim Qx$ as "x is negative."

1. Translate the following statements into symbols.
 (a) Some even integers are positive.
 (b) No odd integers are positive.

(c) All even integers are positive.
(d) Some odd integers are negative.

2. Translate the following symbolic statements into words.

(a) $\exists x(Qx \wedge \sim Px)$ (b) $\forall x(Qx \rightarrow Px)$
(c) $\forall x(\sim Qx \rightarrow \sim Px)$ (d) $\exists x(Qx \wedge Px)$

8.15 TRUTH VALUES OF QUANTIFIED SENTENCES

The set of all natural numbers that are less than 1 contains no members at all, while the set of those that are greater than 1 contains almost all natural numbers. Empty universes sometimes present interesting problems to professional mathematicians or logicians, but to us they would be only a source of distracting special considerations. Let us, therefore, make a special provision eliminating them from our discussion.

RULE ON UNIVERSES In all subsequent work only nonempty universes are considered.

The rules for truth values of quantified sentences are postulates of our logical system like the laws governing truth values of statements. Since they are abstracted from practical experience, however, they should appear reasonable to us. (The numbering begun in Section 8.4 is continued, and so the rules are numbered 6 and 7.)

RULE 6 The quantified sentence $\exists x(Px)$ is true if and only if there is at least one element in the universe of which the open sentence Px is true. The quantified sentence $\exists x(Px)$ is false if and only if the open sentence Px is false of every element in the universe.

RULE 7 The quantified sentence $\forall x(Px)$ is true if and only if the open sentence Px is true of every element in the universe. The quantified sentence $\forall x(Px)$ is false if and only if the open sentence Px is false of at least one element in the universe.

EXERCISE 8.15

1. Find the truth values of the quantified sentences in Exercises 8.14.1 and 8.14.2.

★2. Let Px and Qx represent general open sentences in any universe. Show that in each of the following conditionals whenever the antecedent is true, the consequent is also true.
(a) $\forall x(Px) \rightarrow \exists x(Px)$.
(b) $\forall x(Px) \rightarrow Pa$, where a is the name of any element in the universe.

****3.** Repeat Exercise 2 for the conditionals below.

(a) $(\forall x(Px \rightarrow Qx) \wedge \exists x(Px)) \rightarrow \exists x(Qx)$

(b) $(\forall x(Px \rightarrow Qx) \wedge \forall x(Px)) \rightarrow \forall x(Qx)$

8.16 LOGICAL RELATIONS BETWEEN QUANTIFIED-SENTENCE PATTERNS

Rule 6 sets forth a criterion for the falsity of $\exists x(Px)$, namely, that open sentence Px is false of everything in the universe. Because sentences Px and $\sim Px$ have opposite truth values, the criterion can be restated to say that sentence $\sim Px$ is true of everything in the universe. According to Rule 7, we now have $\forall x(\sim Px)$ true. Hence, $\exists x(Px)$ and $\forall x(\sim Px)$ are logical negations of each other. In the same way, Rule 7 indicates that $\forall x(Px)$ and $\exists x(\sim Px)$ are logical negations of each other. Note that in each case the existential statement presents a counterexample to the universal one, and the universal statement makes it impossible for even one element in the universe to have the characteristics required for the truth of the existential statement.

Now the most obvious logical negation of any statement is its direct denial. Furthermore, two logical negations of the same pattern are logically equivalent. Hence, we have pattern $\sim\exists x(Px)$ equivalent to $\forall x(\sim Px)$ and $\sim\forall x(Px)$ equivalent to $\exists x(\sim Px)$.

Let quantified sentence $\forall x(Px)$ be true. Since the universe is nonempty, there is surely some element of which Px is true; hence, $\exists x(Px)$ is true. In other words, $\forall x(Px)$ logically implies $\exists x(Px)$.

EXERCISE 8.16

In Exercises 1–3 the universe is the five-element system. The dictionary contains four open sentences.

Px: x has an additive inverse
Qx: x has a multiplicative inverse
Rx: $x^* = x'$
Sx: $x' = 5$

1. Translate the four quantified sentences below and thus write a brief composition on the properties of inverses in the five-element system.

$$\forall x(Px) \qquad \sim\forall x(Qx) \qquad \exists x(Rx) \qquad \sim\exists x(Sx)$$

2. Replace each pattern in Exercise 1 with a logical equivalent.

3. Replace each pattern in Exercise 1 with a logical negation. Then translate into words. The result will be a composition different in meaning and wording from that in Exercise 1.

4. For general open sentences Px and Qx in any universe determine whether:
   (a) $\forall x(Px \wedge Qx)$ logically implies $\forall x(Px)$.
   (b) $\exists x(Px \wedge Qx)$ logically implies $\exists x(Px) \wedge \exists x(Qx)$.

8.17 RETROSPECT AND PROSPECT

We have completed our formal study of the algebra of logic. From a few definitions and truth-value rules we have created an impressive system. It possesses a dual nature. On the one hand, it is a self-contained algebraic system, complete with elements, relations, and operations that display not only many expected characteristics but also some peculiarly individual ones. On the other hand, it is a system whose great importance lies in its application to the analysis of the structure and validity of argument. We have been introduced to a powerful tool and have gained a considerable amount of information about its use.

When we direct our attention to the collection of things in the universe of which an open sentence is true, we will discover a new mathematical entity, namely a set, which can serve as the general element in a new algebraic system. The operations of this system are defined in terms of open sentences, and its relations are described by use of quantified sentences. The new system forms a companion algebra to that of logic in the sense that each property of one has a matching property in the other. The correspondence is so complete that it would be quite possible to study the algebra of sets and then develop from it the algebra of logic. In the next chapter we shall investigate the new system.

9
THE ALGEBRA OF SETS

In previous work we have been concerned with the behavior of individual items from a universe. Now we shall consider collections of items and study systems in which such collections are the elements. In Chapter 6 we made the acquaintance of equivalence sets of pairs. In this chapter we shall also discuss aggregates such as the identities of the five-element system or the pairs of the first system having both entries alike. We shall investigate the properties that characterize such collections.

9.1 DEFINITION AND NOTATION

An open sentence is true of or false of each element in its universe of discourse. It is possible that different sentences are true of exactly the same things. For example, in the universe of natural numbers the sentences

$$Px\colon\quad x \text{ is even}$$

$$Qx\colon\quad x + 1 \text{ is odd}$$

$$Rx\colon\quad x \text{ is a multiple of } 2$$

are all true of the even natural numbers and only of the even natural numbers. The importance of an open sentence is, therefore, not so much the words it uses as the selection it makes of items of which it is true. The totality of things so selected is called a *set* of things, the extension of the open sentence. The set of even natural numbers is the extension of any of the open sentences Px, Qx, or Rx above. The notation $\{x \mid Rx\}$ designates the extension of Rx. The braces are standard, and the general symbol is read "the set of all elements x from the universe for which Rx." For our present universe and dictionary we have "the set of all natural numbers that are multiples of 2."

Because an open sentence expresses a property, its extension is the collection of things in the universe that have this property. Given any item of the universe, we can tell whether it belongs in the extension of a given open sentence by testing its fulfillment of the condition expressed in the sentence. The number 4 is in $\{x \mid Rx\}$ because $R4$ is a true statement. The number 3 is not, because $R3$ is false. The symbol $\in$ stands for the phrase *is a member of*. We have $4 \in \{x \mid Rx\}$ and $3 \notin \{x \mid Rx\}$.

When an open sentence is true of everything in the universe, its extension is, of course, the whole universe. This set is called the *universal set* and is represented by the letter U. If the open sentence is true of nothing in the universe, its extension is empty. The symbol $\varnothing$ is used for the empty, or null, set. For example, U is the extension of open sentence $Rx \vee \sim Rx$, and $\varnothing$ of $Rx \wedge \sim Rx$.

A set may be described not only by an open sentence but also, as in the case of equivalence sets, by an enumeration or tabulation of its members. In this case, the membership list is enclosed in braces. The symbol $\{1,2,5,6,7\}$ defines the set consisting of natural numbers 1,2,5,6, and 7. It is customary to list each member of a set only once; hence, $\{1,2,5,6,7\}$ is the same set as $\{1,1,2,2,2,5,6,6,7,7,7,7\}$. The symbol $\{\quad\}$ presents a tabulation of the null set. Capital letters will generally be used to designate sets and lowercase letters to indicate members of a set.

In summarizing our discussion the word *set* is left undefined. No simple definition free of philosophical ambiguities has yet been devised. The nature and characteristics of sets will emerge from the definitions of other terms and from the properties recorded in subsequent theorems.

EXERCISE 9.1

Let the universe be the natural numbers.

1. Describe each of the following sets by determining which elements are in it. Use U and $\varnothing$ to indicate the universal and null sets, respectively.

(a) $\{x \mid x \text{ is less than } 8\}$.
(b) $\{x \mid x \text{ is the identity for multiplication}\}$.
(c) $\{x \mid x \text{ is the identity for addition}\}$.
(d) $\{x \mid x + 2 \text{ is a member of the universe}\}$.

2. Let Gx be the open sentence x *is odd*. Show that each of the statements below is true.

(a) $41 \in \{x \mid Gx\}$.
(b) $22 \in \{x \mid \sim Gx\}$.
(c) $3 \notin \{x \mid Gx \wedge x \text{ is greater than } 10\}$.
(d) $14 \in \{x \mid x = 2 \rightarrow Gx\}$.

3. Show that the following statements are true.

(a) $\{x \mid x \text{ has a multiplicative inverse}\}$ is enumerated by $\{1\}$.
(b) $\{x \mid x = 1 \vee x = 4\}$ is enumerated by $\{1,4\}$.
(c) $\{\quad\}$ is a tabulation of $\{x \mid x = 1 \wedge x = 4\}$.
(d) $\{1,2,5\}$ is a tabulation of $\{x \mid \text{if } x \text{ is greater than 2, then } x = 5\}$.

9.2 RELATIONS BETWEEN SETS

All previous systems possessed a relation of equality or equivalence, and the system of logic had relations of implication and negation as well. What relations exist between sets? Clearly, two sets may be alike or may differ. They are alike if they have exactly the same elements, and there is only one way in which this can occur. If they differ, however, they may be dissimilar in any of several ways. Consider sets of elements of the five-element system. They may have no members whatever in common; such sets are $\{2,3,4\}$ and $\{1,5\}$. They may have some, but not all, members in common; sets $\{2,3,4\}$ and $\{1,3,5\}$ are unlike, but each contains element 3. One set may contain all the members of the other; for example, $\{2,3,4\}$ and $\{1,2,3,4\}$. Each kind of dissimilarity defines a relation. Hence, the system of sets boasts three relations beside the customary one of equality. They are listed below in the order in which we shall study them.

	Relation	*Description*
(1)	Equality	The sets contain exactly the same elements.
(2)	Inclusion	All the elements of one set are elements of the other.
(3)	Meeting	The sets have at least one element in common.
(4)	Disjointness	The sets have no elements in common.

9.3 EQUALITY OF SETS

For equality of sets we use the customary definition of identity. In terms of elements this means that every member of one is a member of the other and conversely. Thus, in the five-element system, $\{2,3,4\}$ is equal to $\{2,3,4\}$, and neither of these sets is equal to $\{1,3,4\}$ or to $\{1,2,3,4\}$.

DEFINITION Two sets are **equal** if and only if they contain exactly the same elements.

What is a symbolic statement of this definition? If the equal sets are A and B, then each element of the universe lies in both A and B or in neither one. That is, open sentences $x \in A$ and $x \in B$ are true together or false together. The biconditional $x \in A \leftrightarrow x \in B$ is, therefore, always true; whence the quantified sentence $\forall x(x \in A \leftrightarrow x \in B)$ is true. On the other hand, when $A \neq B$, there is some element in the universe that lies in one set but not in the other. Of this element, open sentences $x \in A$ and $x \in B$ have opposite truth values. The biconditional above is false, and, by Rule 7, so is the quantified sentence.

THEOREM 9.3-1 For sets A and B, $A = B$ if and only if $\forall x(x \in A \leftrightarrow x \in B)$.

It is not difficult to tell when a given set is the universal set or when it is the null set. In the former case all elements belong to it, and in the latter no elements do. The symbolic statement of these criteria forms the second theorem on sets.

THEOREM 9.3-2 For set A, $A = U \leftrightarrow \forall x(x \in A)$, and $A = \varnothing \leftrightarrow \forall x(x \notin A)$, and $A \neq \varnothing \leftrightarrow \exists x(x \in A)$.

EXERCISE 9.3

1. Let the universe be the natural numbers not exceeding 6. Tabulate the following sets.[1] Then exhibit all couples of equal sets. Show two sets that are not equal.

$A = \{x \mid x \text{ is odd or } x \text{ is greater than } 1\}$ $\quad B = \{x \mid x \text{ has an additive inverse}\}$
$C = \{x \mid x \text{ has a multiplicative inverse}\}$ $\quad D = \{x \mid x \text{ is less than } 2\}$
$E = \{x \mid x \in U\}$ $\quad F = \{x \mid x = x + 1\}$

2. Let the universe be the five-element system. Which of the following sets are equal to each other?

$A = \{1,2,3\}$ $\quad B = \{4,5\}$
$C = U$ $\quad D = \{1^*,5^*\}$
$E = \{1',2',3'\}$ $\quad F = \{2^*,3^*,4^*\}$
$G = \{4,4',5,5^*\}$ $\quad H = \{1,1^*,2,2^*,5\}$
$I = \{1,2,2',3,3'\}$

3. Let the universe be the set of rectangles, and let

A = the set of rectangles having four equal sides
B = the set of rectangles having four equal angles

[1] Instead of tabulating the universal and null sets, use symbols U and $\varnothing$, respectively. Make the further rule that these symbols represent their corresponding sets either in a specifically defined universe (as here) or in an unspecified universe, according as the context of a problem indicates.

C = the set of rectangles having two adjacent sides equal
D = the set of rectangles having opposite sides equal
E = the set of rectangles having equal diagonals
F = the set of rectangles whose diagonals are perpendicular to each other

Which of the above sets are equal to each other?

9.4 INCLUSION OF SETS

Our study of the second of the four relations begins with a definition. A first set is a subset of a second, or the first is contained in the second, if and only if every member of the first is also a member of the second. The relation of being a subset is called *inclusion* and is denoted by the symbol $\subset$. Thus, in the system of integers $\{2,3,4\} \subset \{1,2,3,4,5,7\}$, but $\{2,3,4\} \not\subset \{1,3,5,8\}$. Recall that in describing equality we made the statement that two sets are equal when every member of one is a member of the other, and conversely. Equality could now be defined in terms of inclusion by saying that each set is a subset of the other.

DEFINITION Set A **is a subset of** set B, or $A \subset B$, if and only if every member of A is also a member of B. *Inclusion* is the relation of being a subset.

We digress briefly to describe a second notation that may be used for inclusion of sets. Sometimes the symbol $A \subset B$ is reserved to mean "A is a subset of B but is not equal to B." In this case, A is said to be a *proper* subset of B. When the symbol $A \subset B$ is interpreted in this restricted sense, then the symbol for the relation described in our definition is $A \subseteq B$ and its translation is "A is a proper subset of B or A equals B." The distinction between subset and proper subset introduces notational complications when definitions and theorems on sets are expressed in terms of quantified sentences. Let us therefore use the terminology and notation of the formal definition above.

If $A \subset B$, then for each element in the universe of which $x \in A$ is a true statement, the statement $x \in B$ must also be true. Hence, for all elements in the universe the conditional $x \in A \rightarrow x \in B$ is true; consequently, the quantified sentence $\forall x(x \in A \rightarrow x \in B)$ is true. On the other hand, if $A \not\subset B$, then there exists at least one item in the universe that lies in A but not in B. For such an item, the instances of $x \in A$ and $x \in B$ have truth values true and false, respectively. The conditional above is false and so is the quantified sentence. The criterion for inclusion of set A within set B is, therefore, the truth of the assertion $\forall x(x \in A \rightarrow x \in B)$.

THEOREM 9.4-1 For sets A and B, $A \subset B$ if and only if $\forall x(x \in A \rightarrow x \in B)$.

A conditional is true: first, when antecedent and consequent are identical; second, when the antecedent is false; and third, when the consequent is true. The first case results when we let $B = A$ in the definition of inclusion; then $\forall x(x \in A \rightarrow x \in A)$ is true, and $A \subset A$. The second case occurs when we let $A = \varnothing$ in the definition, because the sentence $x \in \varnothing$ is always false. The true quantified sentence $\forall x(x \in \varnothing \rightarrow x \in B)$ asserts that the null set is a subset of every set. This concept is difficult to grasp but is a correct deduction from the definition. We shall find it so useful that we shall come to accept it with the ease of familiarity. The property presented by the third case places no strain on the imagination. If we let $B = U$ in the definition of subset, the consequent of the conditional, namely, $x \in U$, is always true, and the quantified sentence $\forall x(x \in A \rightarrow x \in U)$ is true. It guarantees that every set is a subset of the universal set.

THEOREM 9.4-2 For set A, $A \subset A$ and $\varnothing \subset A \subset U$.

EXERCISE 9.4

In this exercise, sets A, B, and C are any sets in any universe. For other sets the universe consists of the integers from -4 to 4 inclusive. Their definitions are given below.

$$D = \{1,3\} \qquad E = \{-3,-1\}$$
$$F = \varnothing \qquad G = U$$
$$H = \{-1,1\} \qquad J = \{0,1,4\}$$
$$K = \{0,1\} \qquad L = \{0\}$$
$$M = \{1\}$$

1. Exhibit (a) three pairs of sets in which one set is a subset of the other; (b) three pairs of sets in which neither is a subset of the other.

2. Inclusion of sets is transitive providing that for sets A, B, and C, if $A \subset B$ and $B \subset C$, then $A \subset C$.

(a) Show that the transitive property of inclusion holds for sets L, K, and J.

(b) Show that the transitive property of inclusion holds for general sets in a general universe.

3. (a) Explain why $L \not\subset F$.

(b) What is true of any set A when $A \subset \varnothing$?

(c) How does the definition of inclusion support your answer to (b)?

4. (a) Let set A be nonempty. Can any set containing A be empty? Explain.

(b) How does the definition of inclusion support your answer to (a)?

5. (a) Explain why $G \not\subset E$.

(b) What is true of any set A when $U \subset A$?

(c) How does the definition of inclusion support your answer to (b)?

★6. (a) Explain why the following conditional is true: $(A \subset B \wedge a \in A) \rightarrow a \in B$.

(b) Write the conditional of part (a) in words instead of symbols.

9.5 PROPERTIES OF INCLUSION OF SETS

We have discovered several properties of inclusion of sets, some of which are obvious. Inclusion of sets is transitive. A set contains the universal set if and only if it is the universal set. On the other hand, a set is contained in the empty set if and only if it is the empty set. If a set is empty, all its subsets are empty; equivalently, if a set contains a nonempty subset, then it is nonempty. In the next section we show how diagrams can support our findings.

9.6 USE OF DIAGRAMS

Diagrams can be helpful in both illustrating and suggesting relations between sets. A rectangle may be taken to represent the universe, and in this rectangle circles may be drawn to represent sets. For example, set A in universe U is shown in Figure 9.1. If a set is known to be nonempty, a crossmark is placed inside its circle. Thus, nonempty set A in universe U is pictured in Figure 9.2. If a set is known to be empty, its circle is marked with the symbol $\varnothing$. The absence of crossmark or symbol $\varnothing$ indicates that it is not known whether a set has elements in it.

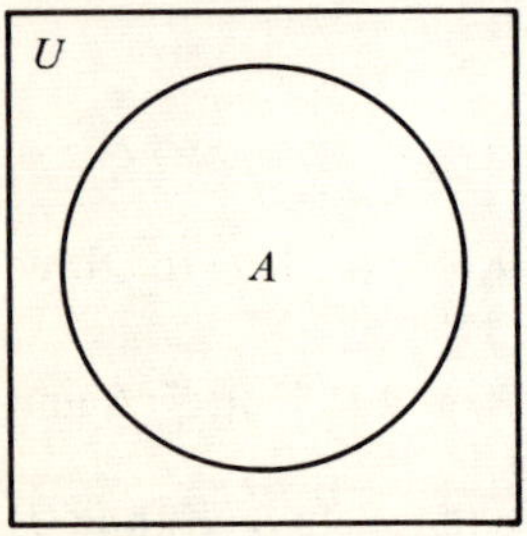

FIGURE 9.1

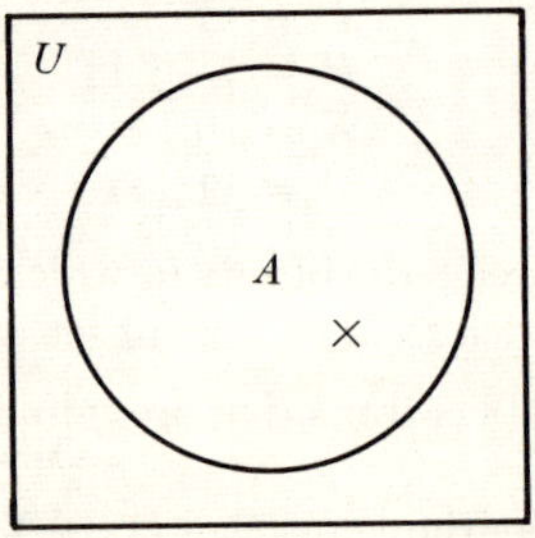

FIGURE 9.2

Now let two sets A and B be equal. Since by definition they are the same set, they must be represented by the same region in the diagram. A single circle is, therefore, marked with the two letters A and B. Thus, the diagram for $A = \varnothing$ in Figure 9.3 is a special case of that for $A = B$.

To show that A is a subset of B, the diagram must make clear that all elements of A are elements of B. It does this when the circle for A is enclosed within that for B as in Figure 9.4. Suppose now that B in Figure 9.5 is known to be empty, then the symbol $\varnothing$ is placed with B in the outer circle. Because there are no elements whatever in the large circle, there cannot be

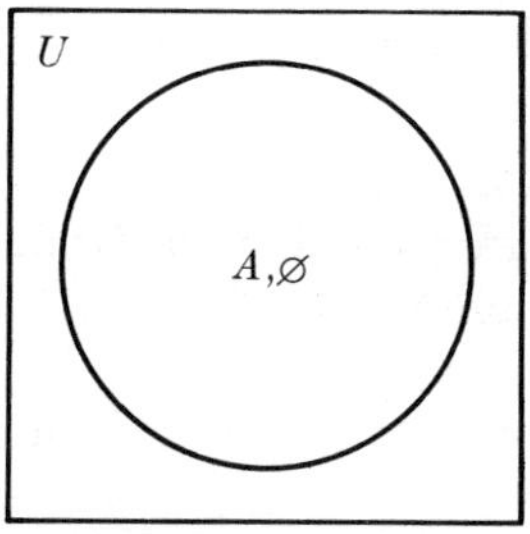

FIGURE 9.3

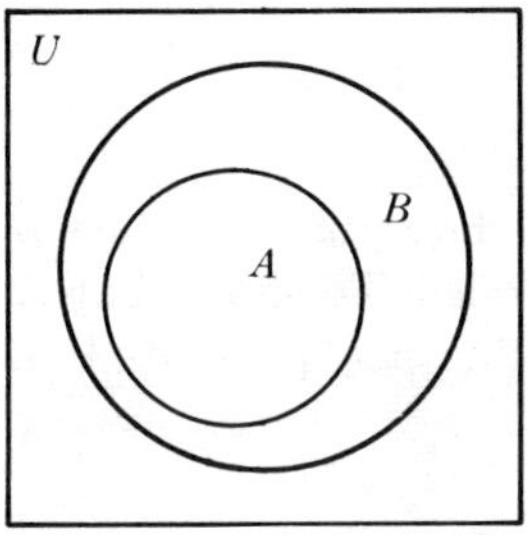

FIGURE 9.4

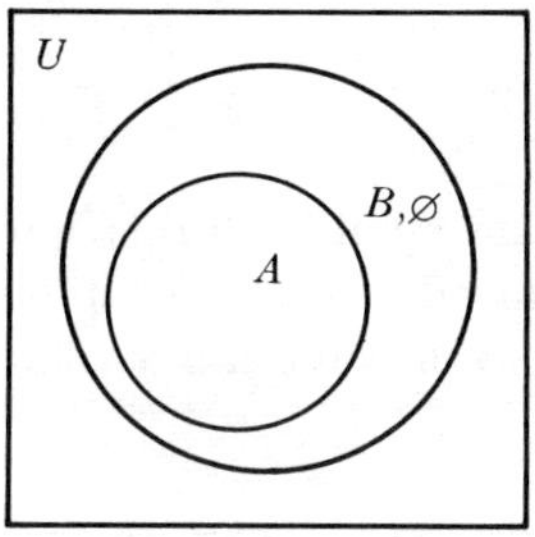

FIGURE 9.5

any within the enclosed circle representing A. Hence, Figure 9.5 shows that any subset of an empty set is empty.

EXERCISE 9.6

1. Draw a diagram to show that set A is nonempty and is a subset of B. How does the diagram show that B also is nonempty?

2. Explain how the diagram for $A = B$ shows that $A \subset B$ and $B \subset A$.

3. Draw a diagram to show that sets A, B, and C have $A \subset B$ and $B \subset C$. How does the diagram confirm the transitive law of inclusion?

9.7 MEETING AND DISJOINTNESS OF SETS

Two sets meet when they have elements in common. In the system of natural numbers let $A = \{2,3,4\}$, $B = \{2,6\}$, and $C = \{1,5,7\}$. Then A meets B, but neither of these sets meets C. There is no symbol for *meets;* we use the word itself. The description of the relation is very simple, whether it is expressed in words or symbols.

DEFINITION Set A **meets** set B if and only if A and B have at least one member in common. The relation is called *meeting.*

THEOREM 9.7-1 Set A meets set B if and only if $\exists x(x \in A \wedge x \in B)$.

The final relation, that of disjointness, holds when sets have no elements in common. For the sets enumerated above, A and B are each disjoint from C. There is no symbol for *is disjoint from;* we use the phrase itself.

DEFINITION Set A **is disjoint from** set B if and only if A and B have no members in common. The relation is called *disjointness.*

THEOREM 9.7-2 Set A is disjoint from set B if and only if

$$\sim\exists x(x \in A \wedge x \in B)$$

The diagramming of both relations is obvious. To show that A meets B a crossmark is placed in the overlap of the A and B circles. To show that A is disjoint from B no overlap is shown. The diagrams are presented in Figure 9.6.

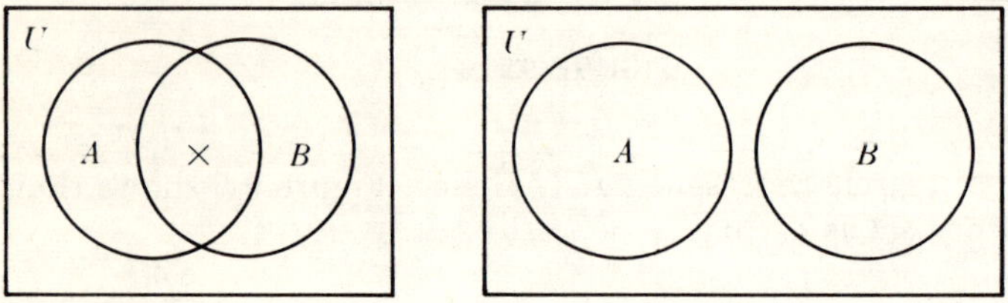

A meets B A is disjoint from B

FIGURE 9.6

The quantified sentences describing disjointness and meeting are logical negations of each other. The two relations are, therefore, *mutually exclusive.* Furthermore, by the law of the excluded middle, $p \vee \sim p$, either the sentence describing disjointness or that describing meeting must be true. In such a case, the two relations are said to be *exhaustive* because between them they exhaust all possibilities.

THEOREM 9.7-3 In the system of sets, the relations of meeting and disjointness are mutually exclusive and exhaustive.

EXERCISE 9.7

1. Study the diagram for A meets B. Explain why it illustrates the two further relations $A \neq \varnothing$ and $B \neq \varnothing$.

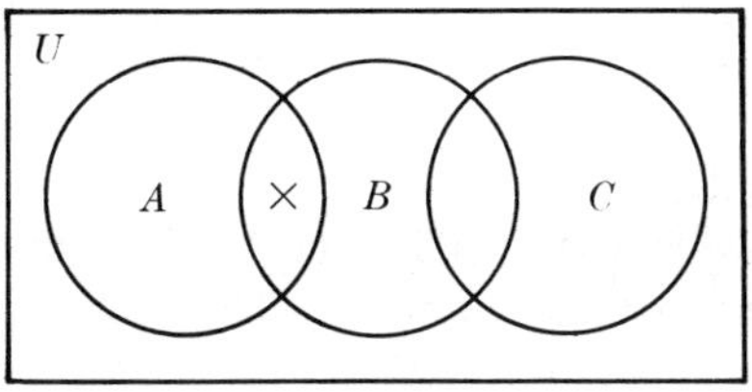

FIGURE 9.7

2. List all the relations shown in the diagram of Figure 9.7.

In Exercises 3–6 the sets of Exercise 9.4 are used.

3. (a) Does F meet M?
(b) Does M meet M?
(c) Does F meet F?

4. (a) Is F disjoint from M?
(b) Is M disjoint from M?
(c) Is F disjoint from F?

5. (a) Show that D meets H and H meets K.
(b) Does D meet K?
(c) Show that E meets H and H meets J.
(d) Does E meet J?

6. (a) Show that D is disjoint from E and E is disjoint from K.
(b) Is D disjoint from K?
(c) Show that E is disjoint from M and M is disjoint from L.
(d) Is E disjoint from L?

7. Using the results of Exercises 3–6 above, give reasonable answers to the following questions.
(a) When does a set meet itself?
(b) When is a set disjoint from itself?
(c) Is meeting of sets transitive? That is, for sets A, B, and C, if A meets B and B meets C, does A meet C?
(d) Is disjointness of sets transitive?

8. Is it true that if sets A and B meet, then
(a) All subsets of A meet all subsets of B?
(b) All sets containing A meet all sets containing B?

9. Is it true that if sets A and B are disjoint, then
(a) All subsets of A are disjoint from all subsets of B?
(b) All sets containing A are disjoint from all sets containing B?

9.8 PROPERTIES OF MEETING AND DISJOINTNESS OF SETS

The existence of the empty set falsifies many of the general statements we would like to make about meeting of sets. For example, it seems plausible that a set should meet itself. The empty set, however, has no elements; hence, it presents a counterexample to the plausible statement. We are tempted to assume that when set A is a subset of set B, then A meets B. Unfortunately, the null set is a subset of every set; yet it has no elements with which to meet its containing set. Appropriately, our first theorem on meeting of sets is the declaration of the intransigence of $\varnothing$.

THEOREM 9.8-1 The null set is disjoint from every set, or, equivalently, if two sets meet, then both are nonempty.

THEOREM 9.8-2 *Law of Meeting for Inclusion.* If two sets meet, then all sets containing one meet all sets containing the other.

THEOREM 9.8-3 *Law of Disjointness for Inclusion.* If two sets are disjoint, all subsets of one are disjoint from all subsets of the other.

9.9 QUANTIFIED SENTENCES AND RELATIONS BETWEEN SETS (OPTIONAL)

Because definitions of the relations between sets are expressed in terms of quantified sentences, it is reasonable to expect that theorems on these relations are similarly expressible. For example, the assertion "$A = B$ if and only if $A \subset B$ and $B \subset A$" has the immediate translation

$$\forall x(x \in A \leftrightarrow x \in B) \leftrightarrow (\forall x(x \in A \rightarrow x \in B) \wedge \forall x(x \in B \rightarrow x \in A))$$

It is interesting to note that the truth of this assertion can be established by use of laws of logic. Replacement of $x \in A$ and $x \in B$ by Px and Qx, respectively, shows that the symbolic sentence is a quantification of the open sentence $(Px \leftrightarrow Qx) \leftrightarrow ((Px \rightarrow Qx) \wedge (Qx \rightarrow Px))$. This, in turn, is a modification of the statement pattern $(p \leftrightarrow q) \leftrightarrow ((p \rightarrow q) \wedge (q \rightarrow p))$. In Exercise 8.7 the two immediate components of the pattern were proved to be logically equivalent. Hence, the pattern always has truth value true and so does the quantified sentence into which it can be transformed.

In Exercise 8.6 it was shown that the conjunction $(p \rightarrow q) \wedge (q \rightarrow r)$ logically implies the conditional $p \rightarrow r$. It follows that the quantified sentence $\forall x(Px \rightarrow Qx) \wedge \forall x(Qx \rightarrow Rx)$ logically implies $\forall x(Px \rightarrow Rx)$. When Px, Qx, and Rx are replaced by $x \in A$, $x \in B$, and $x \in C$, respectively, this implication justifies the truth of the conditional

$$(\forall x(x \in A \rightarrow x \in B) \wedge \forall x(x \in B \rightarrow x \in C)) \rightarrow \forall x(x \in A \rightarrow x \in C)$$

and thus establishes the transitive law of inclusion, namely, if $A \subset B$ and $B \subset C$, then $A \subset C$.

In Exercise 8.5 the pattern $((p \to q) \wedge \sim q) \to \sim p$ was shown to be logically valid. Hence, the quantified sentence

$$(\forall x(x \in A \to x \in B) \wedge \forall x(x \notin B)) \to \forall x(x \notin A)$$

is true. Because the component $\forall x(x \notin B)$ asserts that every element in the universe fails to be a member of B, set B must be empty. The sentence, therefore, translates into the theorem that if A is a subset of B and B is empty, then A is empty; or, more briefly, any subset of an empty set is empty.

EXERCISE 9.9

1. In Section 8.9 it was shown that the conjunction $p \wedge p$ is logically equivalent to p. It follows that the quantified sentence $\exists x(x \in A \wedge x \in A)$ is equivalent to $\exists x(x \in A)$. Explain why this equivalence translates into the assertion that a set meets itself if and only if it is nonempty.

2. The logical equivalence of $p \wedge p$ with p also justifies the equivalence of $\sim\exists x(x \in A \wedge x \in A)$ with $\sim\exists x(x \in A)$. Translate this last equivalence into a law of sets.

9.10 THE USE OF SETS IN LOGICAL ARGUMENT

An interesting application of the system of sets is made to the analysis of arguments involving quantified sentences. Consider the two arguments below, where the universe consists of the oysters in the Tale of the Walrus and the Carpenter.

All little oysters are out of breath.
No thin oysters are out of breath. (1)
Hence, no little oysters are thin.

All little oysters are out of breath.
No little oysters are thin. (2)
Hence, no thin oysters are out of breath.

Let B be the set of oysters that are out of breath, L the set of little oysters, and T the set of thin oysters. Then for argument (1) the premises can be restated as $L \subset B$ and B is disjoint from T. The conclusion is L is disjoint from T, and it follows by the law of disjointness for inclusion.

For argument (2) the premises are $L \subset B$ and L is disjoint from T. The conclusion is T is disjoint from B. There is, however, no law of sets to guarantee this conclusion. A different instance of the same pattern of argument may clarify the situation. All fishes swim. No fishes are dogs. Hence, no dogs swim. The graphical analysis presented in Figure 9.8 may be useful in disclosing the adequacy of the premises in the first argument and their inadequacy in the second.

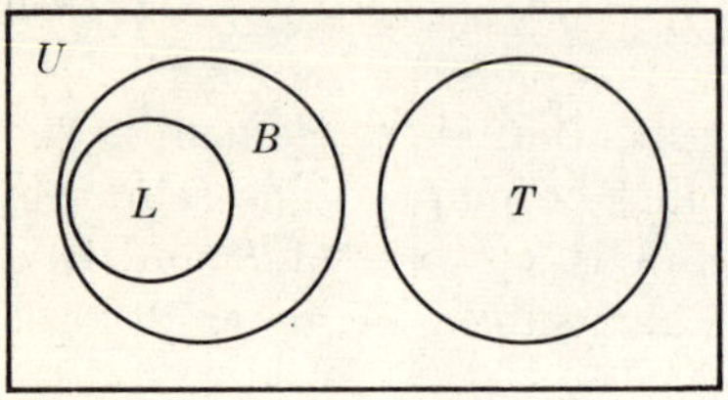

Argument (1)

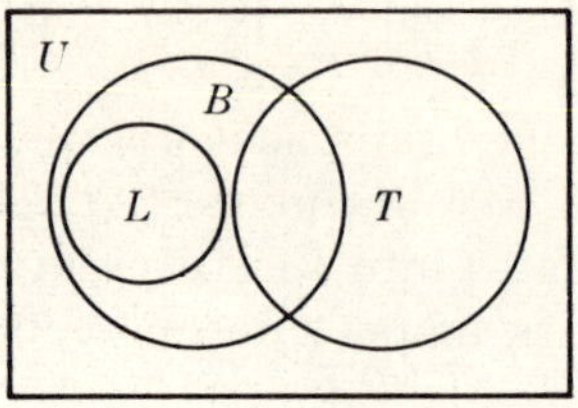

Argument (2)

FIGURE 9.8

EXERCISE 9.10

Use sets to determine whether the following arguments are valid. Let the universe and sets B and L be as described above. Let F be the set of fat oysters.

1. Some oysters are fat. All fat oysters are out of breath. Hence some oysters are out of breath.

2. No little oysters are out of breath. Some fat oysters are out of breath. Hence, no little oysters are fat.

3. Some little oysters are fat. Some little oysters are out of breath. Hence, some fat oysters are out of breath.

4. All oysters are little and fat. Hence, some oysters are little and some are fat.

5. All fat oysters are out of breath. Hence, if no oysters are out of breath, then no oysters are fat.

6. All little oysters are fat. Hence, if no oysters are little, then none are fat.

9.11 OPERATIONS ON SETS

We have found the system of sets to be rich in relations. It also possesses interesting operations. In this section we consider the various ways in which sets can be produced from given sets. Three possible operations come quickly to mind. Given two sets, we can construct a new one by collecting (a) their common elements, (b) all their elements, or (c) their unlike elements. Let us take examples from the universe of the first seven natural numbers.

The process of combining common elements from two sets produces $\{3,5\}$ from $A = \{3,4,5\}$ and $B = \{1,3,5,7\}$. There is an operation that does exactly this. It is known as *intersection* and is denoted by the symbol $\cap$ (read "cap" or "intersection"). Thus $A \cap B = \{3,4,5\} \cap \{1,3,5,7\} = \{3,5\}$.

The set containing all the elements of A and B above is $\{1,3,4,5,7\}$. Because members of the original collections have been united to make the new set, we name this operation *union*. Its symbol is $\cup$ (read "cup" or "union"). We write $A \cup B = \{3,4,5\} \cup \{1,3,5,7\} = \{1,3,4,5,7\}$. Note that elements are listed only once within the braces.

The operation described in (c) constructs a set by choosing all the elements that belong to one or the other of two sets but not to both. For A and B above, this operation results in $\{1,4,7\}$. The operation is known as the *symmetric difference* and is designated by the ordinary symbol $+$. Thus, $A + B = \{3,4,5\} + \{1,3,5,7\} = \{1,4,7\}$. We can also construct the symmetric difference by forming the union of two sets and then removing the elements in their intersection. The correctness of this statement for sets A and B is quickly verified.

The removal of elements suggests the introduction of a binary operation analogous to subtraction. Instead, it is customary to employ a unary transformation that replaces a given set by the collection of all the elements of the universe originally omitted from the given set. This process applied to $A = \{3,4,5\}$ yields $\{1,2,6,7\}$. The operation is known as *complementation* and is denoted by the symbol $'$ (read "prime" or "complement"). Hence, $A' = \{3,4,5\}' = \{1,2,6,7\}$, and $B' = \{1,3,5,7\}' = \{2,4,6\}$.

In the following sections we shall concern ourselves with the three operations of intersection, union, and complementation.

EXERCISE 9.11

Use the universe and sets of the discussion above.

1. Find $(A \cap B)'$. Then show that $A + B = (A \cup B) \cap (A \cap B)'$.
2. Find the following sets.

(a)	$(A \cup B)'$	(b)	$(A + B)'$
(c)	$A' \cap B'$	(d)	$A' \cup B'$
(e)	$A' \cap B$	(f)	$A \cup B'$
(g)	$A' + B'$	(h)	$A' + B$
(i)	$A + B'$		

3. What equalities exist among the sets in Exercises 1 and 2?

9.12 INTERSECTION OF SETS

The intersection of two sets is the set of their common elements. The noun *intersection* thus retains the meaning it has in general mathematical parlance.

DEFINITION The **intersection** of sets A and B, written $A \cap B$, is the set of elements that belong to both A and B. The operation of forming the intersection of two sets is called *intersection.*

An element x of the universe is a member of $A \cap B$ if it is a member of A and also a member of B. In this case it confers truth value true on the open sentence

$$x \in A \wedge x \in B \tag{3}$$

Conversely, any element in the universe of which (3) is true is a member of both A and B and hence of their intersection. It follows that $A \cap B$ is the extension of open sentence (3). We know that such an extension is a unique set. Consequently, every pair of sets has a unique intersection.

THEOREM 9.12-1 For sets A and B and any element x in the universe, $A \cap B = \{x \mid x \in A \wedge x \in B\}$.

THEOREM 9.12-2 The sets of elements in a universe are closed under intersection

When two sets are *disjoint,* they have no elements in common and their intersection is empty. On the other hand, when two sets *meet,* they have common elements. These elements form the membership of the intersection. Conversely, every two sets have an intersection, empty or nonempty. If it is empty, the sets are disjoint; if it is nonempty, the sets meet in its members.

EXERCISE 9.12

In this exercise, the universe consists of the natural numbers, and sets A, B, and C are given by $A = \{1,2,3\}$; $B = \{3,5,7\}$; $C = \{4,5,6,7\}$.

1. Find $A \cap B$, $A \cap C$, $B \cap C$.

2. Determine whether each statement below is true or false.
   (a) $4 \in B \cap C$.
   (b) $A \cap B = \{3\} \wedge A$ meets B.
   (c) $A \cap C = \varnothing \wedge A$ is disjoint from C.
   (d) $A \cap C \subset B \cap C$.
   (e) $B \cap C \subset C$.

The lack of parentheses in Exercise 2 has probably made the statements difficult to read. The notations $4 \in (B \cap C)$ for part (a) and $(A \cap B = \{3\}) \wedge (A$ meets $B)$ for part (b) would be more readily comprehended. Yet, there was no need for parentheses in the comparable statements $4 = 3 \cdot 3$, and $4 = 3 \cdot 3 \wedge 5 = 5^*$ in Chapter 8. Continued use of these new symbols will make their significance clear. Familiarity will eliminate the present necessity for reading slowly and making symbol by symbol translations.

9.13 DIAGRAMMING INTERSECTION OF SETS

Let us use diagrams to study the operation of intersection of sets. When general sets A and B are drawn as interlocking circles, their common elements are represented by the region in which the circles overlap. Because these elements compose the intersection, we can properly write the description $A \cap B$ in this portion of the diagram as in Figure 9.9(a). We can also show the intersection by shading as in Figure 9.9(b). In either case the diagram shows that the whole of $A \cap B$ is contained in A and also in B; hence, the intersection is a subset of each intersecting set. This relation should not come as a surprise to us. An element is a member of an intersection only because it is a member of each intersecting set.

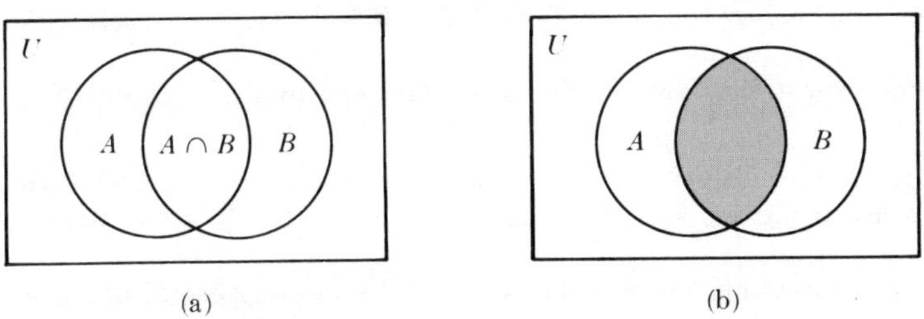

FIGURE 9.9

THEOREM 9.13-1 The intersection of two sets is a subset of each of the intersecting sets. In symbols, for sets A and B, $A \cap B \subset A$ and $A \cap B \subset B$.

If all of the A circle outside its overlap with the B circle is empty in Figure 9.9, then all of A lies in the overlap, and $A = A \cap B$. Furthermore, all of set A now lies in set B so that $A \subset B$. The two relations $A \subset B$ and $A \cap B = A$ always occur together and hence are equivalent. When we apply this equivalence to the inclusions $A \subset A$ and $\varnothing \subset A \subset U$, we get the statements in the corollary below. The first equality in the corollary asserts that intersection is idempotent (see definition in Section 8.9).

THEOREM 9.13-2 One set is a subset of a second if and only if it coincides with their intersection. In symbols, for sets A and B, $A \subset B$ if and only if $A \cap B = A$.

Corollary 9.13-2 For set A, $A = A \cap A = A \cap U$ and $A \cap \varnothing = \varnothing$.

EXERCISE 9.13

For specific sets in Exercises 1–5 the universe consists of the first nine natural numbers.

1. Illustrate Theorems 9.13-1 and 9.13-2 using the following sets:

(a) $A = \{2,3\}$, $B = \{1,2,3,4\}$ (b) $A = \{1,3,5\}$, $B = \{5,7,9\}$

2. The law of intersection for inclusion would say that if $A \subset C$ and $B \subset D$, then $A \cap B \subset C \cap D$.

(a) Test this law using the sets $A = \{1,2,3\}$, $B = \{2,3,4\}$, $C = \{1,2,3,6,7\}$, and $D = \{2,3,4,7,9\}$.

(b) Would you expect this law to hold for general sets in a general universe? Justify your answer.

3. A law of intersection for meeting would say that if A meets C and B meets D, then $A \cap B$ meets $C \cap D$.

(a) Test this law using the sets $A = \{1,2\}$, $B = \{2,5\}$, $C = \{2,3\}$, and $D = \{3,5\}$.

(b) Would you expect this law to hold for general sets in a general universe? Justify your answer.

4. (a) Modify the law of Exercise 3 to be a law of intersection for disjointness.

(b) Test this law using the sets $A = \{2,4,8\}$, $B = \{2,5,9\}$, $C = \{3,7,9\}$, and $D = \{4,6,7,8\}$.

(c) Would you expect this law to hold for general sets in a general universe? Justify your answer.

5. Let sets A, B, and C be enumerated as follows: $A = \{1,2,3,4\}$; $B = \{2,4,8\}$; $C = \{3,4,7,8,9\}$. Find the following sets:

(a) $A \cap B$ (b) $(A \cap B) \cap C$
(c) $B \cap C$ (d) $A \cap (B \cap C)$

6. What do the answers to parts (b) and (d) of Exercise 5 suggest about the operation of intersection of sets?

7. When three general sets are diagrammed, the usual figure is three circles interlocking in such a way that each overlaps the other and all three have a common region as in Figure 9.10. Let sets A, B, and C be general sets in a general universe.

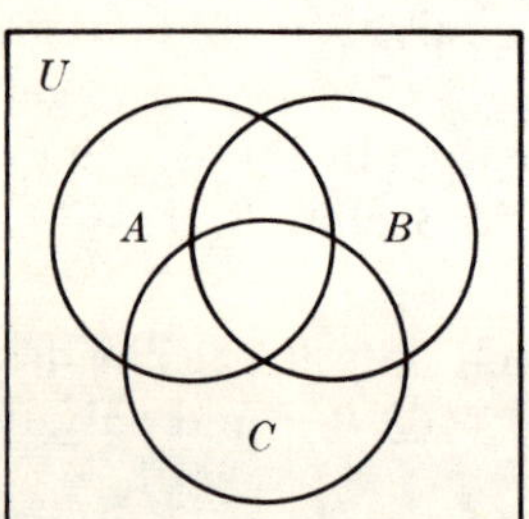

FIGURE 9.10

(a) Draw a diagram showing both $A \cap B$ and $(A \cap B) \cap C$.

(b) Draw a diagram showing both $B \cap C$ and $A \cap (B \cap C)$.

(c) If the diagrams are correctly drawn, then the final set of part (a) occupies exactly the same region in the first drawing as does the final set of part (b) in the second. Hence, these two sets are equal. What property of intersection of sets is thereby established?

8. Explain, by diagrams or other means, why the following equality is correct: $A \cap (A \cap B) = A \cap B$.

9. Show that the universal set acts as an identity set for intersection.

10. Which sets, if any, have inverses under intersection?

9.14 PROPERTIES OF INTERSECTION OF SETS

Let us consider first those properties of intersection of sets that often characterize an operation in an algebraic system. These are commutativity, associativity, and the new property of idempotence. The universal set is the identity, and only this set has an inverse, namely itself. The collection of sets of things in a universe fails, by one criterion, to form a group under intersection.

THEOREM 9.14-1 Intersection of sets is idempotent, commutative, and associative.

THEOREM 9.14-2 The universal set is the identity for intersection.

Because intersection is a binary operation on sets, the law of intersection for equality holds; that is, for sets A, B, C, and D, if $A = C$ and $B = D$, then $A \cap B = C \cap D$. In Exercises 9.13.2 and 9.13.4, similar laws for inclusion and disjointness were discovered. With regard to the latter, it can be shown that the hypotheses are stronger than necessary; either one of them alone would guarantee the conclusion. The equality of Exercise 9.13.8 can be extended to say that the intersection of a finite number of sets depends only on the collection of sets involved and not on the order in which the sets are combined or the number of times any set is represented.

EXERCISE 9.14

1. Explain why one of the hypotheses is redundant in the law of intersection for disjointness.

2. A corollary of the law of intersection for inclusion asserts that for sets A, B, and C, if $A \subset C$ and $B \subset C$, then $A \cap B \subset C$.

(a) Illustrate the statement using specific sets.

(b) Are the hypotheses stronger than necessary or just adequate to guarantee the conclusion?

3. A second corollary to the law of intersection for inclusion asserts that for sets A, C, and D, if $A \subset C$ and $A \subset D$, then $A \subset C \cap D$.

(a) Illustrate this statement using specific sets.

(b) Diagram the statement.

4. (a) Write the converse of the conditional of Exercise 3.

(b) Diagram the hypothesis of the converse in (a).

(c) Does the diagram confirm the conclusions of the converse? Explain.

9.15 UNION OF SETS

The *union of two sets* is a set whose membership consists of all items that belong to the first and all that belong to the second. The phrase, therefore, retains the connotations that it holds in ordinary discourse. The formula $A \cup B$ signifies the set that is the union of A and B; thus in the five-element system $\{1,2,3\} \cup \{2,3,4\} = \{1,2,3,4\}$.

DEFINITION The **union** of set A and set B, written $A \cup B$, is the set of elements that belong to A or to B or to both A and B. The process of forming the union of two sets is called *union*.

Any element of the universe may enter $A \cup B$ by virtue of membership in A or in B or in both. Recall that the logical connective *or* is used in the legal sense of *and/or*; it can, therefore, be used to give an exact description of the criterion for membership, namely,

$$x \in A \vee x \in B \tag{4}$$

Conversely, any element of the universe of which open sentence (4) is true belongs to A or to B or to both sets and is a member of their union. The union is completely described by the assertion that it is the extension of open sentence (4). Furthermore, since the extension of an open sentence is unique, every pair of sets has a unique union.

THEOREM 9.15-1 For sets A and B and any element x in the universe, $A \cup B = \{x \mid x \in A \vee x \in B\}$.

THEOREM 9.15-2 The sets of elements in a universe are closed under union.

Let us diagram the set $A \cup B$. General sets A and B are drawn, of course, as two overlapping circles. Their union now consists of the whole of the interiors of the two circles, shaded in Figure 9.11. From the figure we see that each of sets A and B is contained in $A \cup B$.

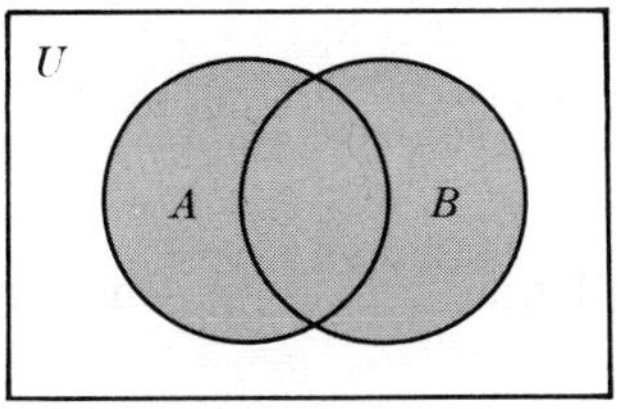

FIGURE 9.11

THEOREM 9.15-3 The union of two sets contains each of the uniting sets as a subset. In symbols, for sets A and B, $A \subset A \cup B$ and $B \subset A \cup B$.

If in Figure 9.11 the portion of A outside its overlap with B is empty, then all of A is contained in the overlap and $A \subset B$. Furthermore, $A \cup B$ now consists entirely of the set B. Hence, the two relations $A \subset B$ and $A \cup B = B$ always occur together and are equivalent. Combining this result with the inclusions $A \subset A$ and $\varnothing \subset A \subset U$ gives the corollary below.

THEOREM 9.15-4 A given set contains another as a subset if and only if it coincides with their union. In symbols, for sets A and B, $A \subset B$ if and only if $A \cup B = B$.

Corollary 9.15-4 For set A, $A = A \cup A = A \cup \varnothing$ and $A \cup U = U$.

EXERCISE 9.15

For sets E through H in this exercise, the universe is the collection of natural numbers not exceeding 9. Descriptions are as follows:

$$E = \{x \mid x \text{ is less than } 6\} \qquad F = \{x \mid x \text{ is even}\}$$
$$G = \{x \mid x \text{ is a multiple of } 4\} \qquad H = \{x \mid x \text{ is a multiple of } 3\}$$

1. Tabulate sets E, F, G, and H.

2. Illustrate Theorem 9.15-3 using sets E and F.

3. Illustrate Theorem 9.15-4 using (a) sets F and G, (b) sets E and H.

4. (a) Write the associative law for union of sets.
(b) Test the law using sets E, F, and H.
(c) Does the answer to part (b) suggest that the law may hold?

5. For general sets A, B, and C in a general universe the associative law may be verified as follows:
(a) Draw a diagram showing both $A \cup B$ and $(A \cup B) \cup C$.
(b) Draw a diagram showing both $B \cup C$ and $A \cup (B \cup C)$.

(c) If the diagrams are correctly drawn, then the final set of part (a) occupies exactly the same region in the first drawing as does the final set of part (b) in the second. Interpret this fact.

6. Explain, by diagrams or other means, why the following equality is true for general sets: $A \cup (A \cup B) = A \cup B$.

7. Is there an identity set for union? If so, which set is it?

8. If you found an identity set in Exercise 7, does E have an inverse under union?

9. (a) Write the law of union for inclusion.
(b) Test the law for the sets in Exercise 9.13.2.
(c) Would you expect this law to hold for general sets in a general universe? Justify your answer.

10. (a) Write the law of union for meeting.
(b) Test the law for the sets in Exercise 9.13.3.
(c) Would you expect this law to hold for general sets in a general universe? Justify your answer.

11. (a) Write the law of union for disjointness.
(b) Test the law for the sets in Exercise 9.13.4.
(c) Would you expect this law to hold for general sets in a general universe? Justify your answer.

9.16 PROPERTIES OF UNION OF SETS

The work of the preceding section has shown that properties of union of sets parallel closely those for intersection. To closure are adjoined further characteristics of an operation of an algebraic system, namely, commutativity, associativity, and idempotence. The null set is an identity set. Because the union of two sets always contains each of the uniting sets as a subset, it cannot omit from its membership any element of either of the uniting sets. Hence, no nonempty set can unite with another set to produce the null set. It follows that only the null set has an inverse. Union, like intersection, fails by one criterion to attain the status of a group operation.

THEOREM 9.16-1 Union of sets is idempotent, commutative, and associative.

THEOREM 9.16-2 The null set is an identity set for union.

In Exercise 9.15.11 above you found a counterexample to the law of union for disjointness. Laws of union for the other relations hold. Note, however, that the law for meeting has a redundant hypothesis. The equality of Exercise 9.15.6 can be extended to say that the union of a finite number of sets depends only on the collection of sets involved and not on the order in which the sets are combined or the number of times any set is represented.

In the next section it is proved that the operations of union and intersection are distributive with respect to each other. In the following theorem, only laws of left distributivity are written; but since both operations are commutative, right distributivity also holds.

THEOREM 9.16-3 *Laws of Distributivity.* Intersection is distributive with respect to union, and union is distributive with respect to intersection. In symbols, for sets A, B, and C,

$$A \cap (B \cup C) = (A \cap B) \cup (A \cap C)$$

and

$$A \cup (B \cap C) = (A \cup B) \cap (A \cup C)$$

EXERCISE 9.16

1. Explain why one of the hypotheses is redundant in the law of union for meeting.

2. A corollary of the law of union for inclusion asserts that for sets A, C, and D, if $A \subset C$ and $A \subset D$, then $A \subset C \cup D$.

(a) Illustrate the statement using specific sets.

(b) Diagram the statement.

(c) Are the hypotheses stronger than necessary or just adequate to guarantee the conclusion?

3. A second corollary to the law of union for inclusion asserts that for sets A, B, and C, if $A \subset C$ and $B \subset C$, then $A \cup B \subset C$. Illustrate this statement using specific sets.

4. Write the converse of the conditional of Exercise 3 and explain why it is true.

5. Using specific sets illustrate Theorem 9.16-1.

6. Write a paragraph on the properties of the null set and the universal set under intersection and union.

7. Illustrate both laws of left distributivity using the following sets of natural numbers:

$$A = \{1,2,3,4\} \qquad B = \{4,5,6,7\} \qquad C = \{1,3,5,8\}$$

★8. Diagram both laws of left distributivity. Use the method employed to verify the associative laws in Exercises 9.13.7 and 9.15.5.

9.17 LOGICAL OPERATIONS AND SET OPERATIONS (OPTIONAL)

Recall that the statement patterns of Chapter 8 frequently occurred in couples, one for conjunction and one for alternation. In the theorems of this chapter, assertions for intersection of sets are frequently matched with assertions that give similar properties for union. Such a pairing of laws is called *duality*. Mathematicians say that laws of conjunction and alternation are duals of each other and that laws of intersection and union are duals of each other. Because criteria for intersection or union of sets A and B are expressed as $x \in A \wedge x \in B$ or $x \in A \vee x \in B$, respectively, it should be clear that this duality must characterize both the logic and the set systems whenever it is a property of either. When two patterns differ only in the interchange of $\wedge$ with $\vee$, they produce analogous statements using $\cap$ and $\cup$, and conversely.

As an example, the commutative law of conjunction asserts that $x \in A \wedge x \in B$ is equivalent to $x \in B \wedge x \in A$. Hence, the two sentences $x \in A \cap B$ and $x \in B \cap A$ have identical truth values, and the quantified sentence $\forall x(x \in A \cap B \leftrightarrow x \in B \cap A)$ is true. This sentence defines the relation $A \cap B = B \cap A$, the commutative law of intersection. Similarly, the commutative law of alternation can be used to establish the commutative law of union, and the associative and idempotent laws for the two logical operations can be made to yield the corresponding laws for the set operations.

In Section 8.9 conjunction was found to be distributive with respect to alternation. Hence, the open sentence

$$(x \in A \wedge (x \in B \vee x \in C)) \leftrightarrow ((x \in A \wedge x \in B) \vee (x \in A \wedge x \subset C)$$

is always true and so is its reformulation in terms of set operations, namely,

$$x \in A \cap (B \cup C) \leftrightarrow x \in (A \cap B) \cup (A \cap C)$$

The universal quantification of this sentence is, therefore, true. The quantified sentence

$$\forall x[x \in A \cap (B \cup C) \leftrightarrow x \in (A \cap B) \cup (A \cap C)]$$

translates into a law of distributivity of intersection with respect to union, $A \cap (B \cup C) = (A \cap B) \cup (A \cap C)$. Because alternation is distributive with respect to conjunction, the second law of Theorem 9.16-3 can be derived.

The duality of these laws is a remarkable and distinguishing characteristic of the logic and the set systems. Recall that in the algebras of the first seven chapters only one operation (multiplication) is distributive with respect to the other (addition). For this reason we can establish a convention whereby the expression $a + bc$ has a clear meaning, the sum of element a with the unique product of b and c, and it cannot be rewritten as $(a + b)c$. In logic, however, as we saw in Section 8.2, the corresponding expression

$p \vee q \wedge r$ can be given either of two equally reasonable interpretations, the alternation $p \vee (q \wedge r)$ or the conjunction $(p \vee q) \wedge r$. Correspondingly, in the system of sets the expression $A \cup B \cap C$ is ambiguous. Insertion of parentheses produces two distinct sets, namely,

$$A \cup (B \cap C) = (A \cup B) \cap (A \cup C) \tag{5}$$
$$(A \cup B) \cap C = (A \cap C) \cup (B \cap C) \tag{6}$$

Formulas must, therefore, be written in such a way that the order of operation is clearly specified.

As a further example of this duality, consider the laws of intersection and union for inclusion. Corollaries to these laws were established in Exercises 9.14.3, 19.14.4, 9.16.3, and 9.16.4:

$$A \subset B \text{ and } A \subset C \text{ if and only if } A \subset B \cap C$$
$$A \subset C \text{ and } B \subset C \text{ if and only if } A \cup B \subset C$$

The corollaries justify the validity of the associated quantified sentences:

$$(\forall x(x \in A \to x \in B) \wedge \forall x(x \in A \to x \in C)) \leftrightarrow \forall x(x \in A \to (x \in B \wedge x \in C))$$
$$(\forall x(x \in A \to x \in C) \wedge \forall x(x \in B \to x \in C)) \leftrightarrow \forall x((x \in A \vee x \in B) \to x \in C)$$

The truth of the sentences now guarantees the validity of the corresponding statement patterns:

$$((p \to q) \wedge (p \to r)) \leftrightarrow (p \to (q \wedge r))$$
$$((p \to r) \wedge (q \to r)) \leftrightarrow ((p \vee q) \to r)$$

Truth tables constructed in Exercises 8.5.5 and 8.5.6 also confirm the logical validity of the patterns. Thus, laws of sets can serve instead of truth tables to determine the truth values of statement patterns.

EXERCISE 9.17

1. Find the sets described by formulas (5) and (6) above using the following sets of natural numbers: $A = \{2,3,4,6\}$, $B = \{1,3,5,7\}$, $C = \{3,5,6,8\}$.

2. In Exercises 8.5.1 and 8.5.4 the conditionals $(p \wedge q) \to p$ and $p \to (p \vee q)$ were shown to be logically valid. Explain why they justify the related laws $A \cap B \subset A$ and $A \subset A \cup B$.

9.18 COMPLEMENTATION OF SETS

It is obvious that whenever we take a set of elements from a universe, we leave behind all the remaining elements. If we take the set $A = \{x \mid Px\}$, the remaining elements are those of which Px is false or, equivalently, of which $\sim Px$ is true. They form a set described by $\{x \mid \sim Px\}$ called the *complement* of A and denoted by A'. To test whether an element in the universe is a member of A', one determines whether it fails to be a member of A. Hence, the complement of A is the extension of the open sentence $x \notin A$.

DEFINITION The **complement** of set A, written A', is the totality of elements in the universe that do not belong to A. The process of taking the complement is called *complementation.*

THEOREM 9.18-1 For set A and any element x in the universe,

$$A' = \{x \mid x \notin A\}$$

THEOREM 9.18-2 The sets of things in a universe are closed under complementation.

The complement of A', written A'', is the set $\{x \mid \sim(\sim Px)\}$ or $\{x \mid Px\}$. Hence, $A'' = A$. Clearly, $A \cap A' = \varnothing$ and $A \cup A' = U$. Conversely, any two sets whose intersection is $\varnothing$ and whose union is U are complements of each other. Among such sets are the universal and null sets.

When a circle representing general set A is drawn within a rectangle representing the universe, set A' coincides with all of the rectangle exterior to the circle. This region is shaded in Figure 9.12.

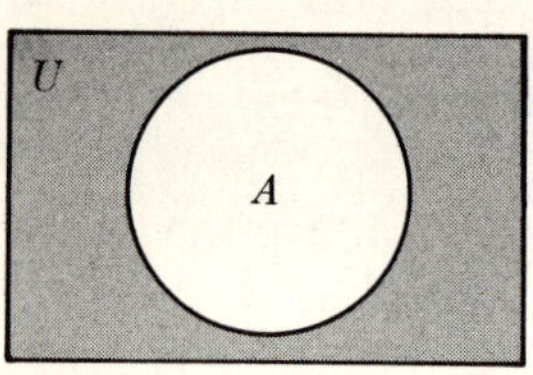

FIGURE 9.12

EXERCISE 9.18

Let sets A, B, C, and D be any sets in any universe. For other sets the universe is the five-element system, and

$$E = \{1,2,3,4\} \qquad F = \{2,3\}$$
$$G = \{1,4,5\} \qquad H = \{1,5\}$$

1. Tabulate E', F', G', and H'.

2. Note that $F \subset E$ and $H \subset G$.
(a) What relation holds between sets F' and E'?
(b) Does a corresponding relation hold between H' and G'?
(c) Write a statement in terms of general sets A and B embodying the property of inclusion suggested in parts (a) and (b).

3. (a) Draw a diagram showing set A as a subset of set B. Mark set A' with horizontal lines and set B' with vertical lines.
(b) Does the diagram support your answer to part (c) of Exercise 2?

4. If sets A and B are equal, what can be said of sets A' and B'? Justify your answer.

5. (a) Show that E meets G. Does E' meet G'?
(b) A law of complementation for meeting would say that if A meets B, then A' meets B'. Does this law hold? Justify your answer.

6. (a) Show that F is disjoint from H. Is F' disjoint from H'?
(b) Write the law of complementation for disjointness.
(c) Does the law of part (b) hold? Justify your answer.

7. Tabulate sets to find out whether any of the following are equal to each other.

$E \cup G$ $\quad (E \cup G)'$ $\quad E' \cup G'$ $\quad E \cap G$ $\quad (E \cap G)'$ $\quad E' \cap G'$
$F \cup H$ $\quad (F \cup H)'$ $\quad F' \cup H'$ $\quad F \cap H$ $\quad (F \cap H)'$ $\quad F' \cap H'$

8. State rules suggested by the results in Exercise 7.

In Exercises 9–12 use diagrams showing sets A and B as two overlapping circles.

9. (a) Diagram set $(A \cap B)'$.
(b) Diagram set $A' \cup B'$.
(c) Do the two diagrams support the rule for $(A \cap B)'$ stated in Exercise 8?

10. (a) Diagram set $(A \cup B)'$.
(b) Diagram set $A' \cap B'$.
(c) Do the two diagrams support the rule for $(A \cup B)'$ stated in Exercise 8?

11. (a) Diagram set $A \cap B'$.
(b) Let $A \subset B$. Then all of A lies in its overlap with B. What can now be said of $A \cap B'$?

12. (a) On a diagram shade the set $A' \cup B$.
(b) Let $A' \cup B = U$. Then a portion of the diagram of (a) is empty; describe this portion. What can now be said of set A?

13. (a) Explain how the diagrams of 11(a) and 12(a) show that $(A \cap B')' = A' \cup B$.
(b) Are the results of 11(b) and 12(b) compatible? Justify your answer.

9.19 PROPERTIES OF COMPLEMENTATION OF SETS

Exercises 9.18.2, 9.18.3, 9.18.11, and 9.18.12 disclosed new criteria for inclusion of sets.

THEOREM 9.19-1 Set A is a subset of set B if and only if one of the following relations holds: $B' \subset A'$, $A \cap B' = \varnothing$, or $A' \cup B = U$.

Corollary 9.19-1 Set A equals set B if and only if $A' = B'$.

The rules discussed and verified in Exercises 7–10 are stated in the theorem below.

THEOREM 9.19-2 For sets A and B the following rules of complementation hold:

$$(A \cap B)' = A' \cup B', \qquad (A \cup B)' = A' \cap B'$$

EXERCISE 9.19

1. In the universe of the five-element system let $A = \{1,3\}$, $B = \{1,2,3,4\}$, and $C = \{2,5\}$.
   (a) Illustrate Theorem 9.19-1 using sets A and B.
   (b) Show that the three following relations hold:

   (i) $A \cap C = \varnothing$ (ii) $A \cap C' = A$
   (iii) $A' \cap C = C$

2. By use of diagrams show that when general sets A and B are disjoint, then $A' \cap B = B$ and $A \cap B' = A$.
3. What can be said of general sets A and B when $A' \cup B = \varnothing$?
4. What can be said of general sets A and B when $A \cup B' = U$?

9.20 LOGICAL OPERATIONS AND COMPLEMENTATION OF SETS (OPTIONAL)

Because $A' = \{x \mid \sim(x \in A)\}$, properties of complementation of sets are analogues of statement patterns using the denial. In Exercise 8.5 the conditional $p \to q$ and its contrapositive $\sim q \to \sim p$ were found to be logically equivalent. It follows that the quantified sentences $\forall x(x \in A \to x \in B)$ and $\forall x(x \notin B \to x \notin A)$ are equivalent and so are their translations into the language of sets: $A \subset B$ and $B' \subset A'$.

Consider next a law of sets stated in Theorem 9.19-1: $A \subset B$ if and only if $A' \cup B = U$. It justifies the logical equivalence of the quantified sentences $\forall x(x \in A \to x \in B)$ and $\forall x(x \in A' \cup B)$. The last sentence can be restated as $\forall x(x \notin A \lor x \in B)$. It follows that the corresponding statement patterns $p \to q$ and $\sim p \lor q$ are equivalent. Hence, a law of sets has been used instead of a truth table to discover a relation between two statement patterns.

EXERCISE 9.20

1. Results in Exercises 8.8.2 and 8.8.3 justify the assertion that the following biconditionals, known as DeMorgan's laws, are logically valid.

$$\sim(p \wedge q) \leftrightarrow (\sim p \vee \sim q) \qquad \sim(p \vee q) \leftrightarrow (\sim p \wedge \sim q)$$

Show that these laws support the rules of Theorem 9.19-2.

2. Using Corollary 9.19-1 justify the validity of the following law of logic.

$$(p \leftrightarrow q) \leftrightarrow (\sim p \leftrightarrow \sim q)$$

9.21 PROBLEMS ON SETS (OPTIONAL)

This section is devoted to interesting problems dealing with both graphic and algebraic aspects of the study of sets.

EXERCISE 9.21

1. Use the diagram of Figure 9.13.

(a) Explain why the set in the shaded portion can be described by $(A \cup B) \cap (A \cap B)'$.

(b) Write another description for the shaded set. [Suggestion: see Exercise 9.18.11(a).]

(c) Recall the discussion in Section 9.11 and show that the shaded set is $A + B$, the symmetric difference of A and B.

(d) For the specific sets in the illustrations in Section 9.11, the relation $A + B = A' + B'$ obtained. Does this equality hold for general sets in a general universe? Draw the diagram for $A' + B'$ and find out.

2. The two sets A and B diagrammed in Figure 9.14 separate the universe into the four regions indicated by Greek letters. Region γ is described by the intersection $A \cap B$. Write descriptions for the remaining regions.

3. Assume that there are 28 items in the universe of Figure 9.14, of which 5 lie in α, 10 in β, and 7 in δ. Find the number of items in set A. In set B.

4. The three sets diagrammed in Figure 9.15 separate the universe into eight regions indicated by Greek letters. Region ζ is described by the intersection $(A \cap B) \cap C$. Write descriptions for the remaining regions.

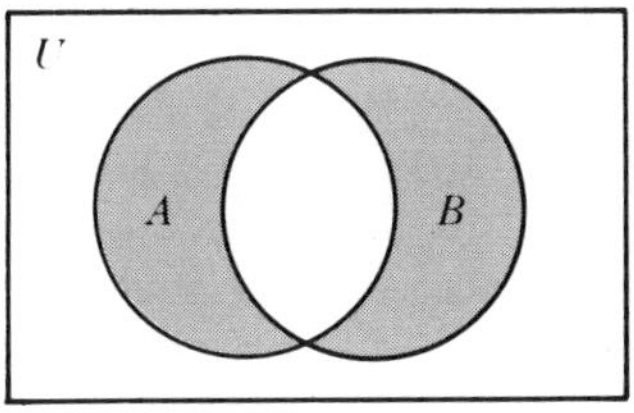

FIGURE 9.13

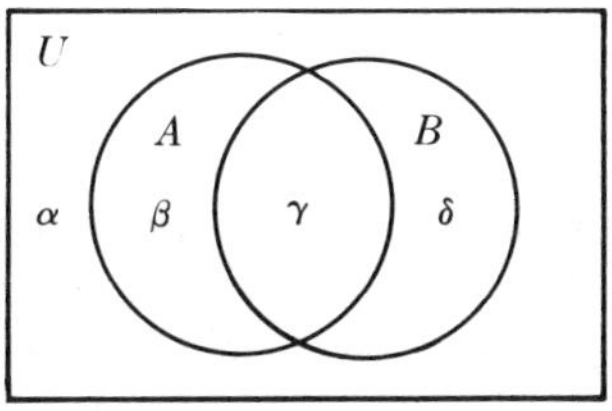

FIGURE 9.14

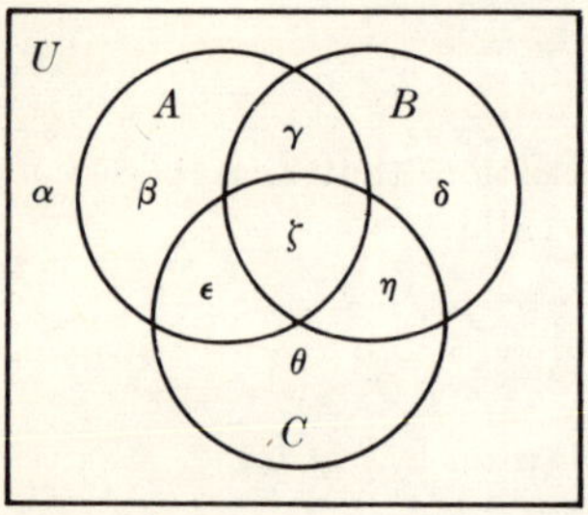

FIGURE 9.15

5. Assume that there are 50 items in the universe of Figure 9.15, of which 25 lie in set A and 34 in set C. Furthermore, certain of the regions contain items as listed below. Find the number of items in set B.

Region	α	β	ζ	η	θ
Number of items	5	6	9	8	11

6. For general sets A, B, and C, let $S = (A \cap B) \cup C'$. Then

$$S' = [(A \cap B) \cup C']'$$

By use of theorems on complementation show that $S' = (A' \cup B') \cap C$.

7. As in Exercise 6, find a formula for the complement of $(A \cup B) \cap C'$.

Using set operations reduce the left member of each of the equalities below to the right member.

8. $[(A \cap B)' \cup C] \cup [(A \cup B) \cup C'] = U$

9. $[(A \cup B)' \cap C] \cap [(A \cap B) \cap C'] = \varnothing$

10. $(A \cup B) \cap (A \cap B)' = (A \cap B') \cup (A' \cap B)$

9.22 BOOLEAN ALGEBRA (OPTIONAL)

In the introduction to Chapter 8 it was noted that during the nineteenth century an Englishman named George Boole had the courage and insight to believe that logic could be formalized and dealt with by methods similar to those of ordinary algebra. As he worked out his idea, he found that certain basic rules would have to be obeyed by the elements in this new mathematical system. In honor of this gifted pioneer, any system governed by these rules is called a Boolean algebra. There are many such algebras, and two of them have been discussed in our study of logic and sets.

Let us now consider an especially simple one. It consists of just two sets U and $\varnothing$, together with their relations and operations. This system is frequently employed without reference to set theory. Such a divorcement from sets is generally accompanied by a change of notation. Symbols U

and $\varnothing$ are replaced by 1 and 0, respectively, and the operations are defined by means of tables which essentially describe the behavior of sets U and $\varnothing$ under the three operations. Formal presentation of this algebra proceeds as follows.

The elements are 1 and 0. They may be represented by letters R, S, ... The relation of equality is defined to be identity. The operations are intersection, union, and complementation, whose symbols are $\cap$, $\cup$, and $'$, respectively, and whose descriptions are presented in Table 9.1.

Table 9.1 Operations Tables for Boolean Algebra of Elements 0 and 1

$\cap$	1	0
1	1	0
0	0	0

$\cup$	1	0
1	1	1
0	1	0

R	R'
1	0
0	1

Let us do some arithmetic in this system. The expression $1 \cap (1 \cup 0')$ is the intersection of two elements, namely, 1 and $1 \cup 0'$. This last element has value $1 \cup 1 = 1$; hence, the whole expression has value $1 \cap 1 = 1$. On the other hand, the union $(1 \cap 0) \cup 1'$ equals $0 \cup 0$ and has value 0. We can construct Boolean tables for these expressions just as we made truth tables for statement patterns, using symbols 1 and 0 in place of T and F and the rules of operation given in Table 9.1. For example, Table 9.2 is a Boolean table for the union $R \cup (R \cap S)$. It shows that the value of the expression is sometimes 0 and sometimes 1.

Table 9.2

R	$\cup$	$(R$	$\cap$	$S)$
1	1		1	1
0	0		0	1
1	1		0	0
0	0		0	0

Table 9.3

R	$\cup$	$(R'$	$\cap$	$S)$
1	1	0	0	1
0	1	1	1	1
1	1	0	0	0
0	0	1	0	0

EXERCISE 9.22

1. In each part below find the single element that results from the operations indicated.

(a) $1 \cap (0 \cup 1)$ (b) $1 \cup (1 \cup 0)'$
(c) $0 \cap (1' \cup 0)'$ (d) $(1 \cap 0') \cup (1' \cap 1)$
(e) $(1 \cap 0)' \cap (1' \cap 0')$ (f) $(0' \cup 1')' \cup (0 \cap 0)$

2. Show that Boolean Table 9.2 is correct.

3. Construct a Boolean table for the union $R \cup (R' \cap S)$. Compare your answer with Table 9.3.

4. Construct Boolean tables for the expressions below.
(a) $(R' \cap S) \cup (R \cap S')$
(b) $(R \cup S) \cap (R' \cup S')$

5. How do the tables of Exercise 4 show that the two expressions are equal?

6. Show that the following equalities hold for any elements R and S.

(a) $R \cap R' = 0$ (b) $R \cup R' = 1$
(c) $(R \cap S)' = R' \cup S'$ (d) $(R \cup S)' = R' \cap S'$

9.23 BOOLEAN ALGEBRA AND THE DESIGN OF ELECTRIC CIRCUITS (OPTIONAL)

One application of our simple Boolean algebra lies in the field of electrical engineering. An electric circuit with one switch is shown in Figure 9.16. When the switch is open, as in Figure 9.16(a), no current flows. When the switch is closed, as in Figure 9.16(b), the current flows. Let us call the switch S and say that when $S = 0$, the switch is open and no current flows; when $S = 1$, the switch is closed and a current flows. Now, consider a circuit with two switches R and S connected *in series* as in Figure 9.17. The current flows if and only if both switches are closed, that is, when $R = S = 1$. Table 9.4 records these conditions.

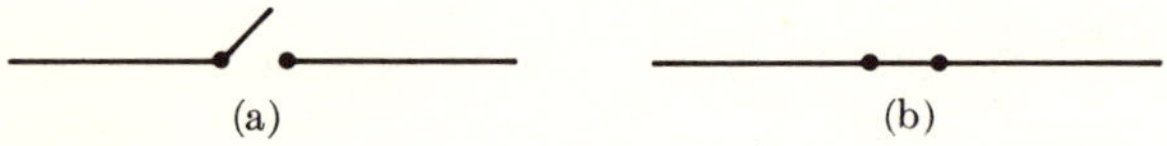

FIGURE 9.16

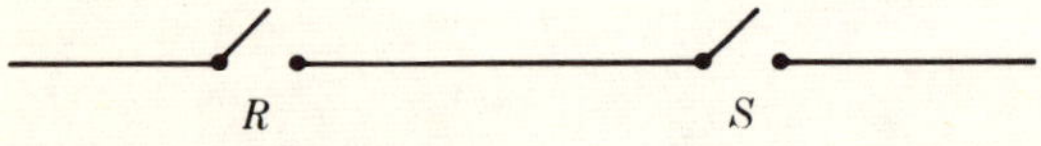

FIGURE 9.17

Table 9.4

Current for switches in series	$S=1$	$S=0$
$R=1$	1	0
$R=0$	0	0

Next, consider a circuit with two switches R and S connected *in parallel* as in Figure 9.18. Now the current flows if either or both switches are closed, that is, when $R = 1$ or $S = 1$. Table 9.5 describes this situation.

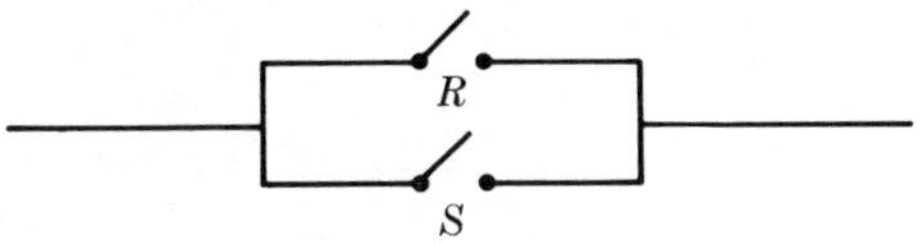

FIGURE 9.18

Table 9.5

Current for switches in parallel	$S=1$	$S=0$
$R=1$	1	1
$R=0$	1	0

It is evident that any switch is either open or closed; "open" and "closed" are opposite and mutually exclusive states for a switch. This is indicated in Table 9.6.

Table 9.6

S = state of switch	S' = opposite state of switch
1	0
0	1

The series, parallel, and state tables are exactly analogous to the intersection, union, and complementation tables, respectively, of the Boolean algebra of elements 1 and 0. Hence, the algebra of switches is a Boolean algebra with the following interpretation for operations.

$\cap$ Switches connected in series

$\cup$ Switches connected in parallel

$'$ Switch in opposite state

To see how this algebra is used in the design of electric circuits, consider the circuit in Figure 9.19(a). In this scheme the two switches R go on or off together. The expression describing this plan is $R \cup (R \cap S)$. We wish to ask two questions about any proposed circuit: (1) When will the current flow and when will it cease flowing? (2) Is the circuit designed to use as few switches as possible? To answer the first, note that Boolean Table 9.2 was constructed for this expression. The analysis makes it clear that the current flows when R is closed and ceases when R is open. To answer the second question, apply theorems on sets. Since $R \cup (R \cap S) = R$, our proposed circuit is too complicated. We could get the same electrical results by using only one switch R as shown in Figure 9.19(b). This result is in accord with our answer to question (1).

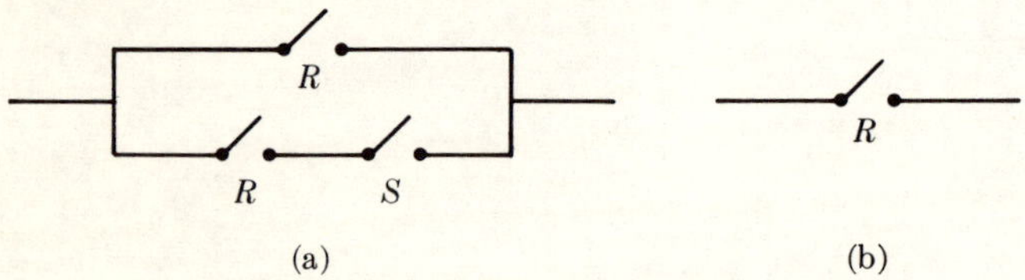

FIGURE 9.19

Next consider the circuit of Figure 9.20(a). Here switches R and R' are in opposite states; that is, when R is closed, R' is open, and vice versa. The expression describing this circuit is $R \cup (R' \cap S)$. Boolean Table 9.3 shows that the current ceases only when R and S are both open. When we simplify the set formula, we get $R \cup S$. This result tells us that, again, our circuit is too complicated. We can get the same electrical service from the simpler circuit shown in Figure 9.20(b). Clearly, in circuits with many switches such analyses can be very useful.

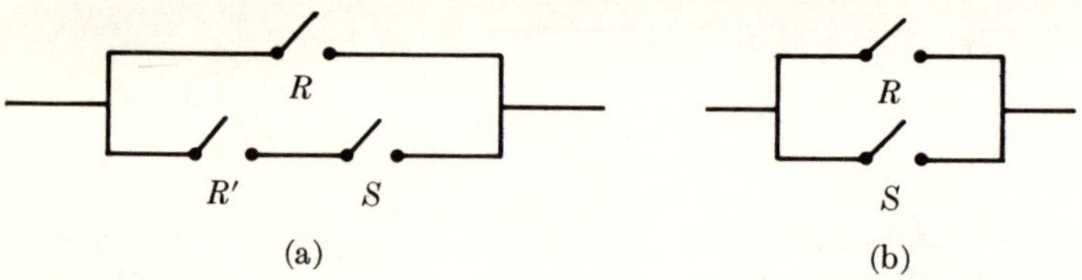

FIGURE 9.20

EXERCISE 9.23

1. Show that the two circuits in Figure 9.21 have exactly the same electrical effect. Under what circumstances does the current flow? (Note Exercises 9.22.4 and 9.22.5.)

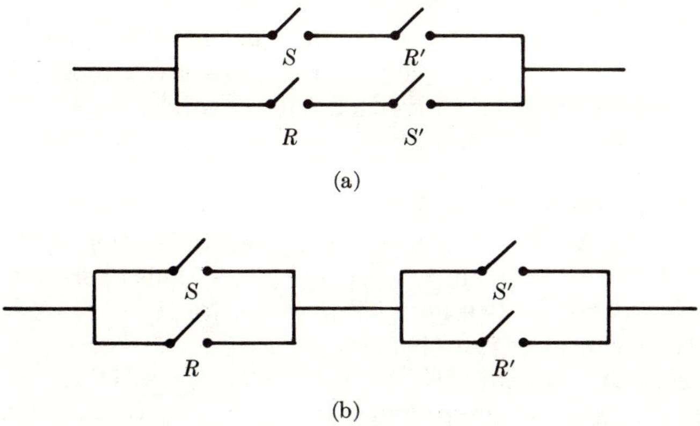

FIGURE 9.21

2. Analyze and, if possible, simplify each of the circuits in Figure 9.22.

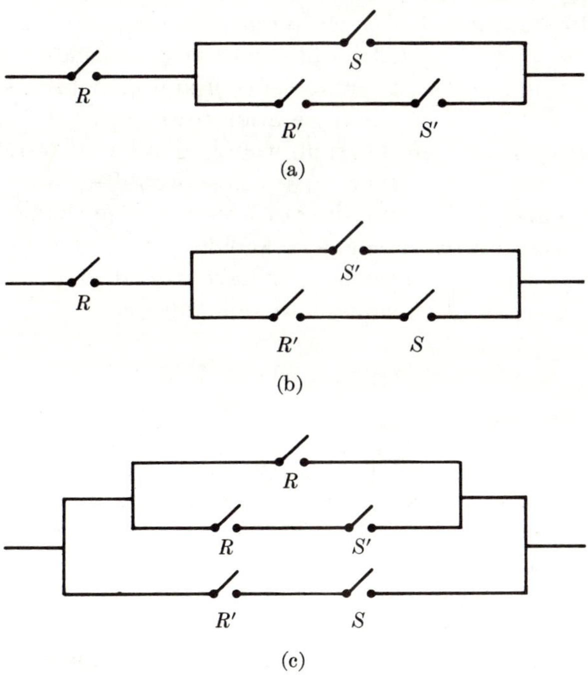

FIGURE 9.22

9.24 THE ALGEBRA OF SETS AND THE ALGEBRA OF LOGIC

The system of statements and the system of sets are so closely allied that theorems for one can be deduced from theorems for the other. Conjunction corresponds with intersection, alternation with union, denial with complementation, and the properties of the associated operations are analogous. The order of our study could have been reversed, and we might have considered first the algebra of sets and then have developed from it the algebra of logic.

There is, however, one operation on sets that has no corresponding operation in logic. The symmetric difference, introduced in Section 9.11, forms a new set out of the unlike elements of two given sets. This operation is commutative and associative, but it is not idempotent since $A + A = \varnothing$. The null set is its identity because all elements of any set A are unlike those of $\varnothing$ and $A + \varnothing = A$. Clearly, each set is its own inverse. Thus, the system of sets forms a commutative group under this operation. Symmetric difference is akin to union and produces a more traditional algebraic system than does union. Why, then, should union be preferred to it? The requirement that the union of two sets include all elements in one or the other or both is related to the definition that by the alternation of two statements one or the other or both are considered. In the system of statements, the connective that corresponds to the symmetric difference is known as *disjunctive alternation.* When it alternates two statements p and q, it means p or q but not both. If we should give this interpretation to the symbol $\vee$, then some properties of the resulting logical system would be different from those that we found. For example, Rule 3 on truth values would say that $p \vee q$ is true if and only if p and q have opposite truth values. The change in this postulate would necessitate other changes, and a system of logic dissimilar to ours would arise. Hence, the systems we have investigated are not the only ones available to us in discussing a two-valued logic.

9.25 RETROSPECT AND PROSPECT

The system of sets is quite different from that of five elements. The fact that it was no less intelligible is due to the mathematical sophistication developed during your long study. You have learned how to read mathematical material and know not only that you must be willing to deal with new concepts, but also that you must satisfy yourselves that proposed properties and theorems develop logically from the previous body of knowledge. In the act of understanding both the relation of the new ideas to earlier work and their significance for further work, you are experiencing the insight that makes the new material truly a creation of your own. To be sure, someone else preceded and guided you to the discovery. Nevertheless, in the illumination of understanding you have become creators of mathematics.

However full a discussion of a given topic may appear to be, the mathematical possibilities inherent in it have been scarcely sampled. Further generalizations, analogous systems, specializations—all these processes and many more make of mathematics an endless series of conquests by the human mind. Theories, polished and elegant in their presumed final form, have become sources from which new and more vigorous theories have sprung.

In the final chapters of the text we shall turn our attention from algebraic to geometric systems. We will find that just as algebraic systems resemble each other in certain respects and differ in others, so there are distinct geometric systems which have some common properties and many dissimilar ones. We will begin with the familiar euclidean geometry and then, by modifying certain of its fundamental postulates, create new and unfamiliar systems. Our work will give ample scope for the play of mathematical imagination.

III

GEOMETRY

In the final chapters of the text we shall turn from algebraic to geometric systems, first recalling the work of Euclid with its many familiar propositions. After considering the postulates on which he founded his treatise, we shall venture to change one of them, thus initiating a new geometry. Inquiry into the qualities of our invention will produce properties that we would never have predicted. We may find ourselves in agreement with an early eighteenth-century investigator, G. Saccheri, who is said to have observed that such a system is untenable *quia repugnantis naturae lineae recta,* because it is repugnant to the nature of a straight line. Let us see what the next chapters hold in the way of unusual mathematical experience.

10

INTRODUCTION TO GEOMETRY

Euclidean geometry is so familiar that many believe that it is the only geometry. Just as there are diverse algebraic systems, however, so there are many different geometries. Although the two systems of pairs were alike in some respects, dissimilarities in definitions of equivalence and multiplication produced such unlike characteristics that the systems bore small resemblance to each other. In the same way, changes in a fundamental postulate of euclidean geometry will create two new geometries similar in certain aspects to the original but, on the whole, so vastly dissimilar that they can be correctly described as noneuclidean.

Euclidean geometry has a venerable history spanning many centuries and involving a distinguished line of scholars. The two new geometries are not yet three hundred years old. Among their originators we shall mention three very young men who were courageous enough to challenge established principle. As we study and understand both the tradition and the dissent, we shall share in one of the great experiences of intellectual history.

10.1 EUCLIDEAN GEOMETRY

The great University of Alexandria was founded about 300 B.C. On its faculty was a scholar named Euclid. During his term of service he wrote a book called *The Elements*, which summarized and extended the mathematical learning of the Greeks. It discussed demonstrative geometry, theory of numbers, and elementary algebra. From its first appearance it was accorded the highest respect, and it has held this respect through the centuries. The section on geometry is its most famous part. It is our good fortune that, having had some experience in the methods of mathematics, we are equipped to study and appreciate it.

We must keep in mind that we are dealing with a *mathematical* system, not a practical or a physical one. Lines and points observed in nature and in architecture may suggest properties to be incorporated in it. The suggested properties cannot become a part of the geometry, however, until they are shown to follow from properties previously established.

Euclid's work is considered the prototype of modern mathematical discourse. If it had been written in this century, it would have begun with a list of undefined terms, such as *point* and *line*, and continued with definitions based on these terms. Euclid attempted to define all his terms, but when he wrote that a point is "that which has no part," he was probably leaving the concept undefined.

Euclid followed the definitions with postulates setting forth the geometric foundation of the work; for example, "a finite straight line can be extended continuously in a straight line." From these initial statements he proceeded to deduce the theorems of the geometry. The formulation of a set of postulates is never easy, and it increases in difficulty as familiarity with the subject matter increases. How many of us would think to include the postulate stated above? Yet many constructions in the geometry require the extending of lines, and we need this permission to do so.

One of the excellences of *The Elements* is the astute recognition of the properties to be stated as postulates and the skillful organization of the properties derived from them. From least assumption, a great edifice of deduction is built. Study through the centuries has turned up only a few omissions. As one example, Euclid failed to state that no matter how far a line is extended, it never repeats itself. When a tacit assumption such as this goes undetected for a long time, it may lead to theorems that do not follow strictly from the declared foundations of the subject. We shall see how the failure to state this property led to error in the work of later geometers.

The postulates are generally brief, neatly stated, and obviously essential. The famous fifth postulate is none of these. It is long and cumbersome in its statement.

> If two straight lines are cut by a transversal so that interior angles on the same side are less than two right angles, then the two straight lines, if extended indefinitely, meet on the side on which the angles are less than two right angles.

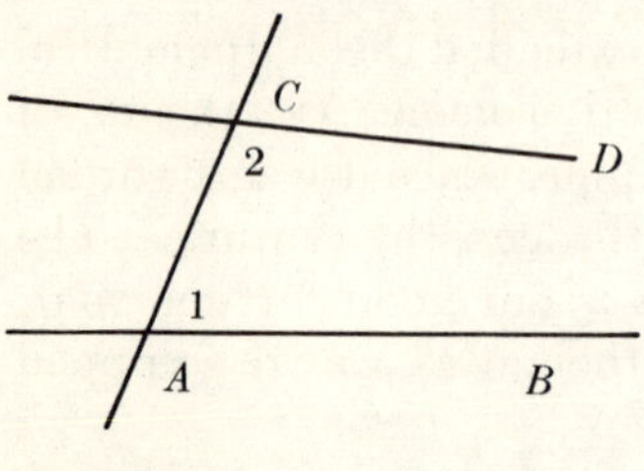

FIGURE 10.1

A diagram will help to make the meaning clear. If angles 1 and 2 in Figure 10.1 add up to less than 180°, lines AB and CD meet on the side toward B and D when they are extended. Euclid does not need this postulate until he proves Proposition 29, which establishes the properties of parallel lines, beginning with the familiar theorem that if parallels are cut by a transversal, alternate interior angles are equal. Furthermore, the postulate is the converse of an early theorem which states that if AB and CD meet so that they form a triangle with AC, then angles 1 and 2 add up to less than 180°.

Early mathematicians were distressed by the lack of elegance of the postulate, by its long delayed use, and by the fact that it is the converse of a theorem of the geometry. If the theorem could be proved, they argued, why not its converse? Of course, no theorems based on the fifth postulate could be used in the proof because that would be using the postulate to prove itself. All such theorems would be forbidden. They begin with Proposition 29; hence, only the first 28 of Euclid's propositions could be used. Every attempt to prove the postulate failed; yet the effort was not abandoned. Mathematicians were determined to remove what seemed to them a flaw in an otherwise perfect jewel.

In the thirteenth century a Persian named Nazir–Eddin attacked the problem by an ingenious indirect method. He constructed a quadrilateral as in Figure 10.2. Base angles A and B are right angles, lateral sides AC and BD are equal. Now, as every good euclidean geometer knows, properties of parallels say that the quadrilateral is a rectangle, that angles C and D, called the summit angles, are right angles, and that $CD = AB$. Nazir–

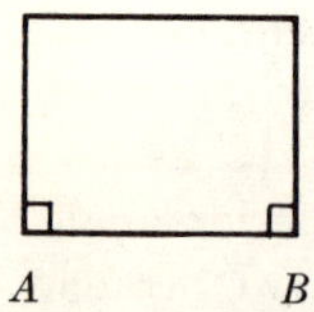

FIGURE 10.2

Eddin, however, went on to establish the remarkable fact that, conversely, if the summit angles are known to be right angles, then the fifth postulate follows as a corollary. Hence, if he could show, without using any of the forbidden theorems, that angles C and D are right angles, he would establish the fifth postulate as an ordinary theorem of the geometry. He sought, therefore, to prove that the angles could be neither acute nor obtuse. He believed that he had succeeded in his effort, but later mathematicians found errors in his work.

In the eighteenth century a Jesuit named Girolamo Saccheri set himself the task of succeeding with Nazir–Eddin's attempt. His plan of attack called for grand strategy. Instead of dealing only with the quadrilateral, as the Persian had done, he would proceed as though the summit angles are actually obtuse instead of right. Using only permitted theorems, he would draw logical conclusions from this assumption until a contradiction appeared. The contradiction would show that the obtuse-angle hypothesis is false. He would then assume that the summit angles are acute. He would argue as though this assumption were correct and would properly investigate its consequences until halted by a contradiction. The contradiction would show that the acute-angle hypothesis also is false. The only remaining possibility, that the summit angles are right angles, would, therefore, be true. From this, the fifth postulate would follow as a corollary. The obtuse-angle hypothesis soon led to contradictory results and was thereby proved untenable. The acute-angle hypothesis, however, did not yield a contradiction. Bizarre properties of lines and polygons appeared, which diverged far from those that a student of Euclid would expect; yet no contradiction arose. In the end, Saccheri could say only that the hypothesis should be rejected because it led to results that were repugnant to the nature of a straight line.

During the remainder of the eighteenth century this method of indirect attack was continued. A German mathematician, Johann Lambert, used a quadrilateral having three right angles and sought to show that the fourth angle must be a right angle. A French analyst, Adrien Marie Legendre, used a right triangle and sought to show that the sum of the two non-right angles must be a right angle. In these attempts the obtuse-angle hypothesis led to a contradiction, but the acute-angle hypothesis did not.

These geometers were, of course, presenting adequate substitutes for the fifth postulate and then seeking to prove the substitutes by using Euclid's first 28 propositions. Many other substitutes have been suggested. The most widely used of these, although known in antiquity, was made popular by the writings of John Playfair, a Scottish mathematician who flourished in the late eighteenth century. Since his time, it has replaced the fifth postulate in most textbooks on geometry. It is called Playfair's axiom and asserts that through a given point not on a given line, only one parallel can be drawn to the line. When we say that it is a substitute for the postulate we mean that (1) the axiom follows as a corollary from the first 28 propositions plus the postulate, and (2) the fifth postulate follows as a corollary from the first 28 propositions plus the axiom. In Section 10.5 it will be shown that both of these requirements are met.

10.2 THREE NONEUCLIDEAN GEOMETERS: LOBACHEVSKI, BOLYAI, RIEMANN

In 1811 a young man of eighteen, Nikolai Ivanovich Lobachevski, was granted a master's degree with distinction by the faculty of the University of Kazan in Russia. For 15 years, although busy with the duties of teacher and administrator, he pondered the problem of the fifth postulate. In 1826, before the Physical–Mathematical Society of Kazan, he read a paper in which he discussed the existence of a geometry different from Euclid's. He argued that the reason for failure to prove the fifth postulate might lie in the fact that the statement, in spite of its form as a theorem, really is a postulate. If it were changed in any way, it would produce a quite different geometry having very little more than the first 28 propositions in common with Euclid's. He argued further that the only way to test this idea would be to alter the postulate and observe the resulting geometry. Because it was not easy to see how to alter the postulate, he decided to change one of the substitutes instead. For this purpose he selected Playfair's axiom and modified it to read, "Through a given point not on a given line *more than one* straight line can be drawn which does not intersect the given line." When he investigated its consequence, he found not only that the summit angles of the Saccheri quadrilateral were acute but that the other strange statements which had shocked Saccheri became reasonable theorems in the new geometry. In 1829 Lobachevski published his results in the *Kazan Bulletin* under the title "Geometric Researches on the Theory of Parallels."

During these same early years of the nineteenth century, a Hungarian mathematician, Wolfgang Bolyai, was conducting the mathematical education of his son John. Wolfgang had been intrigued by the problem of the fifth postulate and passed on his interest to his apt pupil. The boy at first tried to prove the postulate using the method of Saccheri. By the time he reached the age of 18, in 1820, the remarkable results following from the hypothesis of the acute angle began to direct his attention toward the possibility of creating a more general geometry of which not only this new geometry but also euclidean geometry might be special cases. While a student at the Royal College for Engineers in Vienna, he continued his work. In 1823 he wrote with excitement to his father saying that the development of the geometry was progressing so well that he knew it would someday be publishable. The letter has become famous as an expression of the joy of mathematical creation: "Out of nothing I have created a strange new universe." Almost nine years passed, however, before the treatise was completed to his satisfaction. It was published as an appendix to his father's mathematical work called the *Tentamen*.

Both young men developed what is now known as hyperbolic geometry or the geometry of the acute angle. They found many theorems and investigated its trigonometry. In their honor it is sometimes called the Bolyai–Lobachevski geometry.

A third young geometer showed remarkable mathematical insight when he made a criticism that led to the invention of a second noneuclidean geometry. In his doctoral dissertation presented at Göttingen in 1854, Georg Friedrich Bernhard Riemann pointed out that although a straight line may be extended indefinitely, it does not have to be infinite in length. It could be like the arc of a circle that, after a certain amount of extending, begins to retrace itself. Such a line is called *reentrant*. Riemann recognized Euclid's use of the hypothesis that a line is not reentrant. He and other investigators soon discovered that hyperbolic lines are like euclidean lines in this respect. Moreover, they found that the contradictions that had led to the discarding of the obtuse-angle hypothesis were really contradictions to the infinite extent of a line. If a line is permitted to be reentrant, the obtuse-angle hypothesis produces a geometry just as consistent as the others. It is called *elliptic*, or *riemannian*, geometry. In it there are no parallel lines. Each line intersects every other line.

The discovery of noneuclidean geometry was a shock to the mathematicians and philosophers of the day. Scholars and laymen alike had come to believe that Euclid's geometry truly describes the relations of space and that it is a true and permanent part of man's experience of the world. They believed that the postulates of the geometry are self-evident truths and that its theorems are absolute and eternal truths. How could anyone entertain the thought that there might be other geometries besides Euclid's? The demonstration that other geometries actually exist freed, not only geometry, but all of mathematics from almost two thousand years of bondage to "universal truth." No longer is mathematics expected to describe accurately the "real" world. The postulates and axioms no longer have to do with "self-evidence" or "truth." To be sure, a mathematical system frequently serves as a model of a "real" situation and is of help in interpreting and predicting the events of the situation. But neither he who invents the system nor he who applies it is required to believe that the model is a perfect fit. The mathematical world has profited from the courageous intellectual ventures of three young men.

In the remaining sections of this chapter the foundations for our geometric discussion will be laid. Special attention will be given to the quadrilaterals of Saccheri and Lambert and to the right triangle of Legendre. We will discover properties that are independent of the fifth postulate or any of its substitutes or replacements; these properties would be a part of Bolyai's general geometry. We will then see why we need the fifth postulate to establish the well-known characteristics of euclidean parallels, and we will prove that Playfair's axiom is a valid substitute for the postulate.

In Chapter 11 hyperbolic geometry is investigated, modifying Playfair's axiom in the way that Lobachevski did. We will discover for ourselves some of the bizarre results that distressed Saccheri and observe the quadrilaterals and the triangle under these new conditions.

In Chapter 12 Playfair's axiom is replaced with the statement that all straight lines intersect, and it is postulated that lines are reentrant. We will find that we have a further decision to make before proceeding with our investigation. Having made that decision, we will be free to observe the quadrilaterals and the triangle in this distinctive geometric world.

In euclidean geometry exactly one parallel can be drawn to a line through a point external to the line. In the first of the new systems, more than one parallel exists; and, in the second, there is no such parallel. The names given to the geometries arise, in part, from these characteristics. The adjective *hyperbolic* indicates an excess of parallels; it comes from Greek words meaning *to throw beyond the mark*. The name *elliptic* implies a deficiency in parallels; it comes from Greek words meaning *an omission*. In contrast, euclidean geometry may be called *parabolic*, just right, adapted from Greek words meaning *to throw on the mark*.

10.3 THE FIRST PORTION OF *THE ELEMENTS*

The Appendix presents an outline of the first part of *The Elements*, beginning with the definitions and continuing through the important Proposition 32. It is necessary that the student consult this Appendix continuously as he reads and studies this chapter.

Of the many definitions that Euclid gives, only the one on parallel lines is stated. It is followed by the postulates and the common notions. The first of these are geometric in character and are directly concerned with the subject matter. All are easily comprehended except the fifth, which requires study to be understood. The common notions are so called because they apply not only to geometry but also to arithmetic. In modern parlance, the terms *axiom* and *postulate* are used interchangeably to denote all statements that are accepted without proof as the basis of argument.

The outline then lists the first 32 propositions, usually indicating only the topics of which they treat. When a theorem is written in detail, the number by which it is designated in *The Elements* is also given.

Finally, the outline contains four assumptions that Euclid used but did not state. These are all complicated in concept and in general wording, but, as expressed for the specific applications that we shall make, they are simple and clear. The first three are required to establish early theorems. The fourth is the one whose omission was noted above. It is not needed until the justification of Proposition 16 is considered. Let us recall this theorem and then give a modern version of a part of the proof.

PROPOSITION 16 In any triangle, if one of the sides is extended, the exterior angle is greater than either of the interior and opposite angles.

Using the diagram in Figure 10.3 we have the following statements.

Given $\triangle ABC$, with side AB extended to E.

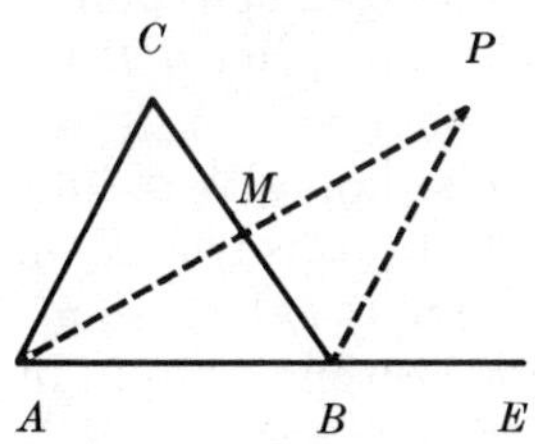

FIGURE 10.3

To Prove Exterior angle CBE is greater than opposite interior angle ACB.

Construction Draw the median from A, cutting side BC at M. Extend the median to P so that $MP = AM$. Draw PB.

Proof

	Arguments		*Reasons*
(1)	$\triangle AMC \cong \triangle PMB$	(1)	Why?
(2)	$\measuredangle MCA = \measuredangle MBP$	(2)	Why?
(3)	$\measuredangle CBE > \measuredangle MBP$	(3)	Why?
(4)	$\measuredangle CBE > \measuredangle ACB$	(4)	Why?

Now this proof appears to be complete, and it makes no special mention of unstated assumption 4. Where was that assumption used? It was used in the reasoning of step 3 when we argued that because PB lies within the angle CBE, it divides the angle into two parts, each of which is smaller than the whole. The question is, how do we know that the line PB lies within CBE? This is true only if, when AM is extended, P lies outside triangle ABC. Just suppose that line AMP is reentrant and that P lies inside the triangle as in Figure 10.4. Then we would have no relation to

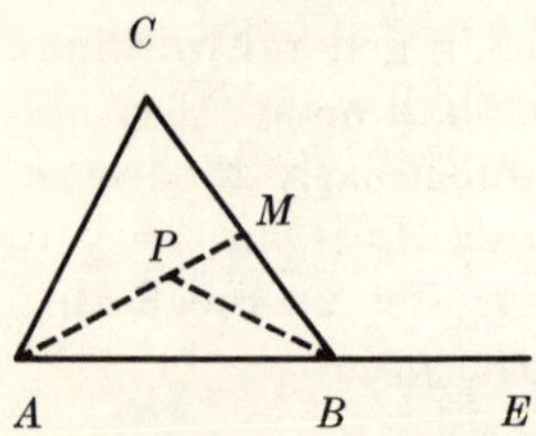

FIGURE 10.4

write down between the exterior angle CBE and the angle MBP. In this case, our proof would fail. We must, therefore, adjoin unstated assumption 4 to the geometry. This new postulate says that straight lines are not reentrant; hence, every time we produce a line segment we get new points. Now we can be sure that point P lies outside triangle ABC, line PB lies within angle CBE, and the proof of Proposition 16 is complete.

In the Appendix a prime follows the number of this proposition, indicating that the proof requires unstated assumption 4. The same is true of the other theorems having their numbers primed. Each of them uses the assumption, either directly or in a theorem quoted as a reason for some step in its proof.

EXERCISE 10.3

1. Supply the reasons in the proof of Proposition 16.

10.4 THREE IMPORTANT POLYGONS

The polygons that were given special mention in Section 10.1 are shown in Figure 10.5. The first is the familiar right triangle. If the right angle is at A, angles B and C are called the *remaining angles.* The second polygon is the Lambert quadrilateral. Three of its angles are right angles; the unrestricted angle at C is called, appropriately, the *fourth angle.* The third polygon is the

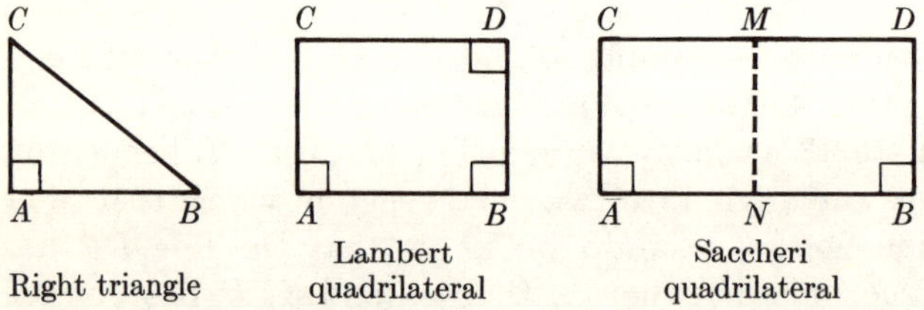

FIGURE 10.5

Saccheri quadrilateral whose base angles are right angles and whose lateral sides are equal. When base angles are at A and B and lateral sides are AC and BD, line CD is called the *summit*, and angles C and D are the *summit angles*. If the summit is bisected at M and the base at N, line MN is called the *median* from summit to base.

We seek to establish some of the relations among these polygons without using any primed propositions. Such relations would be theorems in Bolyai's general geometry. Consultation with the Appendix shows that we are limited to the use of little more than theorems on isosceles triangles, on simple constructions, and on congruence of triangles. Let us see what these theorems will produce.

The Saccheri quadrilateral appears to be symmetric with respect to its median. If it really is, then the summit angles are equal, the angles at M and N are right angles, and each half of the quadrilateral has three right angles and is therefore a Lambert quadrilateral. All these properties can be obtained as a result of congruence of triangles by drawing construction lines and labeling resulting angles as in Figure 10.6.

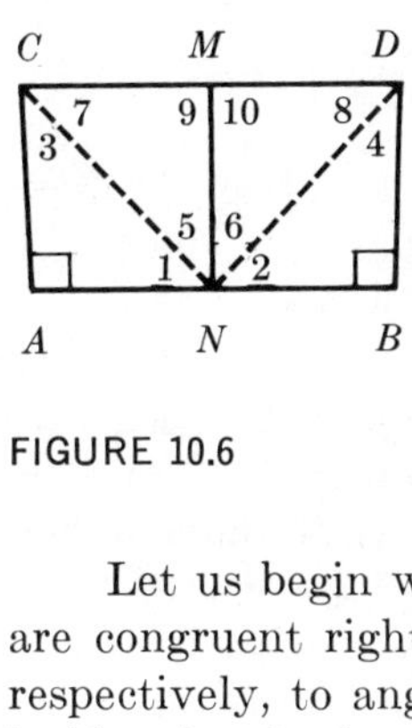

FIGURE 10.6

Let us begin with the outside triangles and move in. CAN and DBN are congruent right triangles; hence, remaining angles 1 and 3 are equal, respectively, to angles 2 and 4. The hypotenuses are also equal. Now, the inside triangles have three sides of one equal to three sides of the other and are congruent. Angles 5, 7, and 9 are equal, respectively, to angles 6, 8, and 10. Hence, summit angles C and D are equal, the angles at M and N are right angles, and all our conjectures are proved true. Furthermore, the two Lambert quadrilaterals into which the median divides the Saccheri quadrilateral have corresponding sides (as well as angles) equal and are congruent. If the original polygon is folded along the median MN, side NB will coincide with NA, side BD with AC, and side MD with MC. We have our first theorem.

THEOREM 10.4-1 In the Saccheri quadrilateral, the summit angles are equal. The median from summit to base is perpendicular to both; it separates the quadrilateral into two congruent Lambert quadrilaterals in such a way that the summit angles of the Saccheri quadrilateral become the fourth angles of the Lambert quadrilaterals.

Since we can separate a Saccheri quadrilateral into two congruent Lambert quadrilaterals, we may be able to join two of the latter to form the former. For the common join, let us use corresponding sides that run between right angles. Thus, in the diagram of Figure 10.7 congruent quadrilaterals $ABDC$ and $A'B'D'C'$ are adjoined along common lines BD and $B'D'$ to form polygon $ABA'C'DCA$. The right angles at D and B guarantee that lines CDC' and ABA' are straight lines. Hence, the polygon is a quadrilateral having right angles at A and A', equal sides AC and $A'C'$, equal summit angles C and C', and median DB perpendicular to summit and base. It satisfies the description and exhibits the properties of a Saccheri quadrilateral.

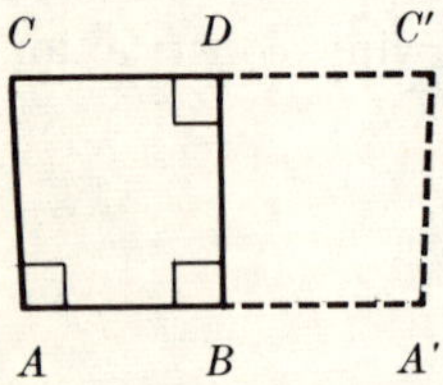

FIGURE 10.7

THEOREM 10.4-2 To any Lambert quadrilateral can be joined a congruent quadrilateral in such a way that the resulting configuration is a Saccheri quadrilateral having as summit angles the fourth angles of the Lambert quadrilaterals.

Let us continue our discussion of the three important polygons by showing that we can start with a right triangle and construct a Lambert quadrilateral whose fourth angle is the sum of the remaining angles of the triangle. At the midpoint P of hypotenuse CB of right triangle ABC draw line PN perpendicular to base AB. Continue NP on the other side of CB to M so that $PM = NP$. Join C and M. The resulting diagram is shown in Figure 10.8. Since triangles NPB and MPC are congruent, angle M is a right angle, and $\measuredangle PCM = \measuredangle B$. Hence, $ANMC$ is a Lambert quadrilateral, and its fourth angle at C is equal to the sum of the remaining angles of triangle ABC.

To reverse the procedure, let us begin with a Lambert quadrilateral and construct a right triangle whose remaining angles are together the fourth angle of the quadrilateral. Let us bisect one of the sides that joins two right angles. In Figure 10.9 take point P at the midpoint of side BD. Draw CP

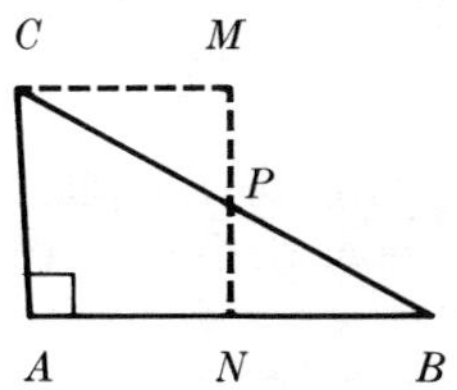

FIGURE 10.8

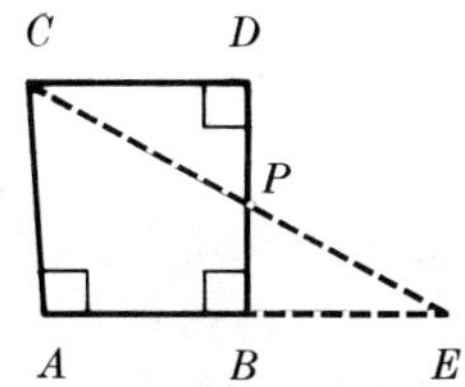

FIGURE 10.9

and extend it until, at E, it meets side AB produced. Triangles DPC and BPE are congruent, and $\measuredangle PCD = \measuredangle E$. Right triangle ACE, therefore, has its remaining angles equal to the fourth angle C of the quadrilateral.

THEOREM 10.4-3 On any right triangle a Lambert quadrilateral can be constructed in such a way that the fourth angle of the quadrilateral is the sum of the remaining angles of the triangle. Conversely, on any Lambert quadrilateral a right triangle can be constructed in such a way that the sum of the remaining angles of the triangle equals the fourth angle of the quadrilateral.

It has now been shown that when any one of the three important polygons is given, each of the other two can be so constructed that the relations stated in the theorems hold between their angles. It follows that whatever can be said about the specified angles of the quadrilaterals can be said about the sum of the remaining angles of the triangle, and vice versa.

THEOREM 10.4-4 Whatever is true of any one of the following is also true of the other two: the sum of the remaining angles of the right triangle, the fourth angle of the Lambert quadrilateral, a summit angle of the Saccheri quadrilateral.

EXERCISE 10.4

1. In the proof of Theorem 10.4-1 explain why
 (a) $\triangle CAN$ is congruent to $\triangle DBN$.
 (b) The angles at C and D are equal.
 (c) The angles at M are right angles.
 (d) The angles at N are right angles.

2. In the proof of Theorem 10.4-3 explain why
 (a) $\triangle NPB \cong \triangle MPC$ in Figure 10.8.
 (b) $\triangle DPC \cong \triangle BPE$ in Figure 10.9.

10.5 THEOREMS ON EUCLIDEAN PARALLELS

Euclid's first proposition concerns the construction of equilateral triangles, thus establishing the existence of lines that meet. The existence of lines that do not meet, on the other hand, is not justified until after a considerable body of geometric knowledge has been accumulated. For the first time, in Proposition 27, the assertion is made that there are lines in the geometry that do not intersect. Then, in Proposition 29, initial properties of such lines are cataloged.

The proofs of these theorems are of interest not only because of the important results that they establish but also because of the method that they employ: a method called indirect proof, proof by contradiction, or proof by *reductio ad absurdum*. To use this method we make an assumption, show that the assumption leads to a contradiction, and conclude that the denial of the assumption is true. One logical justification for the method uses the following implication. Let statement letter a represent the assumption and c the contradiction; then prove that the conditional $a \rightarrow c$ is true. Because $\sim c$ is also true, we have $(a \rightarrow c) \wedge \sim c$, which, by Exercise 8.6.3, logically implies $\sim a$. It is clear that a well-chosen assumption can let us establish the conclusion of our theorem. The method is illustrated in the proof of the first part of Proposition 29.

THEOREM If two parallel lines are cut by a transversal, then alternate interior angles are equal.

The diagram in Figure 10.10 permits us to make a restatement in terms of observed lines and angles.

Given Lines AB and CD are parallel and are cut at points G and H, respectively, by transversal GH.

To prove $\measuredangle 1 = \measuredangle 2$.

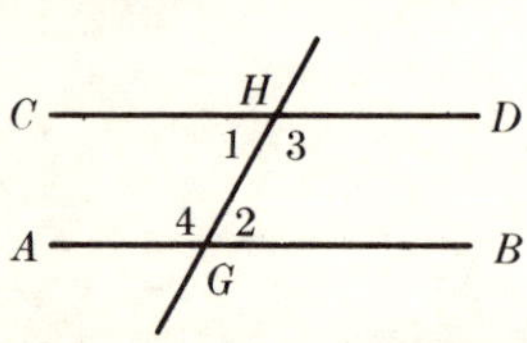

FIGURE 10.10

It is reasonable to begin by recalling criteria for equal angles. There are a plethora of conditions, none of which applies because the angles are not located at vertices of polygons. Let us try, instead, the method of indirect proof and assume the truth of the following statement.

Statement F $\measuredangle BGH$ is less than $\measuredangle CHG$.

Now let us deal with relations among angles. Because $\measuredangle 2$ is less than $\measuredangle 1$, we have $\measuredangle 2 + \measuredangle 3$ less than $\measuredangle 1 + \measuredangle 3$, where angles 1 and 3 are supplementary. Hence, $\measuredangle 2 + \measuredangle 3$ is less than two right angles. In this situation, Euclid's Postulate 5 comes to our rescue. This small angle sum is exactly what the antecedent of its conditional requires. It says that if a transversal cuts two straight lines so that the interior angles on one side of the transversal are less than supplementary, then the two lines meet on that side of the transversal. Lines AB and CD meet at a point on the side toward B and D. This statement is a direct contradiction to the hypothesis that the lines are parallel. Hence, Statement F is false, its denial is true, and $\measuredangle BGH$ is not smaller than $\measuredangle CHG$. By a similar indirect proof it can be shown that $\measuredangle BGH$ is not greater than $\measuredangle CHG$. As a consequence, the two angles are equal, and the theorem is justified. The remaining parts of Proposition 29 can now be established.

The proof could not have been made without using the fifth postulate. For this reason, the number of the proposition is marked with an asterisk in the Appendix. Other propositions similarly marked also use the fifth postulate, either directly or by employing starred theorems in their proofs. Two of these give further characteristics of parallel lines. The last one on the list applies properties of parallels to determine the angle sum of a triangle. A corollary then asserts that the sum of all the angles of a quadrilateral is four right angles. In Exercise 10.5.5 below consequences of this corollary are investigated.

The third starred proposition, number 31, makes two assertions. It says that a parallel can be drawn to a line through a point external to the line. Note that Propositions 27 and 28 present a variety of ways of making this construction. Proposition 31 goes on to assure us that no matter how the construction is made, exactly the same parallel will result. This assertion expresses Playfair's axiom. Its proof begins as an ordinary indirect proof. The contradiction it uncovers, however, does not concern one of the hypotheses but rather is an instance of the pattern $p \wedge \sim p$ which was shown in Exercise 8.5 to be always false. We shall use Figure 10.11 and make an appropriate assumption.

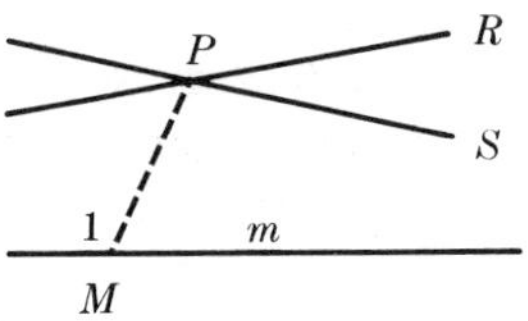

FIGURE 10.11

THE PLAYFAIR AXIOM Through a given point not on a given line there can be drawn only one line that does not intersect the given line.

Given Line m and point P not on line m. Lines PR and PS parallel to line m and extending in the same direction from P.

To Prove Lines PR and PS coincide.

Statement F Lines PR and PS are distinct.

Draw any transversal PM to line m, and mark angle 1 as in the figure. Since PR and PS are distinct, angles MPR and MPS are unequal. On the other hand, since PR and PS are both parallel to line m, angles MPR and MPS are both equal to angle 1 and are equal to each other. But angles cannot be unequal and equal at the same time; hence, Statement F is false, its denial is true, and lines PR and PS are not distinct. In making this proof we have shown that from the first 28 propositions and the fifth postulate there follow logically Proposition 29 and Playfair's axiom.

Let us now show that when we use Euclid's first 28 propositions, Playfair's axiom implies the fifth postulate. We interpret the postulate using the diagram in Figure 10.12.

Given Playfair's axiom. Lines AB and CD cut by transversal GH so that the sum of angles BGH and GHD is less than two right angles.

To Prove Lines AB and CD intersect in the directions toward B and D, respectively.

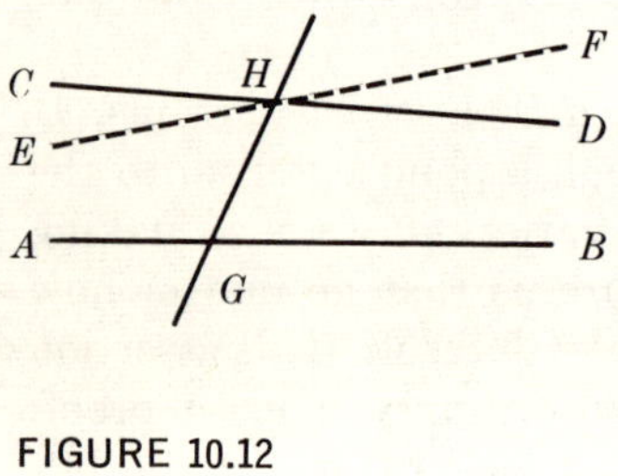

FIGURE 10.12

Through point H draw line EHF so that the interior angles on the same side of the transversal, angles BGH and GHF, are supplementary. By Proposition 28, line EF is parallel to line AB through point H; and, by Playfair's axiom, it is unique in this property. Because line CHD does not coincide with line EHF, it is not parallel to line AB at H. It must, therefore, intersect line AB either in the direction of A and C or in the direction of B and D. By indirect proof it can be shown that the intersection must be in the direction of B and D. For, assume that the intersection is at J in the direction of A and C. Then JGH is a triangle. According to Proposition 17, the sum of its angles JGH and GHJ must be less than two right angles. Hence, the sum of their supplements, BGH and GHD, must be more than two right angles. This contradiction to the hypothesis shows that the assumption is false, its denial is true, and the lines do not meet in the direction of A and C. Thus, by using Playfair's axiom the fifth postulate is proved as a theorem.

EXERCISE 10.5

1. Make the indirect proof to show that when $AB \parallel CD$, then $\measuredangle BGH$ in Figure 10.10 cannot be greater than $\measuredangle CHG$.

2. Given line m and point P not on line m. How can we construct through P a line parallel to m?

3. Prove that in euclidean geometry two lines perpendicular to the same line are parallel.

★4. (a) Prove Proposition 32. (Suggestion: use the diagram of Figure 10.3. Remember that in the proof of Proposition 16 we have shown that $\measuredangle MCA = \measuredangle MBP$.)

(b) Prove that the sum of the remaining angles of a euclidean right triangle is a right angle.

5. Prove that in euclidean geometry the following statements are true.

(a) The Saccheri quadrilateral is a rectangle.

(b) In the Saccheri quadrilateral:

(i) The summit is equal and parallel to the base.

(ii) The median from summit to base is equal and parallel to the lateral sides.

(c) Every Saccheri quadrilateral is a Lambert quadrilateral, and conversely.

10.6 COMMENT ON DIAGRAMS

Diagrams are so helpful in geometry that we may be in danger of placing too much dependence on them. We must remember that drawings are intended chiefly to help us keep in mind details of the argument we are making. Diagrams often suggest properties, but they cannot establish them. For example, two triangles are not congruent simply because they look so. Nor is a straight line defined as the mark we get when we draw a pencil along the edge of a ruler. When we draw such lines, the resulting pictures often look as though the properties we ascribe to them are true. Eye and mind work comfortably together in euclidean geometry; this is our good fortune. Nevertheless, the drawings are only schematic diagrams used to aid the argument.

In the new geometry we are about to develop, the drawings will less faithfully portray the properties that we find. We will continue to draw straight lines with the aid of a ruler because the diagrams so drawn are neat and clear; but lines will intersect or not intersect according as the argument says they do or do not. A drawing will distinguish between the various lines we are discussing, but it may not properly show how the lines behave with relation to each other.

11

HYPERBOLIC GEOMETRY

In this chapter we shall discuss the Lobachevski–Bolyai geometry, which is different from euclidean geometry yet has many properties in common with it. We shall accept Euclid's definitions, axioms, first four postulates, unstated assumptions, and theorems through Proposition 28. Then, instead of using Postulate 5 to get Proposition 29, we shall use a modification of Playfair's axiom and see what sort of geometry results.

11.1 BEGINNING CONCEPTS OF HYPERBOLIC GEOMETRY

The development of the geometry begins with the statement of the modified axiom and requires frequent reference to Figure 11.1. As previously cautioned, the lines in diagrams may not look as though they are obeying the laws that the geometry imposes on them. Nevertheless, drawings are helpful in interpreting arguments.

PARALLEL POSTULATE OF HYPERBOLIC GEOMETRY Through a given point not on a given line, there can be drawn more than one line that does not intersect the given line.

Take line m and point P not on the line. Through P there pass some

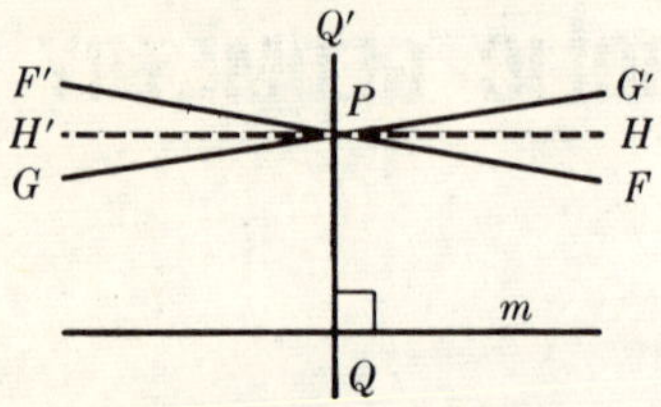

FIGURE 11.1

lines that intersect m and some that do not. Draw the perpendicular $Q'PQ$ through P to m, intersecting m at Q. Draw the line $H'PH$ perpendicular to line PQ at P. By Euclid's Proposition 28, which is a part of our geometry, this line $H'PH$ does not intersect m.

Now consider all lines that pass through P and observe the parts of them that lie on or to the right of the line PQ. Let us say that these halflines or rays represent the various positions of the ray PQ as it rotates in a counterclockwise direction through P. For a while, the rays will cut m; then they will not. All these rays are thus divided into two categories: those that intersect m and those that do not. Both categories are nonempty because the first contains PQ and the second contains PH. Furthermore, the rays that cut line m precede those that do not.

According to the postulate of Dedekind, unstated assumption 3, there must be a specific ray that effects the division of all the rays into the two categories. This dividing ray either intersects line m or it does not. If it does, it is the last one to intersect m. If it does not, it is the first one that does not. Which is it? It will be shown that a definite decision can easily be made because there is no ray that is the last one to cut line m. For suppose that such a last ray does exist, and it cuts m at point S. Then, because m does not stop at point S and does not retrace itself, m must pass through a point T to the right of S. Now, new ray PT cuts m but follows the supposed last ray PS. Hence, there is no last ray to cut line m. The dividing ray must, therefore, be the first of those that do not intersect m. Designate this dividing ray as PF.

To summarize, for all lines through point P the rays on or to the right of PQ are divided into two sets by line PF. The rays that lie within the angle QPF (including PQ but excluding PF) intersect line m. The rays that lie within the angle FPQ' (including PF and PQ') do not intersect m. We note that PH is one of these.

Now let us turn our attention to the left-hand sections of all lines through point P. Again, these halflines represent the various positions of the ray PQ as it rotates in a clockwise direction through point P. For a while, the rays will cut m, and then they will not. All these rays are thus separated into two nonempty sets: those that intersect m and those that do not. There must be a particular ray that separates the rays into the two categories, and this dividing ray must be the first ray that does not intersect m. Letter it PG. It follows that for all lines through P the rays on or to the left of PQ are divided into two sets by line PG. Those that lie within the angle QPG (including PQ but excluding PG) intersect m. Those that lie within the angle GPQ' (including PG and PQ') do not intersect m. Ray PH' is one of these.

Next consider the angles QPF and QPG. Because of the symmetry of the procedure above, it is reasonable to expect that angles QPF and QPG are equal. It can be proved that this is so. The angles are both acute because lines PF and PG must precede lines PH and PH', respectively.

When we study the whole lines through P, not just the parts of them that lie to the right or to the left of PQ, we find that the extension PF' of ray FP back through P lies above PG and that, correspondingly, the extension PG' of ray GP back through P lies above PF. It follows that these extensions do not intersect line m. Furthermore, any lines drawn through P within the vertical angles FPG' and GPF' lie throughout their length above both PF and PG; hence, they do not intersect m. On the other hand, those within the angle $F'PG'$ surely cut m when extended backward through P. The final result is that the sheaf of lines included within the vertical angles $F'PG'$ and GPF intersect m, while those within and including the sides of the vertical angles GPF' and $G'PF$ fail to intersect m.

According to Euclid's definition of parallels, all the last-mentioned lines are parallel to m. In the new hyperbolic geometry, however, it is convenient to make special mention of the limiting lines $F'PF$ and GPG'. They are said to be *parallel to m* and the other lines are said to be *nonintersecting with m*. We name the angles QPF and QPG the *angles of parallelism* of $F'PF$ and $G'PG$, respectively. Because the parallels go in different directions from each other, lines are said to be parallel in a given *direction* or *sense*. Line $F'PF$ is parallel to m in the direction of F and line $G'PG$ in the direction of G; hence, the two lines are parallel to m in *opposite* senses.

The first theorem of hyperbolic geometry is stated using Figure 11.2.

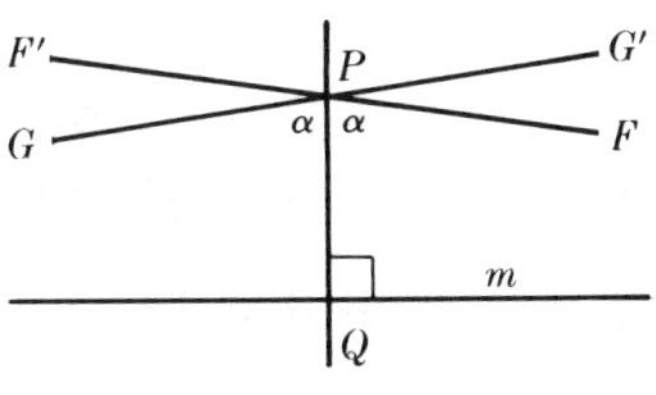

FIGURE 11.2

THEOREM 11.1-1 If m is any line and P is any point not on m, then there exist two lines through P in different directions that do not intersect m and have the following properties:

(a) They make equal acute angles with the perpendicular PQ from point P to line m, one on each side of PQ.

(b) All lines through point P that lie within the angle containing the perpendicular intersect m.

(c) All other lines through P do not intersect m.

DEFINITIONS

The two limiting lines through point P, discussed in Theorem 11.1-1, are called **parallel** to line m, each in a specfic direction or sense as described below.

The equal acute angles of part (a) are called the **angles of parallelism** and are denoted by the Greek letter α.

The **direction of parallelism** of each of the limiting lines is toward the side of PQ on which the angle of parallelism lies.

The lines described in part (b) are called **intersecting** with line m.

The lines described in part (c) are called **nonintersecting** with line m.

11.2 ON NONINTERSECTING LINES IN HYPERBOLIC GEOMETRY

Let us begin with the association between lines m and $H'PH$ of Figure 11.1, because it emphasizes the distinction between concepts of the word *parallel* in euclidean and hyperbolic geometries.

THEOREM 11.2-1 If two lines are perpendicular to a third line, they are nonintersecting with each other. Each is one of the infinite number of lines that do not intersect the other, but neither is the limiting dividing line that is parallel to the other.

In euclidean geometry, parallel lines are equidistant from each other and every transversal that is perpendicular to one of them is perpendicular to the other. We shall see that two nonintersecting lines in hyperbolic geometry have one and only one common perpendicular. They are closest together where they are cut by this common perpendicular and diverge from each other on either side of it.

When two lines are cut by a transversal, certain equalities between the angles formed ensure parallelism for the lines in euclidean geometry. These equalities are sufficient to guarantee only nonintersection in hyperbolic geometry.

THEOREM 11.2-2 If two straight lines are cut by a transversal so that they form equal alternate interior angles, equal corresponding angles, or supplementary interior angles on the same side of the transversal, then the lines are nonintersecting.

This theorem is the hyperbolic restatement of Propositions 27 and 28. It is reasonable to ask whether an analogous modification of Proposition 29 might hold. Such a restatement would say, in part, that if two nonintersecting lines are cut by a transversal, corresponding angles on the same side of the transversal are equal. Unfortunately, this conditional is generally false. In Figure 11.3 two nonintersecting lines m and n are shown diverging from their common perpendicular KH. The diagram is correct when it suggests that for all diagonals labeled RS angle 2 is greater than its corresponding angle 1. Only for diagonal AB which passes through the midpoint O of the common perpendicular are corresponding angles equal.

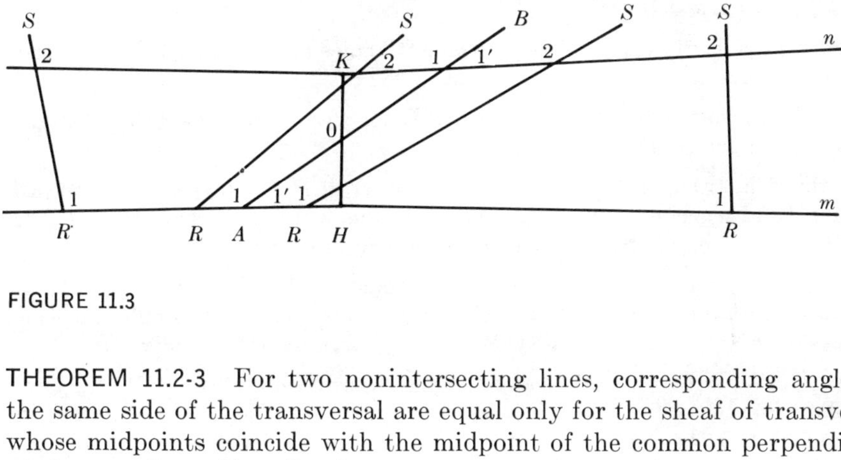

FIGURE 11.3

THEOREM 11.2-3 For two nonintersecting lines, corresponding angles on the same side of the transversal are equal only for the sheaf of transversals whose midpoints coincide with the midpoint of the common perpendicular of the two lines. For all other transversals, the exterior angle on the side toward the midpoint of the common perpendicular is greater than the corresponding interior angle.

EXERCISE 11.2

1. In Figure 11.4, lines m and n are perpendicular to AB. Lines CD, EF, and GH are perpendicular to line m.

(a) Why are the angles marked 1, 2, and 3 all obtuse?

(b) Since line m is a common perpendicular to lines AB and CD, the angle marked 4 is greater than the angle marked 5. By similar analysis of the other lines of the figure, show that for angles as marked, $\measuredangle 4 > \measuredangle 5 > \measuredangle 6 > \measuredangle 7$.

(c) Find the relations between the angles marked 8, 9, 10, and 11 in Figure 11.4.

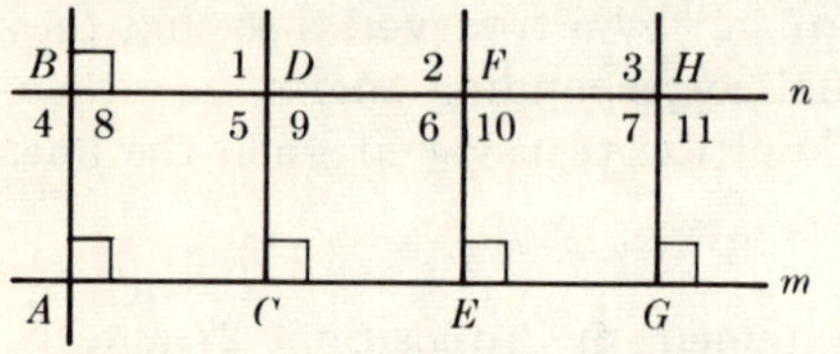

FIGURE 11.4

2. (a) Show that in Figure 11.4 the polygons *ABCD*, *ABEF*, and *AGBH* are all Lambert quadrilaterals.

(b) Which angle is the fourth angle in each quadrilateral of part (a)?

(c) Is *CEDF* a Lambert quadrilateral? Justify your answer.

3. What do Exercises 1 and 2 suggest about the fourth angle of the Lambert quadrilateral?

4. In the Saccheri quadrilateral in Figure 10.5 of Section 10.4, find all pairs of nonintersecting lines.

5. Is it reasonable to expect that the summit angles of the Saccheri quadrilateral are acute? Explain.

6. What does your answer to Exercise 5 imply concerning the sum of the remaining angles of the right triangle?

7. In the right triangle of Figure 11.5 let transversal *DE* be perpendicular to base *AB*. In euclidean geometry, the angles marked 1 and 2 are equal, and triangles *CAB* and *EDB* are similar. What happens in hyperbolic geometry?

8. In right triangle *CAB* of Figure 11.5, let transversal *DE* be drawn so that the angles marked 1 and 2 are equal. In euclidean geometry, $\measuredangle BDE$ is a right angle and triangles *CAB* and *EDB* are similar. What happens in hyperbolic geometry?

★9. Let $\triangle ABC$ in Figure 11.6 be any triangle. Extend side *AB* to *D*. Within angle *CBD* draw line *BE* so that $\measuredangle CBE = \measuredangle C$. In euclidean geometry, lines *AC* and *BE* are parallel, $\measuredangle DBE = \measuredangle A$, as stated in Proposition 32. In hyperbolic geometry:

(a) Show that *AC* and *BE* are nonintersecting. Where does the midpoint of their common perpendicular lie?

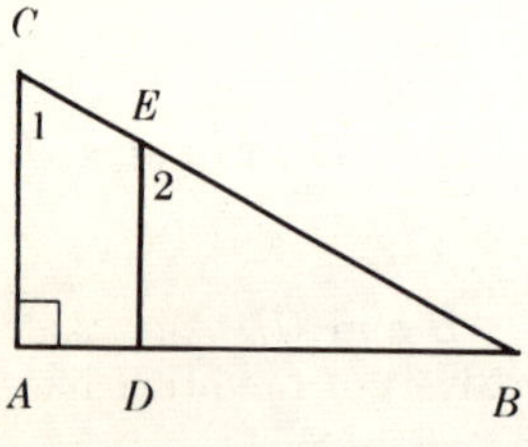

FIGURE 11.5

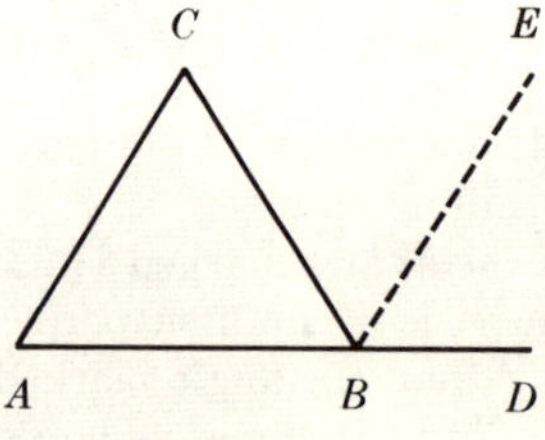

FIGURE 11.6

(b) Prove that $\measuredangle DBE > \measuredangle A$.
(c) State the analogue of Euclid's Proposition 32.

10. The distance from a point to a line (or from a line to a point) is measured in hyperbolic geometry by the same method as in euclidean geometry. Explain how this is done.

11.3 FURTHER PROPERTIES OF NONINTERSECTING LINES IN HYPERBOLIC GEOMETRY

The results in Exercises 11.2.1 and 11.2.7 above emphasize the fact that drawings do us a disservice if we rely on their appearance. The line n in Figure 11.4 and the line CEB drawn in Figure 11.5 do not accord with the laws that govern the angles formed with their transversals. For example, line n leaves line AB at right angles; then it leaves lines CD, EF, and GH at increasingly larger obtuse angles. A more representative drawing would use a broken line for n (Figure 11.7). Similarly, hypotenuse CEB makes a larger angle leaving DE than leaving AC. Here again a broken line would more accurately portray the situation. But a broken line introduces a new distortion to angle relations. Who would guess from Figure 11.7 that angles 1 and 9 are vertical and angles 5 and 9 are supplementary? Clearly, neither unbroken nor broken lines are satisfactory. We might try using curves to represent certain lines. Thus, the drawing in Figure 11.8 portrays the true situation more accurately than any of the preceding diagrams. It makes acute fourth angles seem reasonable for Lambert quadrilaterals $ACBD$, $ABEF$, and $AGBH$. It further suggests the statement made in the preceding section that two nonintersecting lines have only one common perpendicular and that they are closest together at this common perpendicular. This last observation is stated in a theorem even though its proof requires more information than our study of hyperbolic geometry will afford.

THEOREM 11.3-1 Two nonintersecting lines diverge ever farther apart on each side of their common perpendicular, the distance from a point on one

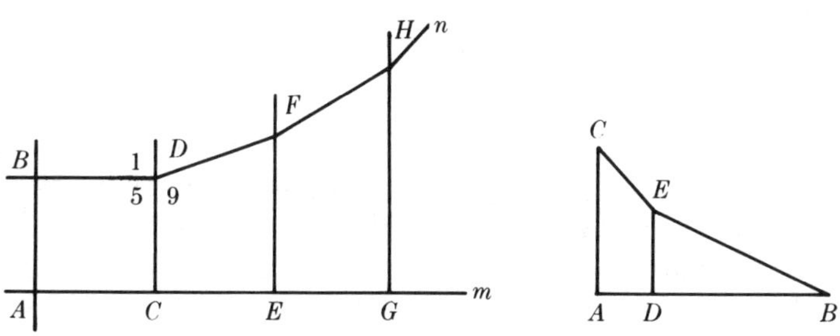

FIGURE 11.7

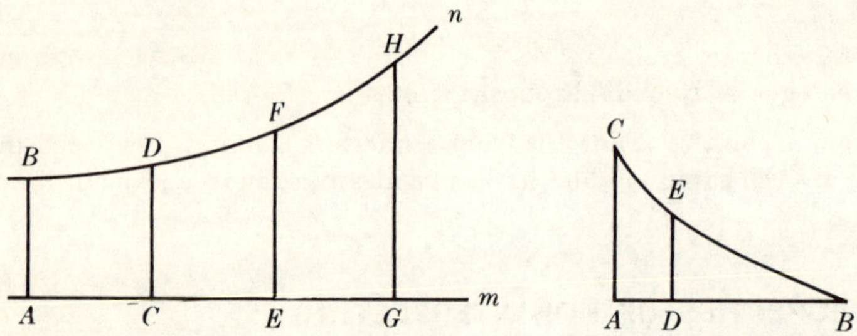

FIGURE 11.8

to the other being shortest when measured along that perpendicular and increasing without limit as the point moves away from the common perpendicular in either direction.

Let us apply this theorem to the Lambert quadrilateral $ACBD$ of Figure 11.4. Because lines m and n have common perpendicular AB, they diverge as they leave AB. The distance from n to m is measured along AB at point B and along CD at point D. (Why?) Hence, $CD > AB$.

EXERCISE 11.3

1. Show that in Lambert quadrilateral $ACBD$ of Figure 11.4 $BD > AC$.

2. Explain why the preceding discussion suggests that for opposite sides of a Lambert quadrilateral the one adjoining the fourth angle is the longer.

3. What does Exercise 2 imply concerning the lengths of the following lines of the Saccheri quadrilateral?
- (a) The lateral sides and median from summit to base.
- (b) The summit and base.

4. In euclidean geometry, every Lambert quadrilateral is a Saccheri quadrilateral, and conversely. What appears to be true in hyperbolic geometry?

5. Show that the results of Exercises 11.2.7 and 11.2.8 imply the following inequality:

$$\text{sum of angles of } \triangle EDB > \text{sum of angles of } \triangle CAB$$

6. Using the results of Exercise 11.2.1, form an inequality that is true of the sum of the angles of the Lambert quadrilaterals $ACBD$, $AEBF$, and $AGBH$.

7. What do Exercises 5 and 6 suggest concerning the relation between size of polygon and sum of angles for each of the polygons below?

(a) Right triangle (b) Lambert quadrilateral
(c) Any triangle (d) Saccheri quadrilateral

8. Let equal line segments AB and CD bisect each other at right angles at point O. Through A and B draw lines a and b, respectively, both perpendicular to AB. Similarly, through C and D draw lines c and d, respectively, both perpendicular to

CD. The four new lines intersect to form a quadrilateral. Find all the properties of this quadrilateral that can be justified by our present knowledge of hyperbolic geometry.

11.4 PROPERTIES OF PARALLELS IN HYPERBOLIC GEOMETRY

In euclidean geometry, two parallel lines remain parallel throughout their length; if a first line is parallel to a second, the second is parallel to the first; and two lines parallel to a third line are parallel to each other. The properties are trivial in euclidean geometry because they say little more than that if two lines do not meet, then they do not meet. Similar characteristics hold in hyperbolic geometry, but their statements are complicated by the necessity to specify the sense of parallelism. Their proofs are not trivial because each line must be established as the limiting line that first fails to meet the other.

Let us inquire next about the relations holding between angles when a transversal cuts two hyperbolic parallels. Recall that in a similar situation, nonintersecting lines afford equal angles only when the transversal passes through the midpoint of their common perpendicular. Since parallel lines have no common perpendiculars, will inequality of angles be the rule? This conjecture is correct and is stated in the following theorem. (You will be asked to prove it in Exercise 11.4.8.)

THEOREM 11.4-1 If two parallel lines are cut by a transversal, then in the direction of parallelism the exterior angle is greater than the corresponding interior angle.

An important consequence of this theorem becomes evident when several parallels are cut by a transversal. The configuration diagrammed in Figure 11.9 shows lines a, b, and c, all parallel to line m and all cut by line PQ

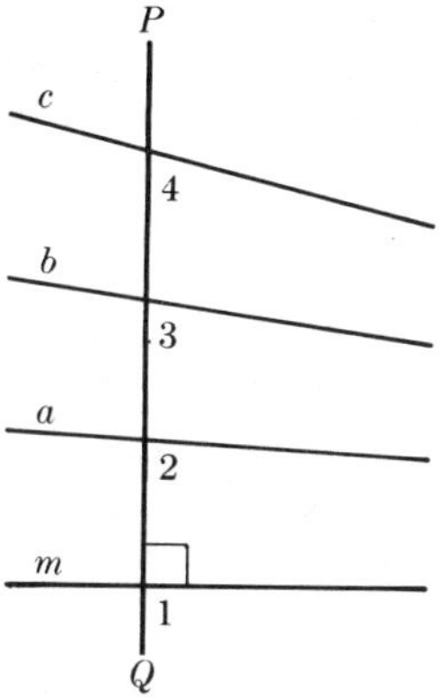

FIGURE 11.9

perpendicular to m. For parallels m and a, exterior angle 1 is greater than interior angle 2. For parallels a and b, exterior angle 2 is greater than interior angle 3. Similarly, $\measuredangle 3$ is greater than $\measuredangle 4$, and so on for as many parallels as we wish to draw. For these angles, which are the angles of parallelism of lines a, b, c, with respect to line m, we have

$$\text{right angle} > \measuredangle 2 > \measuredangle 3 > \measuredangle 4 > \cdots$$

This result is extended in a lemma (a lemma is a subsidiary theorem that is to be used in the proof of a later theorem).

Lemma 1 As the length of the perpendicular PQ (Figure 11.9) increases, the angle of parallelism decreases, approaching zero as P recedes from Q and approaching a right angle as P approaches Q.

A reasonable question now arises. Does each line have its own sequence of relations between length of perpendicular and size of angle of parallelism? Or, for all lines, is a given angle associated with a given length, and conversely? We have the answer in the following lemma, which we shall here accept without proof, although the interested student can easily prove it by using the methods that are effective for Theorem 11.4-1.

Lemma 2 Let two pairs of parallels be cut by transversals. If, in each pair, one line is perpendicular to the transversal, then angles of parallelism of the second lines are equal if and only if lengths of perpendicular are equal.

For a given length of perpendicular, the angle of parallelism is fixed, and for a given angle of parallelism, the length of perpendicular is fixed. The formula connecting the two measures is expressed in terms that would have little meaning for us because they belong to the field of analysis, which we have not studied. The theorem below results from a combination of the two lemmas.

THEOREM 11.4-2 Given line m. To every distance away from m there corresponds an angle of parallelism, and, conversely, to every acute angle there corresponds a distance for which the angle is the angle of parallelism. The angle increases as the distance decreases, approaching a right angle as the distance approaches zero. The angle decreases as the distance increases, approaching zero as the distance increases without limit.

To construct a line through point P parallel to line m, measure the distance from P to m, find the proper angle of parallelism, and draw the line. But not so fast! The new line is approaching m. Hence, at a point S, as shown in Figure 11.10, the distance from S to m is shorter than the distance

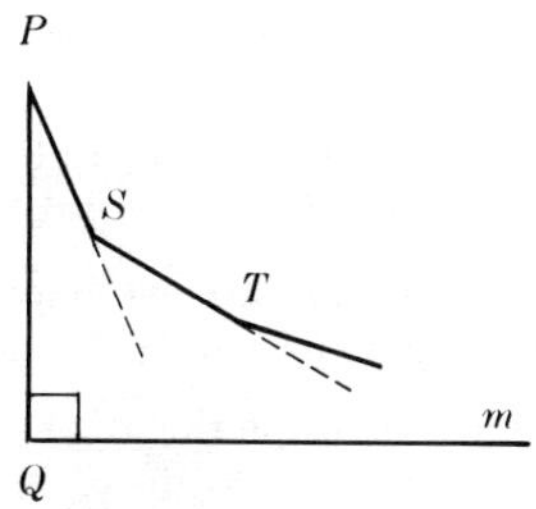

FIGURE 11.10

from P to m. The angle of parallelism at S should be a little larger than it was at P. Make this correction. Similarly, at point T the distance to m is shorter than it was at S. Another correction is required, and so on. Obviously, as the parallels draw closer together, they take more nearly the same direction and actually look more like euclidean parallels. The drawing also supports the suggestion that if we represent some of the lines of this geometry by curves, the diagrams may more faithfully interpret the theory. The following theorem is stated without proof.

THEOREM 11.4-3 Two parallel lines draw ever closer together in the direction of parallelism and diverge ever farther apart in the opposite direction; the perpendicular distance from a point on one of them to the other decreases without limit as the point moves in the direction of parallelism and increases without limit as it moves in the opposite direction.

This theorem points up one of the striking dissimilarities between euclidean and hyperbolic parallels. Every pair of parallel lines in hyperbolic geometry is congruent. Consider the pairs (m,a) and (m,b) of Figure 11.11. If P should move to the right in pair (m,b), the length of perpendicular PQ would decrease and would finally reach the present size of $P'Q$. Then, the angle of parallelism would be the same as angle 1. If P' should move to the left in pair (m,a), perpendicular $P'Q$ would increase in size. When it reached the present length of PQ, the angle of parallelism would be identical with angle 2.

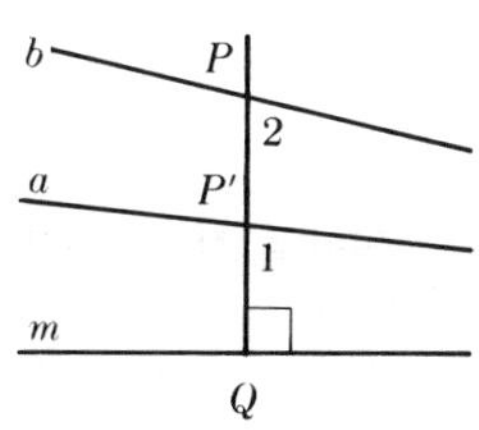

FIGURE 11.11

EXERCISE 11.4

1. At S, the midpoint of line segment RT, erect perpendicular SV. Draw RU and TW parallel to SV in the direction of SV. Show that $\measuredangle SRU = \measuredangle STW$.

2. Let lines r and s through point P be, respectively, parallel to and nonintersecting with line m. Is it possible for lines r and s to coincide? Justify your answer.

3. In a Lambert quadrilateral is it possible for the fourth angle at C to be the angle of parallelism for length AC? Justify your answer.

4. Let line CE through vertex C of $\triangle CAB$ be parallel to base AB in the direction of AB. Show that $\measuredangle BCE < \measuredangle B$.

5. Through midpoint M of the summit of a Saccheri quadrilateral, lines are drawn in both directions parallel to base AB. They cut lateral sides AC and BD in points G and H, respectively. From the symmetry of the figure we would expect that $CG = DH$ and $MG = MH$. Prove that this conjecture is correct. Show further that line MN is the perpendicular bisector of GH.

6. In the Saccheri quadrilateral $ABCD$ of Figure 11.12, lines AE and BF are drawn from A and B parallel to NM in the direction of NM. Prove that the diagram is correct when it makes the following visual implications:

(a) Point E lies between C and M; point F lies between D and M.
(b) $\triangle ACE \cong \triangle BDF$.
(c) $ME = MF$.
(d) Quadrilateral $ABEF$ has
 (i) Equal lateral sides AE and BF.
 (ii) Equal acute base angles at A and B.
 (iii) Equal obtuse summit angles at E and F.
 (iv) MN as median from summit to base.

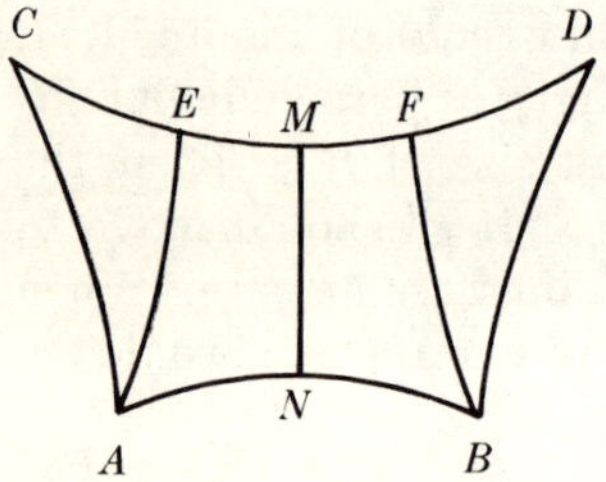

FIGURE 11.12

7. In euclidean geometry what is the relation between quadrilaterals $ABCD$ and $ABEF$ of Exercise 6?

8. When two pairs of parallel lines intersect, the polygon formed is called a parallelogram. In euclidean geometry opposite sides are not only parallel but also equal. What appears to be true in hyperbolic geometry?

9. Given two intersecting lines:

(a) How can a line be drawn that is parallel to both?

(b) How can a line be drawn that is parallel to one and perpendicular to the other?

★10. In this problem use an indirect proof to establish Theorem 11.4-1. Let parallel lines AB and CD be cut by transversal GH forming interior and exterior angles 1 and 2, respectively, in the direction of parallelism as in Figure 11.13.

(a) Show that $\measuredangle 2$ cannot equal $\measuredangle 1$ because of Theorem 11.2-2.

(b) Assume that $\measuredangle 2$ is less than $\measuredangle 1$ and draw line HN so that $\measuredangle JHN = \measuredangle 1$. Now show that line HN has two contradictory properties.

(c) Complete the proof.

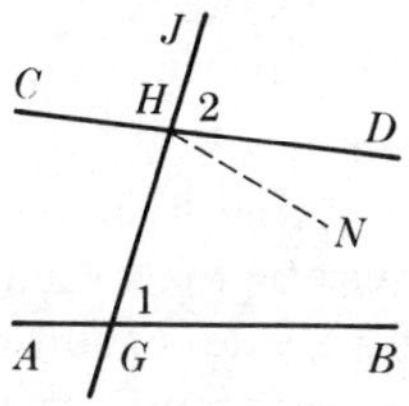

FIGURE 11.13

11.5 THE THREE POLYGONS IN HYPERBOLIC GEOMETRY

In Exercises 11.2 and 11.3, characteristics of nonintersecting lines, most of which were accepted without proof, were used to demonstrate properties of lines and angles in the three important polygons. In this section these same results will be shown to be derivable from properties of parallel lines, many of which we have proved. We will then deduce the uniqueness of the common perpendicular of two nonintersecting lines.

Let us begin with the proposition that the summit angles of the Saccheri quadrilateral are acute. This is neatly derived from the theorem on parallels cut by a transversal, Theorem 11.4-1. Lateral sides AC and BD of the quadrilateral are equal and are perpendicular to base AB (Figure 11.14). Using the correct angle α, draw lines CF and DG parallel to AB

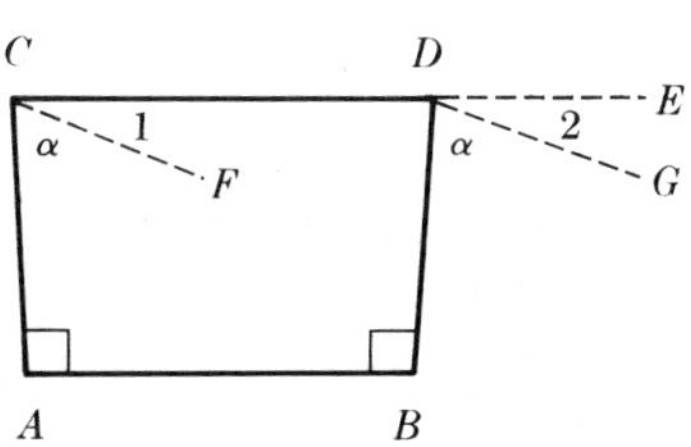

FIGURE 11.14

in the direction of AB. If the summit is produced to E as in Figure 11.14, then CE is a transversal of the two parallels. The exterior angle marked 2 must, therefore, be larger than the corresponding interior angle marked 1. Hence, $\measuredangle 2 + \alpha$ or $\measuredangle BDE$ is greater than $\measuredangle 1 + \alpha$ or $\measuredangle ACD$. Now summit angle BDC is equal to $\measuredangle ACD$ and, because CDE is a straight line, it is supplementary to $\measuredangle BDE$. It is, therefore, the smaller of two supplementary angles and must be acute.

Two immediate consequences of this result, derived by using Theorem 10.4-4, assert that the fourth angle of a Lambert quadrilateral is acute and that the sum of the remaining angles of a right triangle is less than a right angle. Now consider any nonright triangle. The perpendicular from the vertex of the largest angle to the side opposite separates the figure into two adjacent right triangles (see Figure 11.15). Each of the two parts has angle sum less than a straight angle; hence, the total configuration has angle sum less than two straight angles. Since V is not a vertex of the original triangle, the straight angle formed by the sum of angles 1 and 2 must be discarded. It follows that the sum of all the angles of a general triangle is less than a straight angle. Finally, since any quadrilateral can be separated into two triangles by a diagonal joining opposite vertices, there is the further result that the sum of all the angles of a quadrilateral is less than two straight angles.

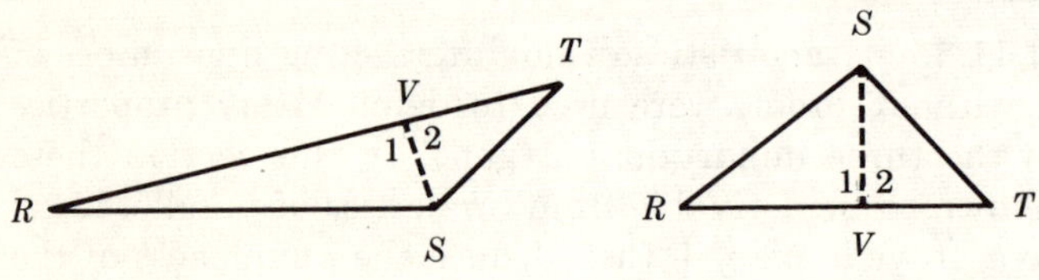

FIGURE 11.15

THEOREM 11.5-1 The summit angles of the Saccheri quadrilateral are acute. The fourth angle of the Lambert quadrilateral is acute. The sum of the remaining angles of the right triangle is less than a right angle. The sum of all the angles of any triangle is less than two right angles, and the sum of all the angles of any quadrilateral is less than four right angles.

We can now make an easy indirect proof of the uniqueness of the common perpendicular of two nonintersecting lines. Let the lines have two common perpendiculars QR and ST. Then a quadrilateral $QRST$ is formed whose angle sum is exactly two straight angles and therefore exceeds the permissible limit. As a corollary to the theorem let us write a relationship suggested by Exercise 11.3.7. Its justification is beyond the scope of this work.

Corollary 11.5-1 Of two Saccheri quadrilaterals, the one with the larger area has the smaller summit angles. Of two Lambert quadrilaterals, the one with the larger area has the smaller fourth angle. Of two triangles, the one with the larger area has the smaller angle sum.

The acuteness of the summit angles of the Saccheri quadrilateral permits us to establish by indirect proof that in the Lambert quadrilateral the side adjacent to the fourth angle is longer than the side opposite. Let us begin with contrary assumptions, using Figure 11.16.

STATEMENT 1 Side AC equals side BD.

STATEMENT 2 Side AC is smaller than side BD.

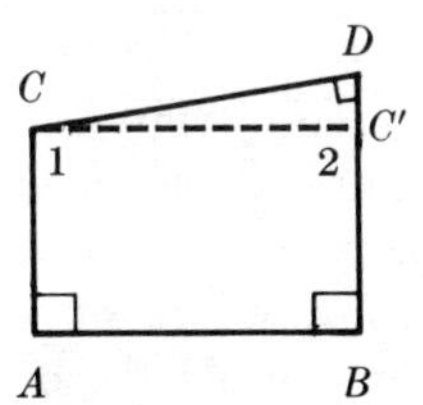

FIGURE 11.16

According to Statement 1, polygon $ABCD$ satisfies the definition of a Saccheri quadrilateral and its summit angles C and D must both be acute. But angle D of the Lambert quadrilateral is a right angle. Hence, Statement 1 is false and side AC is not equal to side BD. To investigate Statement 2, mark off on BD the distance $BC' = AC$. Polygon $ABCC'$ is a Saccheri quadrilateral with summit angles 1 and 2 both acute. Then, triangle CDC' has exterior angle 2 smaller than opposite interior angle D. This contradiction to Proposition 16 shows that Statement 2 is false and AC is not shorter than BD. Side AC is, therefore, longer than BD.

THEOREM 11.5-2 In the Lambert quadrilateral opposite sides are nonintersecting lines and each side adjacent to the fourth angle is greater than the corresponding opposite side.

In the Saccheri quadrilateral the summit and base are nonintersecting lines having the median as common perpendicular; the lateral sides and median are nonintersecting lines having the base as common perpendicular. The summit is greater than the base, and each of the lateral sides is greater than the median.

These characteristics are so much unlike those that hold in euclidean geometry that it is difficult to visualize them. Let us take up the earlier suggestion that curved lines may serve well to represent the straight lines of this geometry. In Figure 11.17 are diagrams that are visually consistent with the hyperbolic qualities of the polygons.

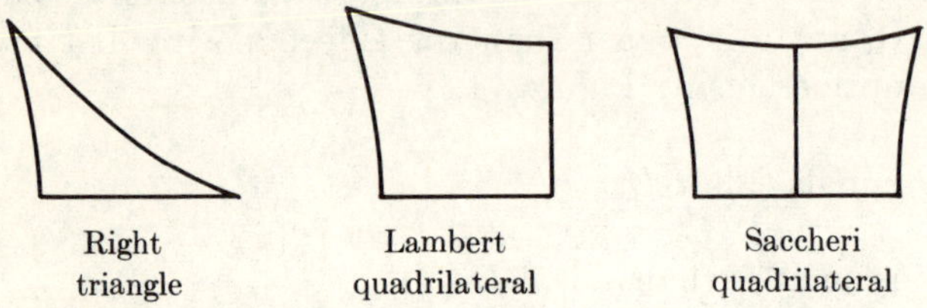

FIGURE 11.17

This section closes with mention of some properties of general triangles. In euclidean geometry an exterior angle is equal to the sum of the two opposite interior angles, and when we know two angles of a triangle we can reckon the third. According to Exercise 11.2.9, the exterior angle of a hyperbolic triangle is greater than the sum of the two opposite interior angles. Moreover, when we know two angles we can give only an upper limit to the size of the third. Two euclidean triangles having equal corresponding angles may be only similar, with corresponding sides proportional. In hyperbolic geometry, two such triangles are congruent and their corresponding sides are equal. In euclidean geometry, the line joining the midpoints of two sides of a triangle is parallel to the base and equal to half of it. In hyperbolic geometry, the median is nonintersecting with the base and less than half its length, and the common perpendicular bisects the base.

EXERCISE 11.5

1. If, in the Lambert quadrilateral with fourth angle at C, we have $AB = BD$, does $AC = DC$?

2. Given a Lambert quadrilateral with fourth angle at C. Draw diagonal CB. In euclidean geometry this diagonal separates the quadrilateral into two congruent right triangles. What happens in hyperbolic geometry?

3. Let $ABCD$ be a Saccheri quadrilateral. Let $MNDB$ be one of the two congruent Lambert quadrilaterals into which median MN separates $ABCD$. The area of $MNDB$ is half the area of $ABCD$. How do the angle sums of the two quadrilaterals compare?

4. In the Saccheri quadrilateral draw the line joining midpoints J and K of the lateral sides AC and BD, respectively. In euclidean geometry, the two polygons

$ABJK$ and $JKCD$ are (a) Saccheri quadrilaterals and (b) congruent. Do the same properties hold in hyperbolic geometry? Explain.

5. Given isosceles right triangle ABC with right angle at A. Draw the median from A to hypotenuse BC. In euclidean geometry this median divides the triangle into two right triangles which are (a) congruent and (b) isosceles. What happens in hyperbolic geometry?

6. At any point P on base AB of $\triangle ABC$ draw transversal PQ to side BC so that $\measuredangle BPQ = \measuredangle A$. In euclidean geometry, $\triangle ABC$ is similar to $\triangle PBQ$, and $\measuredangle C = \measuredangle PQB$. What can be said of these triangles and angles in hyperbolic geometry?

11.6 REPRESENTATIONS OF HYPERBOLIC GEOMETRY

We have seen that diagrams using curved lines may give more faithful representations of properties of polygons in hyperbolic geometry than diagrams using only straight lines. The efficacy of this method leads to the hope that there may be a curved surface whose properties are as well described by hyperbolic geometry as those of a plane surface are described by euclidean geometry. There are, indeed, solids on small portions of whose surfaces we find a congenial drawing medium for hyperbolic figures. One, in particular, permits the properties of hyperbolic lines and polygons to be clearly displayed. It is a trumpet-shaped solid called a *pseudosphere*. The Italian mathematician Beltrami first suggested its use to show that the geometry of Bolyai and Lobachevski is as consistent as that of Euclid.

To create a pseudosphere, first draw a curve that may be described by the following story. A rabbit runs along the top of a straight stone wall that borders a field. A dog in the field chases the rabbit, constantly changing direction so that his body always points toward the rabbit [see Figure 11.18(a)]. The rabbit is too frightened to jump off the wall. Furthermore, both it and the dog are putting forth maximum effort so that neither can increase its speed and neither slackens speed. The path of the dog is called

(a) (b)

FIGURE 11.18

a *curve of pursuit*, or *tractrix*. The path of the rabbit is called the *axis* of the tractrix. If the curve is rotated around the axis, the horn-shaped pseudosphere of Figure 11.18(b) results.

On the surface of this solid many curves can be drawn joining two given points. We are interested in the one along which the distance is shortest; such a curve is called a *geodesic*. We designate the geodesic as the straight line joining the points. With these points and lines, the properties of the geometry are well displayed. Suppose we have access to a pseudosphere and hold a rubber band stretched against it between two points. The physical laws of the tensions involved guarantee that the band lies so that the distance between the two points is the shortest possible. The rubber band, therefore, lies along a geodesic. With help from nimble-fingered friends, we can convince ourselves that the properties of hyperbolic points and lines are the reasonable and natural ones for this surface. In Figure 11.19, on the surface of the pseudosphere, two sets of parallels to line m are constructed at points P and P'. Below line m is a Saccheri quadrilateral with one summit vertex D at point Q. The fact that summit angles are acute is evidenced by the position of summit CD below line m. Because lateral side BDP is perpendicular to line m, it is oblique to summit CD.

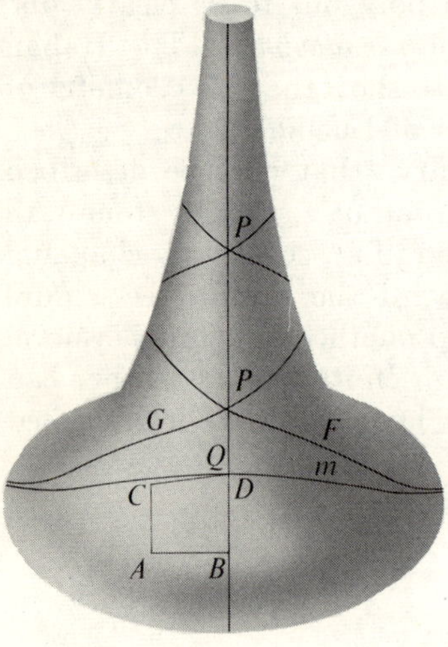

FIGURE 11.19

If no pseudosphere is available, the geometry can still be seen to work in a completely natural way by using the following scheme: take for the hyperbolic surface the inside of a circle not including its circumference. Call this fundamental circle k. The points of our geometry are ordinary euclidean points inside k. The lines of our geometry are the diameters of k

and arcs (inside k) of ordinary euclidean circles that meet k at right angles. For example, in Figure 11.20 circle a has radius AR perpendicular to the radius KR of k at their point of intersection R. The circles are perpendicular; hence, the short arc of a between R and S is a line of the geometry. The remainder of circle a is of no interest to us since it lies outside the hyperbolic plane.

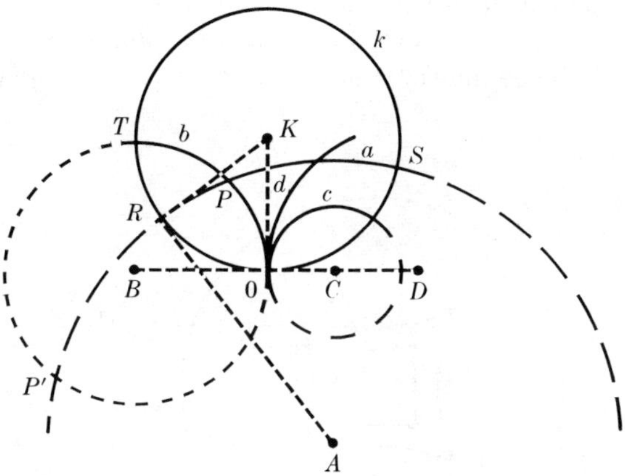

FIGURE 11.20

Draw circle b meeting circle k at right angles. The small arc of b between points O and T is another line of the geometry. Now, for a moment, consider the whole circles a and b. They intersect in two points, one inside circle k and one outside. The inside intersection P is a point of the hyperbolic plane; the outside intersection is not. Hence, hyperbolic lines a and b intersect in exactly one hyperbolic point; similarly, any two intersecting hyperbolic lines have a unique hyperbolic point in common. Dually it can be shown that two hyperbolic points determine a unique hyperbolic line. In these properties the lines satisfy the conditions expected of straight lines.

Two hyperbolic lines are parallel if their circles are tangent to each other on the circumference of k. In Figure 11.20 lines c and d are both parallel to b. The drawing accords with the theorem that parallel lines converge in the direction of parallelism and diverge in the opposite direction. Two hyperbolic lines are nonintersecting when their circles do not meet. For example, lines a and c are nonintersecting. According to the drawing, the lines have closest points from which they diverge. This situation illustrates the theorem that nonintersecting lines diverge on either side of their common perpendicular.

In Figure 11.21(a) are shown the construction of Theorem 11.1-1 and a Saccheri quadrilateral separated by its median into two Lambert quadrilaterals. In (b) are two examples of the hyperbolic polygon that corresponds to the square in euclidean geometry. The drawing illustrates the relationship that as the area of a polygon increases, the angle sum decreases.

This second model of hyperbolic geometry uses euclidean points, lines, and circles, and euclidean relations between them. If any contradiction were found in the theorems of hyperbolic geometry, it would be immediately translatable into a corresponding contradiction in euclidean theorems. Hence, hyperbolic geometry is as consistent as euclidean geometry. It was this inescapable fact that convinced the nineteenth-century philosophers and mathematicians that hyperbolic geometry must be accepted as a proper and independent geometry.

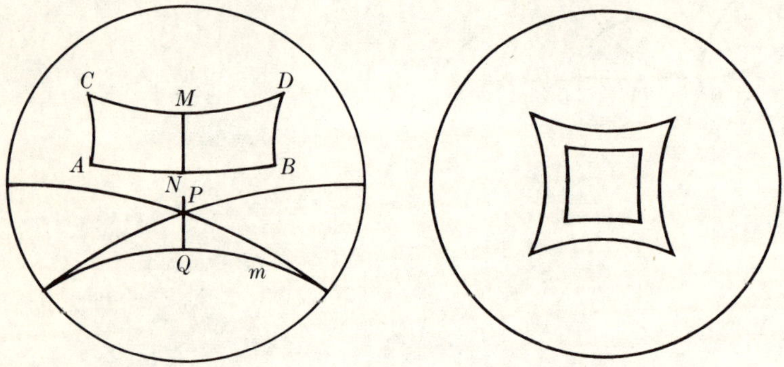

FIGURE 11.21

11.7 COMMENT

This introduction has permitted only a glimpse of the field of hyperbolic geometry. Lobachevski and Bolyai had many further problems to investigate. We mention only two of them: Since parallel lines and nonintersecting lines do not remain equidistant, what sort of configuration would have all of its points a fixed distance from a given straight line? Since any straight line perpendicular to one of a family of parallels is oblique to all the rest, what sort of curve would cross each member of the family at right angles? Regrettably, the study of these and similar questions lies beyond the intended scope of our work.

12
ELLIPTIC GEOMETRY

In this chapter we shall discuss elliptic geometry, which is distinct from the euclidean and hyperbolic geometries, although it has properties in common with each. We shall begin by making the remaining change in Playfair's axiom; because of this change we must also modify the postulate on extent. As we proceed with our study, situations will arise that will call for further decision. In investigating elliptic geometry we shall experience both restriction and freedom.

12.1 MODIFICATION OF AXIOMS

Playfair's axiom is modified by changing the phrase "only one" to "not any." The new axiom says that through a given point not on a given line no straight line can be drawn that does not intersect the given line. In other words, all straight lines intersect.

This postulate directly contradicts Euclid's Propositions 27 and 28, which set forth conditions under which straight lines do not meet. The proofs of these propositions require unstated assumption 4, which asserts that the extension of a line always results in new points. Hence, properties of infinite extent and universal intersection are incompatible for lines. The change in Playfair's axiom must be accompanied by a change in assumption 4. The only modification possible in the latter is its denial. Let us write the new foundation statements:

PARALLEL POSTULATE OF ELLIPTIC GEOMETRY All straight lines intersect.

POSTULATE ON EXTENT OF ELLIPTIC GEOMETRY A straight line is reentrant.

From the discussion in Section 10.2 recall that Riemann was the first to suggest that these two postulates might be consistent. In his honor, geometries that use the new postulate on extent are frequently designated as riemannian. In contrast to the two young inventors of hyperbolic geometry, Riemann became a very productive mathematician, and the name "riemannian geometry" is applied to a wide class of geometries, in not all of which lines are reentrant.

12.2 BEGINNING CONCEPTS OF ELLIPTIC GEOMETRY

Let us first inquire concerning the parts of euclidean geometry that hold in elliptic geometry. There are, of course, the definitions, although those concerning parallel lines no longer have significance. There are also all the postulates, except the fifth, and all unstated assumptions, except the fourth. The common notions still hold. We have also the first 15 propositions and Proposition 26, which can be proved by superposition. Finally, there are the properties of lines, of right triangles, and of Lambert and Saccheri quadrilaterals listed in the appropriate theorems of Section 10.4.

Which parts of this material will be most useful in developing the new geometry? When straight lines intersect, they sometimes intersect at right angles. Moreover, we have theorems to tell us when two straight lines are perpendicular. It would be reasonable, therefore, to use the relation of perpendicularity and see what develops from it. Let us begin with straight line m. At two points A and B on it erect perpendiculars (see Figure 12.1); the perpendiculars will, of course, intersect, say at point M. Triangle AMB has equal base angles and is therefore isosceles. Use the letter d to indicate the length of each of the equal sides AM and BM. Then, $AM = BM = d$.

Now take point C on line m so that $BC = AB$, and erect a perpendicular to m at C. This perpendicular must meet line BM, say at E. Triangles ABM and BCE have equal bases and equal base angles; hence, they are congruent. Corresponding sides satisfy the relation $AM = BE = BM = CE$; therefore point E coincides with M, and the distance from C to M is d. By arguments using limits, such as those employed by Euclid in finding properties of circular arcs, we can show that if D is any point whatever on line m, then

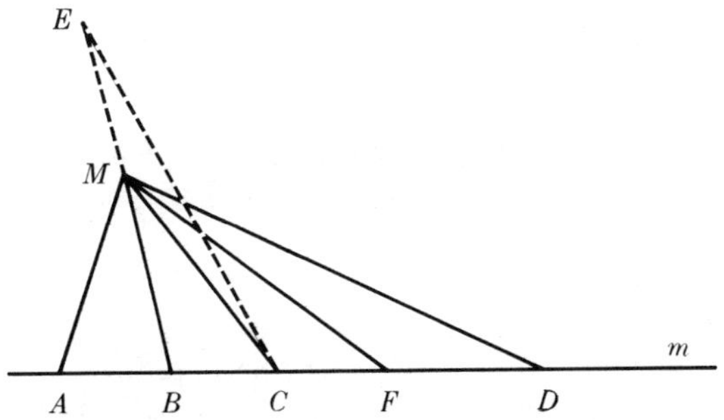

FIGURE 12.1

the perpendicular to m at D passes through point M and has length d. Thus, all perpendiculars to line m meet at point M and have length d. We call M the pole of line m.

Finally, take point F on line m so that $CF = BC$. Join F and M. Triangles MCB and MCF have a common side, equal bases, and equal included right angles. They are congruent; hence, the angle at F is a right angle, and $FM = d$. For any point D on line m we can show, by arguments using limits, that line DM is perpendicular to m and has length d. It follows that all lines joining line m to its pole M are perpendicular to m and have length d. We have our first theorem of elliptic geometry.

THEOREM 12.2-1 With every line there are associated a point M, called the *pole*, and a distance d. All perpendiculars to m pass through M, and conversely, all lines through M are perpendicular to m. The rays from M to m all have length d.

EXERCISE 12.2

1. The distance from a line to a point (or from a point to a line) is measured in elliptic geometry by the same method as in euclidean geometry. Explain how this is done.

2. At any points A, B, and C on line m erect perpendiculars to m.
   (a) Why are triangles formed in this construction?
   (b) What can be said about the angle sum of each of these triangles?

3. Join any points A, B, and C on line m to pole M. What properties have the triangles so formed?

4. Let line n intersect line m at right angles at point N. On n take segment $NT = d$. Is point T the pole of line m? Explain.

5. Let line n intersect line m at point N, not at right angles. On n take segment $NT = d$. Is point T the pole of line m? Explain.

12.3 LINES, POLES, AND DISTANCES IN ELLIPTIC GEOMETRY

We have seen that line m has a pole and a distance d associated with it. Any other line will, by the same reasoning, have a pole and a distance associated with it. We ask whether two different lines, m and n, have the same or different distances to their poles.

At points A and B on line m erect perpendiculars to m. The perpendiculars meet at M and have length d. On line n take points A' and B', so that $A'B' = AB$, and erect perpendiculars to n at these points. The perpendiculars meet at the pole of n, call it N, and have a given length, say e. The resulting diagrams are shown in Figure 12.2. Now, triangles MAB and $NA'B'$ have two angles and the included side of one equal to two angles and the included side of the other. They are congruent; corresponding sides AM and $A'N$ are equal; and distance d equals distance e. In elliptic geometry, the distance from a line to its pole is fixed.

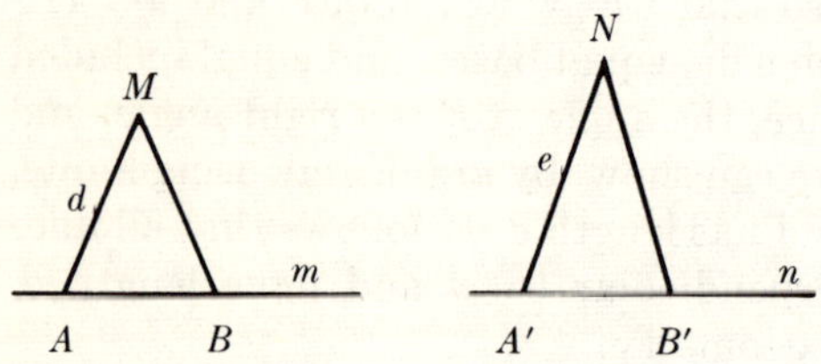

FIGURE 12.2

THEOREM 12.3-1 The distance d is the same for all lines in elliptic geometry.

Next, we ask whether two different lines have the same or different points for poles. Let us assume that lines m and n have the same pole and see what happens. Because the lines are elliptic, they intersect. Call their point of intersection A and their common pole B. Draw AB and indicate angles 1 and 2 as in Figure 12.3. Now angles 1 and 2 are each right angles. They are, therefore, adjacent supplementary angles, and, according to Proposition 14, their exterior sides lie in a straight line. Hence, lines m and n are the same straight line. We have thus shown that if two straight lines have the same pole, they are the same straight line. By the law of contraposition, if two lines are not the same straight line, they do not have the same pole.

THEOREM 12.3-2 Distinct lines have distinct poles, and two lines having the same pole coincide.

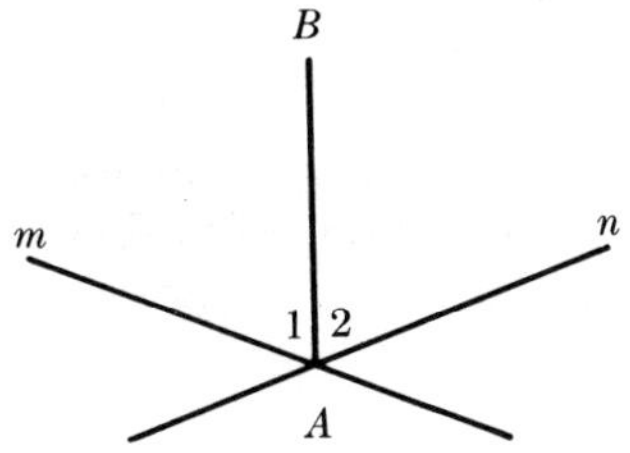

FIGURE 12.3

The association between point and line in this geometry suggests that there may be a dual relationship between *point* and *line* whereby the theorem above would be associated with a corresponding proposition found by interchanging the concepts of pole and line. Such a theorem would assert that distinct poles are associated with distinct lines and that two poles associated with the same line coincide. Let us assign two poles to a given line and find out. In Figure 12.4 each of points P and Q is a pole of line m, and each is joined to point A of m. Lines PA and QA are both perpendicular to line m at A. They must coincide because at any point of a given line only one perpendicular can be erected to the line. Since $AP = d = AQ$, points P and Q also coincide. But not so fast! This conclusion is true only if P and Q lie on the same side of m. If they lie on opposite sides, then line PAQ remains perpendicular to m, but P and Q are distinct. A valid theorem would have to say that if two points lying *on the same side* of a given line are both poles of the line, then they coincide.

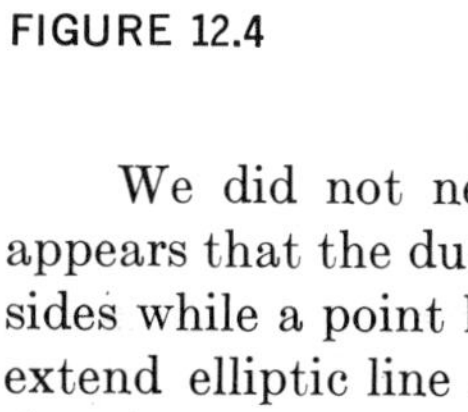

FIGURE 12.4

We did not need any such qualifying clause in Theorem 12.3-2. It appears that the duality between point and line is not perfect. A line has two sides while a point has only one. But must a line have two sides? We could extend elliptic line PA by bouncing back along AP from A. Line m would then have only one side. We are free to make our own decision. Shall a line have only one side? In this case we have a complete duality as far as Theorem 12.3-2 is concerned. Or shall a line have two sides and, in this respect, be like the lines to which we are accustomed? Depending on our decision, we will be discussing one or the other of two distinct geometries. We make the arbitrary choice of two-sided lines. The resulting system is called *double elliptic geometry*. Let us state the new assumption and write the modified dual of Theorem 12.3-2, emphasizing it by use of the contrapositive. Let us state two further results that have also been established.

POSTULATE ON LINES Every line has two sides.

THEOREM 12.3-3 If two points lying on the same side of a line are both poles of the line, they coincide; if they do not coincide, they are not both poles of the given line.

THEOREM 12.3-4 A line has exactly two poles, one on each of its sides.

THEOREM 12.3-5 All points a distance d from line m and on the same side as pole M coincide with M. All points a distance d from pole M lie on line m.

EXERCISE 12.3

1. Given line m. Explain how to construct a pole of m.

2. Given a point M. Explain how to construct a line of which M is a pole.

3. Could two sides of a triangle have the same pole? Explain.

4. Show that in any triangle the altitude to the base (when extended) passes through the pole of the base.

5. If a triangle has only one right angle, must one side have length d? Explain.

6. If a triangle has one side of length d, must it have one right angle? Explain.

7. When a triangle contains two right angles, it is called *birectangular*. Show that if a triangle is birectangular, it has two sides equal to d.

8. On line m take points R and S a distance d apart. Join each point with the pole M. Show that triangle RSM has three right angles. Such a triangle is called *trirectangular*.

9. Let the lines forming a Saccheri quadrilateral be extended.

(a) Show that the lateral sides and median from summit to base all meet at the pole of the base.

(b) Find other line and pole relations in the quadrilateral.

10. In a Lambert quadrilateral each pair of opposite sides, when extended, meets at the pole of one of the included sides. Explain.

11. For a Saccheri quadrilateral show that:

(a) If a lateral side has length d, then the quadrilateral collapses into a birectangular triangle.

(b) If the base has length d, then each of the diagonals also has length d.

12. For a Lambert quadrilateral with no side greater than d show that:

(a) If side AC is of length d, then C is the pole of AB, whence sides BD and DC lie in one straight line.

(b) If side AB is of length d, then A is the pole of BD, whence sides AC and CD lie in one straight line.

13. On line m take points R and S a distance d apart. Complete the equilateral triangle RSV having RS as base. Must vertex V fall at the pole M of line m? Justify your answer.

14. On line m take points R and S a distance less than d apart. Complete the equilateral triangle RSV having RS as base.

(a) Draw the median from vertex V to base RS. Show that when this median is extended backward through V, it passes through M, the pole of line m.

(b) Show that the vertex V does not fall at M.

15. Join point A on line m with pole M. Continue AM to A' so that $MA' = d$. Does point A' lie on line m? Explain.

12.4 FURTHER PROPERTIES OF LINES IN ELLIPTIC GEOMETRY

Let us return to the simple configuration of a line m, its pole M, and two perpendiculars drawn from M intersecting m at points A and B. Let us extend line MA its own length through A to point M'. Join B and M'. The resulting diagram is shown in Figure 12.5(a). Triangles MAB and $M'AB$ have two sides and the included angle of one equal to two sides and the included angle of the other. They are congruent. Corresponding sides MB and $M'B$ are equal as are corresponding angles MBA and $M'BA$. At B, therefore, we have two adjacent right angles whose exterior sides MB and $M'B$ must lie in a straight line. Hence, we have two straight lines MAM' and MBM', each of length $2d$, each perpendicular to line m, and intersecting

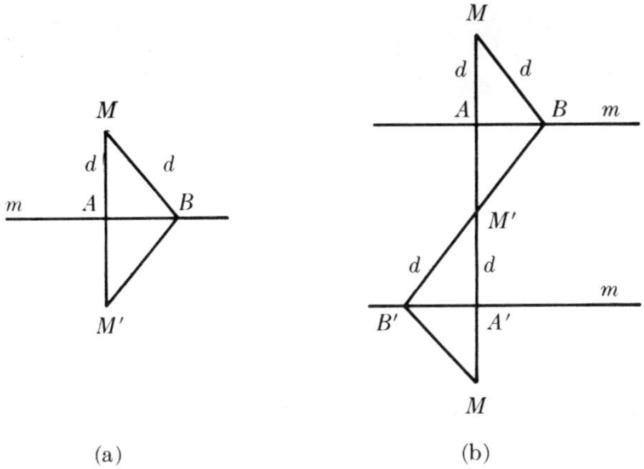

FIGURE 12.5

each other in two points M and M', the poles of m. The angles at M and M' are equal.

Produce line MAM' a distance d to A', and MBM' a distance d to B'. Points A, B, A', and B' all lie on a single straight line having M' for pole. (Why?) This line must be line m already determined by A and B, and $M'A'$ and $M'B'$ must be perpendicular to it. (Why?) Continue $M'A'$ and $M'B'$ on the other side of m [Figure 12.5(b)]. After a distance d they meet at the pole on that side of m. Since a line has only two poles, MA and MB have returned to their starting point.

Each of $MAM'A'M$ and $MBM'B'M$ is a reentrant straight line of length $4d$. The lines meet in two distinct points M and M' at which the angles of intersection are equal. They have two common perpendiculars AB and $A'B'$ that are equal in length. These same qualities characterize every line and pair of lines in the geometry.

Clearly, these lines differ so much from euclidean lines that drawings made with a ruler do not portray their properties. Open or closed curves might give a better representation. In Figure 12.6 we experiment with this suggestion. The drawings support the results stated in the theorems below and in the corollary that you will prove in Exercise 12.5.

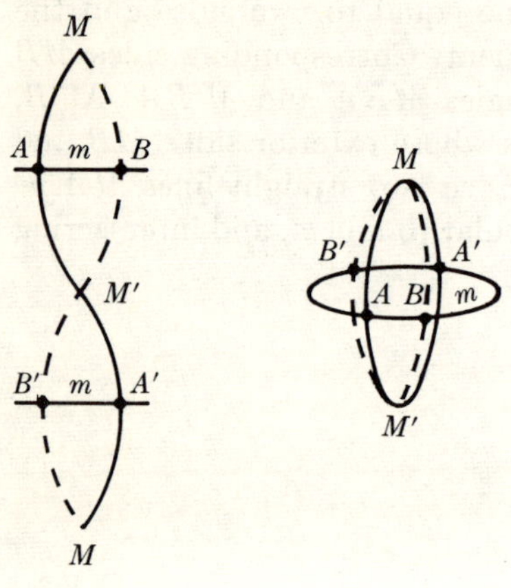

FIGURE 12.6

THEOREM 12.4-1 Every straight line is of length $4d$ and is reentrant.

THEOREM 12.4-2 Every two straight lines meet in two points, enclose a portion of the plane, and have two common perpendiculars. The common perpendiculars are equal in length, and the angles of intersection at the two common points are equal.

Corollary 12.4-2 For two pairs of straight lines, the pair having the greater angles of intersection has the longer common perpendiculars, and conversely.

In Euclid's first postulate lies the implication that two distinct points determine a straight line. Consider two elliptic points M and M' that lie a distance $2d$ apart. We have already drawn two distinct straight lines joining them, and there is no reason why we should not draw many more. All these lines are perpendicular to line m because they pass through the poles of m. On the other hand, if two points M and N are a distance less than $2d$ apart, they cannot be poles of the same straight line. Hence, N cannot fall at M', and only one line of the multitude joining M and M' can pass through N.

THEOREM 12.4-3 Two points a distance less than $2d$ apart determine one and only one straight line. Through two points a distance $2d$ apart there pass an infinite number of straight lines all perpendicular to a fixed straight line of which the two given points are poles.

In euclidean geometry, two lines having a common perpendicular are parallel, and for every transversal corresponding angles on the same side are equal. In hyperbolic geometry, two lines having a common perpendicular are nonintersecting, and the angles they form with a transversal may or may not be equal as specified in Theorem 11.2-3. We have just seen that in elliptic geometry every pair of lines has a common perpendicular. The laws that govern the angles they form with a transversal are analogous to those for hyperbolic lines.

THEOREM 12.4-4 For any two elliptic lines, corresponding angles on the same side of a transversal are equal only for the sheaf of transversals whose midpoints coincide with the midpoint of one of the common perpendiculars of the lines. For all other transversals, the exterior angle on the side toward the midpoint of a common perpendicular is smaller than the corresponding interior angle.

EXERCISE 12.4

1. In both the euclidean and hyperbolic geometries there is the theorem that one and only one perpendicular can be drawn to a line from a point external to the line. Under what circumstances does an analogous theorem hold in elliptic geometry?

2. Show that Euclid's Proposition 16 holds in elliptic geometry for a triangle in which all sides and all medians are of length less than d.

3. Show that when base AB has length $2d$, the Saccheri quadrilateral collapses into two straight lines meeting at right angles at A and B.

4. Refer to Figure 11.4 in Exercise 11.2. Let lengths of AG, AB, and BH all be less than d.

(a) Why are the angles marked 1, 2, and 3 all acute?

(b) Show that $\measuredangle 4 < \measuredangle 5 < \measuredangle 6 < \measuredangle 7$.

(c) Find the relations between angles 8, 9, 10, and 11.

5. The observations in Exercise 11.2.2 remain valid. What do the results obtained in Exercise 4 above suggest in elliptic geometry about:

(a) The size of the fourth angle of the Lambert quadrilateral?

(b) The sum of the angles of the Lambert quadrilaterals $ACBD$, $AEBF$, and $AGBH$?

6. Modify and answer Exercise 11.2.9 for elliptic geometry when all line segments have length less than d.

7. and **8.** Answer Exercises 11.2.7 and 11.2.8, respectively, for elliptic geometry when side CA is less than d.

9. For the triangles of Exercises 7 and 8 above, show that the following inequality holds:

$$\text{sum of angles of } \triangle EDB < \text{sum of angles of } \triangle CAB$$

10. If two lines have a common perpendicular, they remain equidistant in euclidean geometry; in hyperbolic geometry, they diverge from the common perpendicular. What appears to be the case in elliptic geometry?

11. Using the answer to Exercise 10, make an informed guess about the relative lengths of opposite sides for:

(a) The Lambert quadrilateral.

(b) The Saccheri quadrilateral.

12.5 RIGHT TRIANGLES IN ELLIPTIC GEOMETRY

In this section a formal investigation of the relation between length of side and size of angle in an elliptic polygon is begun. In particular, let us consider a right triangle ABC with right angle at A. When AB is restricted to be less than $2d$, the size of each of the remaining angles is determined by the length of the side opposite it, and conversely. In Figure 12.7 consider side AC and the angle at B. First, take $AC = d$. Then, C is the pole of line AB; angle CBA is a right angle; and side BC equals d. Next, take AC' less than d. Now, line $C'B$ lies within the angle CBA; hence, angle $C'BA$ is acute. Finally, if AC'' is greater than d, line $C''B$ lies above CB, and angle $C''BA$ is obtuse. It is clear that the size of the angle at B depends on the length of side AC. In the exercise below you are asked to show that the converse of this relation also holds. We here accept without proof the final statement in the following theorem.

THEOREM 12.5-1 If one angle of a triangle is a right angle and the side adjacent is less than $2d$, then each of the remaining angles is less than

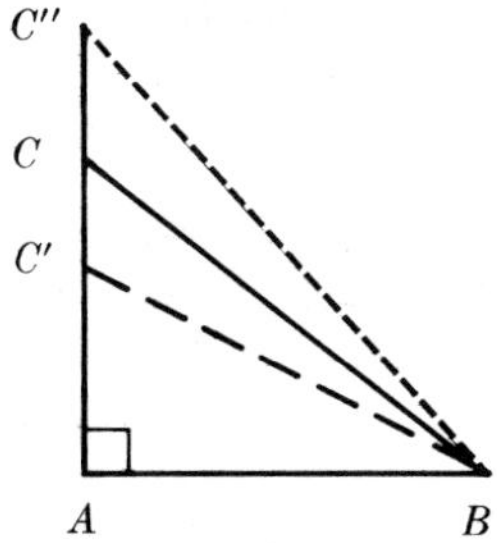

FIGURE 12.7

(equal to, or greater than) a right angle if and only if the side opposite is less than (equal to, or greater than) d. If the side adjacent to the right angle is greater than $2d$, the results are reversed.

EXERCISE 12.5

1. In triangle ABC of Figure 12.7 let $AB = AC = d$. What can be said about the angles of triangle ABC? Those of $\triangle ABC'$? Those of $\triangle ABC''$?

2. Prove that in right triangle ABC with right angle at A (and side AB less than $2d$) if the angle at B is a right angle, the side opposite equals d; if the angle at B is acute, the side opposite is less than d; if the angle at B is obtuse, the side opposite is greater than d.

3. Prove Corollary 12.4-2.

4. Let Lambert quadrilateral $ABCD$ have each side less than d. Extend AC and BD until they meet at M, the pole of AB.

(a) Explain why triangle MCD is a right triangle with adjacent side MD less than d.

(b) What can be said of the size of $\measuredangle MCD$?

(c) What can be said of the size of the fourth angle of the quadrilateral?

5. In euclidean geometry every Lambert quadrilateral is a Saccheri quadrilateral, and conversely. What appears to be true in elliptic geometry?

12.6 THE THREE POLYGONS IN ELLIPTIC GEOMETRY

In previous chapters we considered the Lambert and Saccheri quadrilaterals to be *convex*, meaning that a straight line drawn through any interior point has exactly two intersections with the sides. To maintain this condition of convexity for elliptic quadrilaterals, restrictions must be placed on the lengths of the sides. Some of these have been anticipated in Exercises 12.3.11 and 12.3.12. Let us consider one additional case. When sides AB and CD of the Lambert quadrilateral have length between d and $2d$, then the pole P of side BD lies between A and B, and side DC must pass through P on its way from D to C (see Figure 12.8). Investigation shows that for convexity each side of the Lambert quadrilateral must be less than d. This rule imposes the restrictions expressed below on the other special polygons.

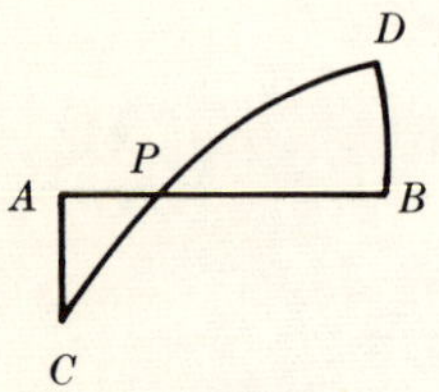

FIGURE 12.8

RESTRICTION ON SIDES OF POLYGONS For the Lambert quadrilateral, each side is less than d. For the Saccheri quadrilateral, each lateral side is less than d, and summit and base are each less than $2d$. For the right triangle, each side is less than $2d$, and of the sides adjoining a right angle, at least one is less than d.

Exercise 12.5.4 above showed that a Lambert quadrilateral with sides less than d has its fourth angle obtuse. Relations between the angles of the three polygons now give the theorem below.

THEOREM 12.6-1 The fourth angle of the Lambert quadrilateral is obtuse. The summit angles of the Saccheri quadrilateral are obtuse. The sum of the remaining angles of the right triangle is greater than a right angle, and the sum of all the angles of any triangle is greater than two right angles. The sum of the angles of any quadrilateral is greater than four right angles.

A corollary to the theorem presents a set of relations consistent with those found in Exercise 12.4. Their justification requires more information about the geometry than shall be discussed in this text.

Corollary 12.6-1 Of two Saccheri quadrilaterals, the one with larger area has the larger summit angles. Of two Lambert quadrilaterals, the one with larger area has the larger fourth angle. Of two triangles, the one with larger area has the larger angle sum.

Elliptic geometry resembles hyperbolic geometry in that the Saccheri and Lambert quadrilaterals are distinct polygons. One has equal summit angles, while the other has one right and one obtuse angle. The Lambert quadrilateral must, therefore, have unequal lateral sides. It would seem reasonable for the larger side to be opposite the obtuse fourth angle. The truth of this conjecture can be established by an indirect proof similar to that made for the corresponding hyperbolic property in Section 11.5. Now the lateral sides of the Saccheri quadrilateral must be shorter than the median, and the summit shorter than the base. Here again, diagrams with curved lines may give a good visual representation of the polygons, as in Figure 12.9.

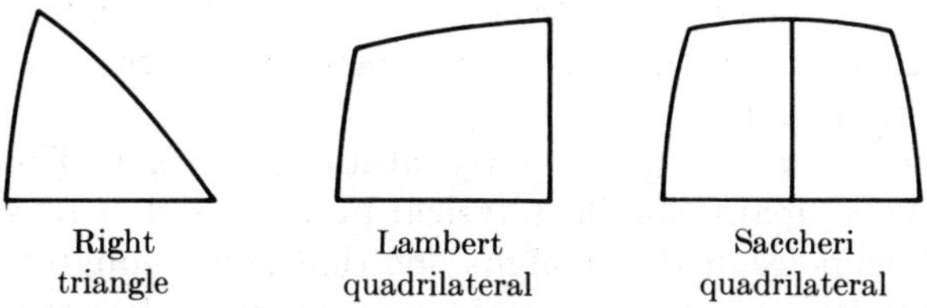

FIGURE 12.9

THEOREM 12.6-2 In the Lambert quadrilateral each side adjacent to the fourth angle is shorter than the corresponding opposite side. In the Saccheri quadrilateral the summit is shorter than the base and each of the lateral sides is shorter than the median.

We have found an important property of elliptic right triangles that has no analogue in either of the other two geometries. However, many characteristics of general elliptic triangles *resemble* those of hyperbolic triangles. For example, when we know the sum of two angles, we can give only a lower limit for the size of the third angle. The exterior angle of a triangle is less than the sum of the two opposite interior angles. When two triangles have all corresponding angles equal, they are congruent and their corresponding sides are equal. Finally, in any triangle the median to two sides is greater than half the base, and one of its common perpendiculars with the base bisects the base.

EXERCISE 12.6

Exercises **1–6.** Adapt Exercises 11.5.1–11.5.6 for elliptic geometry.

7. Given birectangular triangle ABC with right angles at A and B. Join the midpoints D and E of sides AC and BC, respectively. Show that CDE is an isosceles triangle. Are its base angles acute, right, or obtuse? Justify your answer.

12.7 A REPRESENTATION OF ELLIPTIC GEOMETRY

When we seek a surface on which to portray this geometry, we find one immediately at hand, namely, the surface of a sphere. Its points are ordinary euclidean points. A plane through the center of a sphere intersects the surface in a circle whose center and radius are the center and radius of the sphere. The circles are called great circles and are the geodesics. They are the straight lines of the elliptic geometry. Let us consider the earth to be a giant sphere. The equator is a great circle on its surface, and so are the meridians of longitude. The length d is the distance from pole to equator or half the distance measured on the equator from the Greenwich Meridian to the International Date Line. Using this model it is easy to illustrate the properties developed in Section 12.4.

The earth, however, may not be a good representational surface. The North and South Poles have so many special physical properties that it is hard to consider them only as poles of the straight line that is the equator. It is difficult to remember that each of the meridians of longitude has two poles. Furthermore, for east–west directions we are accustomed to the coordinate systems of maps, which use parallels of latitude rather than great-circle arcs. It might be better to use a ball for a model of the surface and rubber bands stretched over it for the lines. With these tools we can diagram all the properties of the geometry.

This representation employs euclidean points and circles on a euclidean sphere. If any contradiction should arise in the theorems of elliptic geometry, it would be translatable immediately into a corresponding contradiction in euclidean theorems. Hence, elliptic geometry is as consistent as euclidean. We have left many of its properties unmentioned and many obvious questions unanswered; for example, what curve would be formed by all points that are a fixed distance, less than d, above a given line? We could also investigate the single elliptic geometry that we abandoned when we decided to deal with two-sided lines.

12.8 THE THREE GEOMETRIES

Euclidean geometry is known and used far more widely than either of the others we have discussed because of its ancient origin and the simple and easily applied model it presents for confined space relations. It satisfies most engineering needs because the distinction between the three geometries lies within the limits of measuring error when distances are small. For example, the geodesics of most engineering projects are short when compared with the distance from pole to equator. Hence, triangles drawn to extend over even sizable portions of the earth's surface have such small sides in comparison with the distance d that, as we saw in Exercise 12.4, Euclid's Proposition 16 holds for them. Moreover, their angle sum is experimentally indistinguishable from two right angles. Even if the earth were fashioned from giant pseudospheres, the situation would not change. A triangle on a limited portion of the earth would have an angle sum so close to two right angles that measurement could not detect the discrepancy. A mathematician would say that the three geometries are locally almost indistinguishable. Euclidean geometry is by far the simplest with respect to the formulas required when it is applied in engineering or physics or mathematics. It is used whenever a situation permits. For example, in theoretical physics where ordinary distances and velocities are encountered, euclidean geometry is applicable. Great distances and velocities approximating the speed of light, however, require the concepts of Einstein's relativity theory, which use a riemannian geometry of a type different from any of the three geometries that we have studied.

EXERCISE 12.8

1. Which of the theorems discussed in Chapters 10–12 are true in all three geometries? These would form a part of Bolyai's "general geometry."

2. Which of the properties discussed in Chapters 10–12 hold in hyperbolic and elliptic geometries but not in euclidean geometry?

12.9 RETROSPECT

We have observed how a change of postulate can modify results in a mathematical system and how dissimilar developments are possible for a given type of material. The resulting systems may vary in difficulty, in importance, and in the interest that they hold; yet, as long as each develops logically from its foundations, it is acceptable as mathematics.

The three geometries have been the subject of long study and vigorous debate. They have stirred men's imaginations and have been of inestimable value in demonstrating the freedom of the mathematical art. Now that you have been introduced to them and have gained confidence in dealing with geometric material, you may wish to investigate other geometric concepts. You may even be emboldened to learn how the geometry of Euclid accords with the algebra of real numbers in the harmony and competence of analytic geometry.

As you gain further experience with the vividly imaginative subject of mathematics, you will discover still other concepts and systems in whose order and elegance resides a lasting beauty.

APPENDIX

GEOMETRY

A. FROM EUCLID'S *THE ELEMENTS*

DEFINITIONS

23. Parallel straight lines are straight lines which, being in the same plane and being produced indefinitely in both directions, do not meet one another in either direction.

POSTULATES

1. Two points determine one and only one straight line.

2. A finite straight line can be produced continuously in a straight line.

3. A circle can be drawn with any center and radius.

4. All right angles are equal.

5. If two straight lines are cut by a transversal so that interior angles on the same side are less than two right angles, then the two straight lines, if produced indefinitely, meet on the side on which the angles are less than two right angles.

COMMON NOTIONS

1. Things that are equal to the same thing are also equal to one another.

2. If equals are added to equals, the wholes are equal.

3. If equals are subtracted from equals, the remainders are equal.

4. Things that coincide with one another are equal to one another.

5. The whole is greater than the part.

PROPOSITIONS

On constructions

1. Of an equilateral triangle.

On congruence of triangles

4. *SAS*

8. *SSS*

26. *ASA*

26′.[1] *SAA*

On isosceles triangles

On the intersection of lines

13. and **15.** If two lines intersect, adjacent angles are supplementary and vertical angles are equal.

14. If two adjacent angles are supplementary, their exterior sides lie in a straight line.

On inequalities

16′. In any triangle, if one of the sides is produced, the exterior angle is greater than either of the interior and opposite angles.

17′. The sum of two angles of a triangle is less than two right angles.

18′. and **19′.** In any triangle, the greater side subtends the greater angle, and conversely.

On parallels

27′. and **28′.** If two straight lines are cut by a transversal so that they form equal alternate interior angles, equal corresponding angles, or supplementary interior angles on the same side of the transversal, the lines are parallel.

29*.[2] If a transversal cuts two parallels, it makes equal alternate interior angles, equal corresponding angles, and supplementary interior angles on the same side of the transversal.

[1] The meaning of a prime following the number of a proposition is explained in Section 10.3.

[2] The meaning of an asterisk following the number of a proposition is explained in Section 10.5.

30*′. Straight lines parallel to the same straight line are parallel to one another.

31*′. Through a given point not on a given straight line one and only one straight line can be drawn parallel to the given line.

32*′. The exterior angle of a triangle is equal to the sum of the two opposite interior angles, and the sum of the three angles of a triangle is equal to two right angles.

B. UNSTATED ASSUMPTIONS (STATEMENTS ADEQUATE FOR WORK IN THIS TEXT)

POSTULATE OF ARCHIMEDES For two unequal line segments, there is always some finite multiple of the smaller that exceeds the greater.

AXIOM OF PASCH If a line cuts one side of a triangle, it cuts another side also; or if a line enters a triangle at a vertex, it cuts the opposite side.

POSTULATE OF DEDEKIND If all straight lines through a point are divided into two nonempty sets such that all lines in the first precede all lines in the second, then there is one particular line through the point that brings about this division of the lines into two sets.

POSTULATE ON EXTENT A straight line is not reentrant.

REFERENCES

PART I. CHAPTERS 1–7

J. H. Banks, *Elements of Mathematics*. New York: Allyn and Bacon, 1956.
Chapter 5. What is Algebra?

J. E. Hafstrom, *Basic Concepts in Modern Mathematics*. Reading, Mass.: Addison-Wesley, 1961.
Chapter 5. Groups.

B. W. Jones, *Elementary Concepts of Mathematics*. New York: Macmillan, 1970 (3rd edition).
Chapter 3. The Positive Integers and Zero.

B. W. Jones, *Modular Arithmetic*. New York: Blaisdell, 1964 (paperback edition).

M. L. Keedy, *A Modern Introduction to Basic Mathematics*. Reading, Mass.: Addison-Wesley, 1963.
Chapter 3. Some Small Mathematical Systems.

C. D. Miller and V. E. Heeren, *Mathematical Ideas, An Introduction*. Glenview, Ill.: Scott, Foresman, 1969.
Chapter 3. Mathematical Systems; Chapter 4. Group Theory.

L. C. Peck, *Secret Codes, Remainder Arithmetic, and Matrices*. Washington, D.C.: National Council of Teachers of Mathematics, 1961 (paperback edition).

W. L. Schaaf, *Basic Concepts of Elementary Mathematics*. New York: Wiley, 1960.
Chapter 4. The Number Concept; Chapter 5. Extending the Number System.

M. F. Willerding, *Elementary Mathematics, Its Structure and Concepts*. New York: Wiley, 1966.

Chapter 14. Mathematical Systems.

PART II. CHAPTERS 8, 9

S. M. Dowdy, *Mathematics, Art and Science*. New York: Wiley, 1971.

Chapter 5. Mathematical Logic.

R. M. Exner and M. F. Rosskopf, *Logic in Elementary Mathematics*. New York: McGraw-Hill, 1959.

H. Meschkowski, *Ways of Thought of Great Mathematicians*. San Francisco: Holden-Day, 1964.

Essay on George Boole.

W. L. Schaaf, *Basic Concepts of Elementary Mathematics*. New York: Wiley, 1960.

Chapter 2. Elementary Logic.

PART III. CHAPTERS 10–12

J. W. Armstrong, *Elements of Mathematics*. New York: Macmillan, 1970.

Chapter 5. Two Geometries.

E. T. Bell, *Men of Mathematics*. New York: Simon and Schuster, 1937.

Chapter 16. The Copernicus of Geometry.

C. J. Keyser, *Mathematical Philosophy*. New York: Dutton, 1922.

Lecture XVII. Non-Euclidean Geometries.

M. Kline, *Mathematics in Western Culture*. New York: Oxford, 1953.

Chapter XXVI. New Geometries, New Worlds.

M. Kline, *Mathematics and the Physical World*. New York: Crowell, 1959.

Chapter 26. Non-Euclidean Geometries; Chapter 27. Mathematics and Nature.

L. R. Lieber, *Non-Euclidean Geometry*. New York: Academy, 1931.

H. Meschkowski, *Evolution of Mathematical Thought*. San Francisco: Holden-Day, 1965.

Chapter III. The Road to non-Euclidean Geometry.

G. Saccheri, *Euclides Vindicatus* (translated by G. B. Halsted). Chicago: Open Court, 1920.

H. E. Wolfe, *Introduction to non-Euclidean Geometry*. New York: Dryden, 1945.

ANSWERS AND HINTS

EXERCISE 1.3

1. The entries in the 3 row are the sums $3 + 1$, $3 + 2$, etc. The entries in the 3 column are the sums $1 + 3$, $2 + 3$, etc.

3. (a) 3 (b) 1 (c) 3 (d) 2 (e) 4 (f) 4 (g) 5 (h) 4 (i) 3

5. Table A1.3-5

+	1	2	3	4	5
1	2	3	4	5	6
2	3	4	5	6	7
3	4	5	6	7	8
4	5	6	7	8	9
5	6	7	8	9	10

EXERCISE 1.4

1. No. **3.** Yes. **5.** 1 **7.** Yes.

EXERCISE 1.5

1. See discussion preceding Theorem 1.4-1.

3. By commutativity, $2 + 3 = 3 + 2$. Hence, $4 + (2 + 3) = 4 + (3 + 2)$.

5. (a) By closure the sum $a + b$ is a single element. By commutativity the elements $(a + b)$ and c can be interchanged; hence, $(a + b) + c = c + (a + b)$. By commutativity $a + b = b + a$; hence, $c + (a + b) = c + (b + a)$.

(b) By closure the sum $b + c$ is a single element. By commutativity the elements $(b + c)$ and a can be interchanged. A second use of commutativity gives the desired result.

7. (a) Yes. (b) No.

9. No. Closure requires that each pair of elements shall have a unique sum; it is not concerned with the number of times a given element appears as a sum.

EXERCISE 1.6

3. (a) By associativity $1 + (2 + 3) = (1 + 2) + 3$. By commutativity $1 + 2 = 2 + 1$; hence, $(1 + 2) + 3 = (2 + 1) + 3$.

(b) By associativity $(3 + 4) + 2 = 3 + (4 + 2)$. By commutativity $4 + 2 = 2 + 4$; hence, $3 + (4 + 2) = 3 + (2 + 4)$. By associativity $3 + (2 + 4) = (3 + 2) + 4$.

(c) Use commutativity to get $2 + 3 = 3 + 2$. Then use associativity.

(d) See Exercise 1.5.5(b).

5. Yes.

7. No. Check triple 9, 8, 7.

9. (a) $d + (d + b) = (d + d) + b = a + b = b$; hence, $d + b = e$.

(b) $(c + c) + b = c + (c + b) = c + a = c$; hence, $c + c = b$.

(c) $(c + f) + f = c + (f + f) = c + a = c$; hence, $c + f = d$.

11. No.

EXERCISE 1.7

1. (a) 3 (b) 3 (c) 3 (d) 4

3. Prove that if $b + a = c + a$, then $b = c$.

(a) Let the sum $b + a$ be represented by d. Element d appears in the body of the addition table where the b row crosses the a column. Because d also equals $c + a$, it appears where the c row crosses the a column. But entry d occurs only once in the a column. Hence, the b row and the c row are the same row, and $b = c$.

(b) By commutativity of addition, $b + a = a + b$ and $c + a = a + c$. The "if" clause can be rewritten "if $a + b = a + c$," and by the law of left cancellation $b = c$.

5. (a) Make an argument similar to that preceding Theorem 1.7-1.

(b) Make an argument similar to that for Exercise 3(a).

EXERCISE 1.8

1. $1 + 4 = 3 + 2 = 5$. The sum is 1.

3. (a) 0 (b) None. (c) None.

5. No.

EXERCISE 1.9

1. Enter the addition table in the 2 column. (Why?) Find the sum 5 and look across to the caption of its row.

3. Yes. Each element is its own inverse.

5. Number $-n$ is the negative of number n.

7. Element 5 is its own inverse.

9. Yes. By definition of inverse, the equality $a + b = b + a = 5$ describes both of the relations $a^* = b$ and $b^* = a$.

EXERCISE 1.10

1. When $a = b$, then a and b are the same element. Because each element has a unique additive inverse, a^* is the same element as b^*.

3. $(a + b + c)^* = [(a + b) + c]^* = (a + b)^* + c^*$.

5. (a) Yes. $(a + b)^* = a + b = a^* + b^*$.
(b) Every element is its own additive inverse.

EXERCISE 1.11

1. (a) 3 (b) 3 (c) 4 (d) 3 (e) 5 (f) 1 (g) 4 (h) 5

3. (a) (i) 1 (ii) 4 (iii) 7
(b) There are no additive inverses in system S.

5. Addition is not commutative in system T.

7. (a) $x = a + b^*$. (b) Yes.

9. (a) It gives the same information for system T as Theorem 1.11-1 does for the five-element system.

(b) In system T each of the equations $a = a + x$ and $a = x + a$ has the unique solution $x = 1$.

EXERCISE 1.13

1. $-2, 5, -10, 6, 0, -1, -m, m$

3. For integers m and n:

(i) If $-m$ is the negative of m, then m is the negative of $-m$. In particular, $-(-m) = m$.

(ii) $-(m+n) = (-m) + (-n)$
(iii) $0 = -0$, and 0 is the only integer that is its own negative
(iv) For every integer m, there exists an integer $-m$ such that $m = -(-m)$.

EXERCISE 2.1

1. $1 \cdot 3 = 3$; $2 \cdot 3 = 1 \cdot 3 + 3 = 6$; etc.

EXERCISE 2.2

1. 3, 1, 4, 2, 5

3. 1, 4, 5

5. (a) 2, 3, 4, 5 (b) All. (c) 1, 2, 3, 5 (d) 5 (e) 1, 2, 3, 4 (f) All.

7. (a) is false.

9. (a) 2 (b) 3 (c) 3

EXERCISE 2.3

1. If and only if each pair of numbers has a unique product that is a number in the system.

3. All closed except (d) and (e).

5. Use the method of the discussion preceding Theorem 1.5-1 or of Exercise 1.5.10.

7. No.

EXERCISE 2.4

1. For any three elements a, b, and c of the five-element system, $abc = (ab)c$.

5. Use method analogous to that which established $(a+b)+c = (c+b)+a$ in Exercise 1.6.4.

7. (a) 4 (b) 5

EXERCISE 2.5

1. No.

3. The equality $ab = ba = 1$ describes both of the relations $a = b'$ and $a' = b$.

5. For elements having inverses, an element is its own inverse or else a pair of elements are inverses of each other. Furthermore, an odd number of elements have additive inverses while an even number have multiplicative inverses.

7. (a) Yes. 1 (b) 1, -1

EXERCISE 2.6

1. (a) 4 (b) 1 (c) 1 (d) 1 (e) 2 (f) 2 (g) 4 (h) 1 (i) 1 (j) 1 (k) 2 (l) 3

3. Consider two cases: $a = 5$ and $a \neq 5$.

5. Use $(abc)' = [(ab)c]' = (ab)'c'$.

EXERCISE 2.7

1. (a) 3 (b) 1 (c) 3 (d) 2 (e) 3 (f) 2 (g) All. (h) 5 (i) 5

EXERCISE 2.9

1. Operate with $a^\dagger$ on the right on both members of the equality.

3. Assume that $e^\dagger$ and $e^\sim$ are both inverses of e under $\circ$. By definition $e \circ e^\dagger = e^\dagger \circ e = i$ and $e \circ e^\sim = e^\sim \circ e = i$. Hence, $e^\dagger \circ e = e^\sim \circ e$, and a law of cancellation applies.

5. Use $(a \circ b) \circ (b^\dagger \circ a^\dagger) = a \circ (b \circ b^\dagger) \circ a^\dagger = a \circ i \circ a^\dagger$.

EXERCISE 2.10

1. No. **3.** Yes. **5.** Yes. **7.** Yes. **9.** No. **11.** No.

EXERCISE 2.11

1. Yes.

3. In a group every element is either its own inverse or is one of a pair of elements that are inverses of each other. (Why?) The identity which is its own inverse and all pairs of elements which are mutual inverses account for an odd number of elements. An odd number of elements remains.

EXERCISE 2.12

1. (a) 5 (b) 1, 2, 3, 4 (c) 5 (d) None. (e) All. (f) All. (g) All. (h) None. (i) 5

3. The product is 5.

5. The product is different from 5.

7. (a) Use Exercises 4 and 5.
(b) See discussion preceding Theorem 1.4-1.

9. (*b*) Use $a + 4a = 5a = 5$.

EXERCISE 2.13

3. (a) 1 (b) 2

5. $(a + b)(c + d)$

7. Use $a'b'(a + b) = (a'b')a + (a'b')b = (a'a)b' + a'(b'b) = b' + a'$.

9. Use triples 4, 4, 4 and 7, 7, 7.

Table A2.13-9

×	1	4	7
1	4	7	1
4	7	1	4
7	1	4	7

EXERCISE 2.14

1. Note that $a^*(b^*) = [a(b^*)]^* = [(ab)^*]^*$.

EXERCISE 2.15

1. $1/14$, n, 27, n/m, $4/3$, $-1/6$, $9/11$, $-5/4$, $1/n$, $-1/n$

3. (a) 0 (b) None. (c) 0 (d) All. (e) 0 (f) All. (g) 0 (h) 0

5. (a) Use the distributive law and the definition of zero as the additive identity.

EXERCISE 2.16

1.

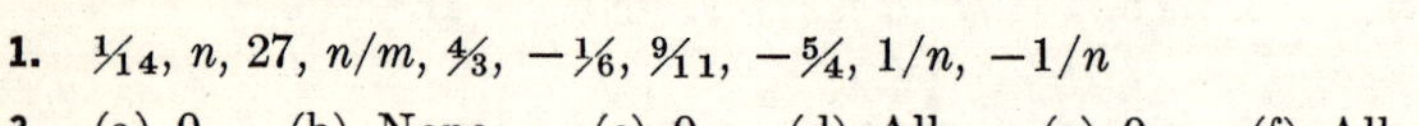

FIGURE A.1

3.

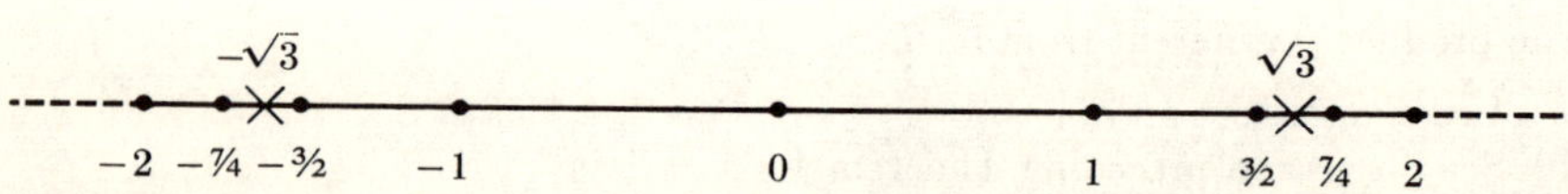

FIGURE A.2

5.

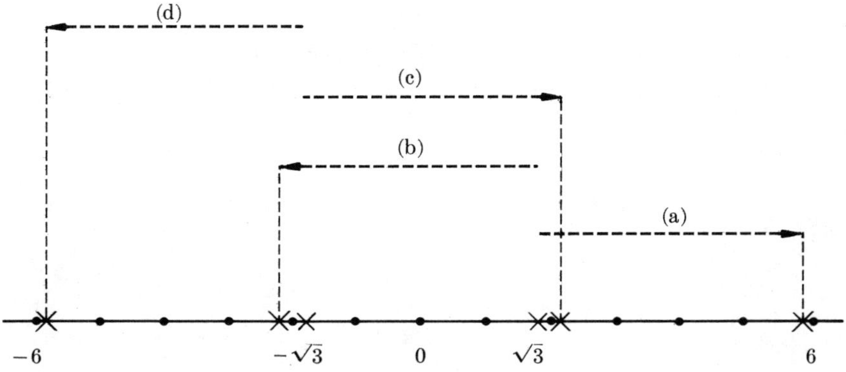

FIGURE A.3

EXERCISE 3.1-A

1. (a) 1 (b) 3 (c) 3 (d) 4 (e) 1 (f) 3 (g) 3 (h) $a'b^*$

EXERCISE 3.1-B

3. $a \neq 5$ and $c \neq 5$; hence, a' and c' both exist.

5. (a) None. (b) All. (c) All. (d) None. (e) All. (f) 5

7. (a) $-7/3$ (b) $-\pi/9$ (c) None. (d) 0 (e) All. (f) -35 (g) $1/12$ (h) $-\sqrt{2}/\pi$ (i) $24/17$, $-\pi/5$ (j) $-20/3$, $3/4$

EXERCISE 3.2

1. Solve equations $1 = 4 + x$ and $1 = 2 + x$. Note that $4 + 2 = 2 + 4$.

3. (a) 4 (b) $3 - 4 = 4$ (c) 3 (d) No. $3 - 5 \neq 5$

5. (a) 4 (b) 1 (c) 2 (d) 2 (e) 4 (f) $a - b$

7. (a) (i) 12 (ii) 3 (iii) 5 (iv) 1 (v) None. (vi) None. (vii) None. (viii) 2
(b) Note answers to (a).
(c) Note answers to (a).
(d) Yes. Adapt discussion preceding Theorem 3.2-3.

9. The real numbers are an additive group.
(a) Use Theorem G.4 of Section 2.10.
(b) Use criteria 1 and 4 of the definition of group in Section 2.10.
(c) Yes. Adapt discussion preceding Theorem 3.2-3.

EXERCISE 3.3

1. No. Construct a counterexample.

3. If $a - b = a - c$, then $a + b^* = a + c^*$; whence $b^* = c^*$.

5. (a) Use Theorem 3.2-2 and associative relation $(a + b) + c^* = a + (b + c^*)$.
(b) Use $(a + b^*) + c = a + (b^* + c) = a + (b^* + c^{**}) = a + (b + c^*)^*$.

7. $(b - c)a = ba - ca$.

EXERCISE 3.4

1.

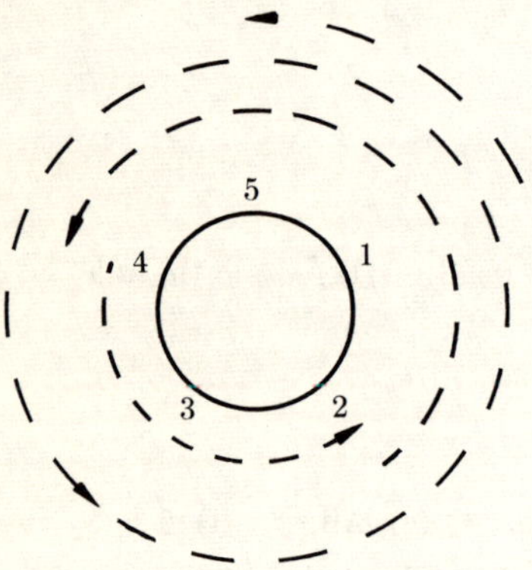

FIGURE A.4

5. For elements a, b, c, and d of the five-element system, $a - b = c - d$ if and only if $a + d = c + b$.

7. Use $(a - b)(c - d) = (a + b^*)(c + d^*)$. Apply Corollary 2.13-1 and Theorem 2.14-1.

EXERCISE 3.5

1. Table A3.5-1 $3c = 4$

×	c
3	4

3. Table A3.5-3 $2c = 1$

$\times$	c
2	1

5. (a) None. (b) All.

EXERCISE 3.6

1. 4, 2, $3 = 2 \cdot 4 = 4 \cdot 2$

3. (a) 1 (b) None. (c) 5 (d) 3 (e) 3 (f) 1, 2, 3, 4 (g) 3 (h) 1, 2, 3, 4 (i) 3

5. 5

7. Whenever b and c are different from 5, then $a \neq 5$.

9. With the omission of division by zero, the rational number system is closed under division. The quotient $c = a \div b$ is found from the equation $a = bc$. This has a unique solution for c when $b \neq 0$, has no solution when $b = 0$ but $a \neq 0$, and has all rational numbers for solution when $b = a = 0$.

11. (a) For nonzero rationals a, b, and c, $a \div b = c$ if and only if $a \div c = b$.
(b) $3 \div 5 = 3/5$ if and only if $3 \div 3/5 = 5$.

EXERCISE 3.7

3. (b) No. Eliminate each element in turn.

5. Use Theorem 3.6-2.

7. Use Theorem 2.6-1(b) and Theorem 3.6-2.

EXERCISE 3.8

1. Use Theorems 3.6-2, 2.6-1(b), and 2.4-1.

3. Use suggestions for Exercise 1.

5. For elements a, b, c, and d, with $bd \neq 5$, $ad = cb$ if and only if $a \div b = c \div d$.

7. Use suggestion for Exercise 1.

EXERCISE 3.10

1. 2134 2145 1334 4142

5. (a) Hans has hash
(b) Nan has Anna's sash

7. (a) an ash sash has Hannah
(b) Sasha has an ass

9. (a) Anna has Hans' hash
(b) Shah Hassan has a sash

11. (a) Hans has hash
(b) Nan has an ash sash

EXERCISE 4.2

1. Table A4.2-1

+	1	2	3	4	5	6
1	2	3	4	5	6	1
2	3	4	5	6	1	2
3	4	5	6	1	2	3
4	5	6	1	2	3	4
5	6	1	2	3	4	5
6	1	2	3	4	5	6

3. 3 **5.** 4 **7.** 5 **9.** All. **11.** All.

13. None. **15.** 1, 2, 3, 4, 6 **17.** (b) Identity.

19. (a) Yes. (b) Yes.

21. Yes. Every space in the table contains one and only one entry that is an element.

EXERCISE 4.3

1. (a) $(6 + b) + c = b + c$ and $6 + (b + c) = b + c$.

3. See Exercise 1.6.4.

5. (a) 5 (b) 4 (c) 3, 6 (d) 2 (e) 1, 4 (f) 2, 5

7. (a) 6 (b) 1

EXERCISE 4.4

1. $1 \cdot 2 = 2$; $2 \cdot 2 = 1 \cdot 2 + 2 = 4$; etc.

3. The entries in the 2 column are the products $1 \cdot 2$, $2 \cdot 2$, $3 \cdot 2$, etc., while the entries in the 2 row are $2 \cdot 1$, $2 \cdot 2$, $2 \cdot 3$, etc.

5. 4. **7.** 2, 5 **9.** 2, 4, 6 **11.** All. **13.** All. **15.** 5

17. The number of solutions is not predictable.

19. (a) 6 (b) Yes.

EXERCISE 4.5

1. (a) 5 (b) 5 (c) 3, 6 (d) All. (e) 6 (f) None. (g) 4 (h) None. (i) None. (j) 2, 4, 6 (k) 3, 6 (l) 2, 5 (m) 4 (n) 2 (o) 2

3. When a, b or c equals 6, then $(ab)c = a(bc) = 6$.

5. (c) and (d). **7.** 125

EXERCISE 4.6

1. $1 \cdot x = 6$ and $5x = 6$ only when $x = 6$.

3. (a) 2 (b) 3 (c) 2

5. Use $2(b + d) = 2b + 2d$.
(a) $2d = 6$ when $d = 3$ or 6.
(b) $2d \neq 6$ when $d = 1, 2, 4, 5$.
(c) By (a), $2b = 2(b + d)$ when $d = 3$ or 6; by (b), $2b \neq 2(b + d)$ when $d = 1, 2, 4, 5$.

7. Use $3(b + d) = 3b + 3d$. Proceed as in Exercise 5.

EXERCISE 4.7

1. Use the method of Section 2.14 and the equality $6a + ba = (6 + b)a = ba$.

3. See Exercise 2.14.1.

EXERCISE 4.8

1. (b) 3, 6 **3.** None.

EXERCISE 5.2

1. (a) $p_1 = 4, p_2 = 2$ (b) $p_1 = 2, p_2 = 3$ (c) $p_1 = 2', p_2 = 5$
(d) $q_1 = 5, q_2 = 5$ (e) $q_1 = 4, q_2 = 1$ (f) $q_1 = 4^*, q_2 = 3'$

3. $P = T, Q = U$.

5. (a) (1,1), (2,2), (3,3), (4,4), (5,5) (b) (1,4), (2,3), (3,2), (4,1), (5,5)
(c) (1,3), (2,1), (3,4), (4,2), (5,5) (d) (1,1), (2,5), (3,4), (4,3), (5,2)

EXERCISE 5.3

1. (a) (4,1) (b) (1,2) (c) (5,5) (d) (2,3) (e) (2,4) (f) (2,1)

3. (a) $x = 4$ (b) $y = 4$ (c) None. (d) $x = 5$ (e) All. (f) None. (g) None. (h) $x = 3, y = 1$ (i) All. (j) None.

5. $(p_1,p_2)^* = (p_1^*,p_2^*)$

7. Commutativity.

EXERCISE 5.4

1. (a) (3,6) (b) (3,5) (c) (4,6) (d) (4,6) (e) (4,5) (f) (6,6)

3. Step 2.

5. By closure the sum $(4,2) + (5,1)$ is a single pair.

EXERCISE 5.5

3. (a) $x = 1, y = 2$ (b) $x = 1, y = 4$ (c) $x = 3, y = 1$ (d) $x = y = 4$ (e) $x = 1, y = 2$ (f) $x = y = 4$

5. By the rule for addition of cross-pairs, the first (second) entry in $P + Q$ is the second (first) entry in $Q + P$.

EXERCISE 5.6

1. (a) (1,1) (b) F (c) (2,2) (d) (4,1) (e) F (f) (2,1) (g) (3,4) (h) $(1,2)^* = (4,3)$ (i) $(4,2)^* = (1,3)$ (j) (1,4)

3. (a) Yes. (b) Yes. (c) Yes. (d) No.

EXERCISE 5.8

1. A table of 625 entries is too cumbersome. It is better to argue from multiplicative properties of the five-element system.

3. (3,3) **5.** (4,5) **7.** F **9.** $x = y = 1$

11. None. **13.** All. **15.** $x = 1, 2, 3, 4;\ y = 1, 2, 3, 4$

17. All. **19.** None. **21.** (4,1)

23. Associativity. **25.** (2,1) **27.** (3,4)

EXERCISE 5.9

1. (a) $x = y = 3$ (b) $x = 4,\ y = 1$ (c) $x = y = 5$ (d) $x = 3,\ y = 5$ (e) $x = 1, y = 4$

3. Steps in the proof:
(1) p_1, etc., are elements of the five-element system.
(2) $p_1q_1 = q_1p_1$ and $p_2q_2 = q_2p_2$.
(3) $(p_1q_1,p_2q_2) = (q_1p_1,q_2p_2)$.
(4) $(p_1q_1,p_2q_2) = PQ$ and $(q_1p_1,q_2p_2) = QP$.
(5) $PQ = QP$.

5. For pairs P, Q, and R, $P(Q + R) = PQ + PR$ and $(Q + R)P = QP + RP$.

7. Proceed as in Section 2.14. (i) Use $PQ = P(Q + F) = PQ + PF$. (ii) Use $PQ + PQ^* = P(Q + Q^*) = PF = F$. (iii) See Exercise 2.14.

EXERCISE 5.10

1. (1,5), (2,5), (3,5), (4,5), (5,1), (5,2), (5,3), (5,4)

3. (a) $(2,5)(x,y) = (2x,5y) = (2x,5)$ for all elements y.
(b) $(5,4)(x,y) = (5x,4y) = (5,4y)$ for all elements x.

5. (a) $PQ = (2q_1,5)$ and $PR = (2r_1,5)$ for all elements q_2 and r_2. If $PQ = PR$, then $2q_1 = 2r_1$.
(b) Make an analysis similar to that for (a).

EXERCISE 5.11

1. (a) (3,4) (b) F

3. By Exercise 5.8.20(b) we know that I is an identity for the whole system. Because no other pair can serve as identity for the group, no other pair can serve as identity for the whole system.

5. Multiply both members of the equality on the left by P'.

7. (a) (1,1), (1,2), (1,3), (1,4)
(b) Use Theorem 2.11-1.

9. The group consists of the identity and 15 remaining pairs. Each of these is either its own inverse or, by Corollary 5.11-2, is coupled with another pair in such a way that each is the inverse of the other. All such couples comprise an even number of pairs. Hence, there remains in the group an odd number of pairs each of which must be its own inverse. There is a total of four such pairs.

EXERCISE 6.1

1. $A \sim J$, $B \sim D \sim G$, $C \sim H$.

EXERCISE 6.2

1. See the discussion preceding this Exercise.

3. Use relation $x + 5 = y + 3$ or $y = x + 2$.

5. No.

EXERCISE 6.3

1. (a) The entries in the pairs are not identical.
(b) $1 + 3 = 2 + 2$.
(c) A and B belong to the same equivalence set.

3. $[1,5] = \{(1,5),(2,1),(3,2),(4,3),(5,4)\}$
$[3,5] = \{(1,3),(2,4),(3,5),(4,1),(5,2)\}$
$[4,5] = \{(1,2),(2,3),(3,4),(4,5),(5,1)\}$

EXERCISE 6.4

1. (a) True. (b) True. (c) True. (d) False. (e) True. (f) False.

5. (a) False. (b) True. (c) False. (d) True. (e) True. (f) True.

EXERCISE 6.5

1. Table A6.5-1

+	[1,5]	[2,5]	[3,5]	[4,5]	[5,5]
[1,5]	[2,5]	[3,5]	[4,5]	[5,5]	[1,5]
[2,5]	[3,5]	[4,5]	[5,5]	[1,5]	[2,5]
[3,5]	[4,5]	[5,5]	[1,5]	[2,5]	[3,5]
[4,5]	[5,5]	[1,5]	[2,5]	[3,5]	[4,5]
[5,5]	[1,5]	[2,5]	[3,5]	[4,5]	[5,5]

EXERCISE 6.6

1. (a) [4,5] (b) [3,5] (c) [4,5] (d) [2,5] (e) [3,5] (f) [4,5]

EXERCISE 6.7

1. (a) True. (b) True. (c) False. (d) False. (e) True. (f) True.

5. (a) True. (b) False. (c) False. (d) True. (e) True. (f) True.

EXERCISE 6.8

1. Table A6.8-1

×	[1,5]	[2,5]	[3,5]	[4,5]	[5,5]
[1,5]	[1,5]	[2,5]	[3,5]	[4,5]	[5,5]
[2,5]	[2,5]	[4,5]	[1,5]	[3,5]	[5,5]
[3,5]	[3,5]	[1,5]	[4,5]	[2,5]	[5,5]
[4,5]	[4,5]	[3,5]	[2,5]	[1,5]	[5,5]
[5,5]	[5,5]	[5,5]	[5,5]	[5,5]	[5,5]

5. For all equivalence sets $[a,5]$ and $[b,5]$, the product $[a,5][b,5] = [5,5]$ if and only if $[a,5] = [5,5]$ or $[b,5] = [5,5]$.

7. (i) All sets except [5,5] have unique multiplicative inverses.
(ii) If $[a,5]' = [b,5]$, then $[a,5] = [b,5]'$.
(iii) $[a,5]'' = [a,5]$.
(iv) $([a,5][b,5])' = [a,5]'[b,5]'$.

9. (a) [2,5] (b) [3,5] (c) [4,5] (d) [2,5] (e) [3,5] (f) [5,5]

EXERCISE 6.11

1. (a) Adapt the proof of Theorem 5.4-1.
(b) Adapt the proof in Exercise 5.5.2.

3. Adapt the proof in Exercise 5.9.6.

5. The definition of additive identity is satisfied.

7. (a) Note that $[2n_1 + 2n_2, n_1 + 2n_2] = [n_1 + (n_1 + n_2), n_2 + (n_1 + n_2)]$.

9. The definition of multiplicative inverse is satisfied.

EXERCISE 7.2

1. (a) $a_{11} = 3, a_{12} = 4, a_{21} = 1, a_{22} = 5$
(b) $a_{11} = 5, a_{12} = 3, a_{21} = 2, a_{22} = 4$

3. (a) C, M (b) E

EXERCISE 7.3

1. (a) $\begin{pmatrix} 3 & 5 \\ 3 & 5 \end{pmatrix}$ (b) $\begin{pmatrix} 2 & 4 \\ 5 & 2 \end{pmatrix}$

3. $x =$ All.

5. $x = 1, y = 5, z = 2, w = 5$.

7. None.

9. $x = 2^* = 3, y = 3^* = 2, z = 4^* = 1, w = 1^* = 4$.

11. Correct. **13.** Correct.

EXERCISE 7.4

1. (a) $\begin{pmatrix} 3 & 2 \\ 5 & 1 \end{pmatrix}$ (b) $\begin{pmatrix} 3 & 5 \\ 4 & 4 \end{pmatrix}$ (c) $\begin{pmatrix} 4 & 1 \\ 2 & 5 \end{pmatrix}$ (d) $\begin{pmatrix} 2 & 5 \\ 4 & 3 \end{pmatrix}$

3. For uniqueness, $\begin{pmatrix} 2x & 2y \\ 2z & 2w \end{pmatrix}$ must equal $\begin{pmatrix} 5 & 5 \\ 5 & 5 \end{pmatrix}$.

EXERCISE 7.6

1. 5. Yes.

EXERCISE 7.7

3. True. **5.** False. **7.** True. **9.** True. **11.** True.

EXERCISE 7.8

1. Table A7.8-1

	Urban	Polis
Index A	2	1
Index B	1	3

3. Table A7.8-3

	Wool	Cotton
Urban	1	4
Polis	1	4

5. Table A7.8-5

	(a)		(b)	
	Paper	Cord	Qualo	Supro
Martus	4	2	3	1
Tradus	2	1	4	3

EXERCISE 7.9

1. (a) $\begin{pmatrix} 4 & 4 \\ 4 & 1 \end{pmatrix}$ (b) $\begin{pmatrix} 4 & 3 \\ 1 & 2 \end{pmatrix}$

3. (b) $AI = IA = A$ (c) I is an identity for multiplication.

5. (b) The matrices are multiplicative inverses of each other.

9. Note definition of multiplication of matrices.

11. (a) For matrices A, B, and C, $(B + C)A = BA + CA$.
(b) Partial answer:

$$(B + C)A = \begin{pmatrix} b_{11} + c_{11} & b_{12} + c_{12} \\ - & - \end{pmatrix} \begin{pmatrix} a_{11} & - \\ a_{21} & - \end{pmatrix}$$

$$= \begin{pmatrix} (b_{11} + c_{11})a_{11} + (b_{12} + c_{12})a_{21} & - \\ - & - \end{pmatrix}$$

$$= \begin{pmatrix} (b_{11}a_{11} + c_{11}a_{11}) + (b_{12}a_{21} + c_{12}a_{21}) & - \\ - & - \end{pmatrix}$$

$$BA + CA = \begin{pmatrix} b_{11} & b_{12} \\ - & - \end{pmatrix} \begin{pmatrix} a_{11} & - \\ a_{21} & - \end{pmatrix} + \begin{pmatrix} c_{11} & c_{12} \\ - & - \end{pmatrix} \begin{pmatrix} a_{11} & - \\ a_{21} & - \end{pmatrix}$$

$$\begin{pmatrix} b_{11}a_{11} + b_{12}a_{21} & - \\ - & - \end{pmatrix} + \begin{pmatrix} c_{11}a_{11} + c_{12}a_{21} & - \\ - & - \end{pmatrix}$$

$$= \begin{pmatrix} (b_{11}a_{11} + b_{12}a_{21}) + (c_{11}a_{11} + c_{12}a_{21}) & - \\ - & - \end{pmatrix}$$

(c) No.

EXERCISE 7.10

7. $x = 2, y = 3, z = 2, w = 5$ **9.** Yes.

EXERCISE 7.11

3. (a) $|A| = 3, |B| = 4.$ (b) $A + B = \begin{pmatrix} 3 & 3 \\ 3 & 5 \end{pmatrix}, |A + B| = 1.$

(c) No. (d) $AB = \begin{pmatrix} 2 & 3 \\ 1 & 5 \end{pmatrix}, |AB| = 2.$

5. (a) 1. (b) Use Theorem 7.11-2.
(c) $|A|$ and $|A'|$ are multiplicative inverses of each other.
(d) Yes. (e) No. $|A| = 5$; hence, $|A| \cdot |A'| \neq 1.$

7. (a) 5. (b) Use Theorem 7.11-2.
(d) No. $|A| \neq 5$ and $|B| \neq 5$; hence, $|A| \cdot |B| \neq 5.$

9. Use Theorem 2.14-1.

EXERCISE 7.12

3. (a) $\tilde{A} = \begin{pmatrix} 5 & 2 \\ 1 & 2 \end{pmatrix}, A\tilde{A} = \tilde{A}A = \begin{pmatrix} 3 & 5 \\ 5 & 3 \end{pmatrix}$

(b) $\tilde{A} = \begin{pmatrix} 1 & 2 \\ 3 & 1 \end{pmatrix}, A\tilde{A} = \tilde{A}A = F$

(c) $\tilde{A} = \begin{pmatrix} 2 & 2 \\ 4 & 1 \end{pmatrix}, A\tilde{A} = \tilde{A}A = \begin{pmatrix} 4 & 5 \\ 5 & 4 \end{pmatrix}$

5. (a) $\begin{pmatrix} 5 & 4 \\ 2 & 4 \end{pmatrix}$

7. (a) 3 (b) $\begin{pmatrix} 4 & 1 \\ 1 & 2 \end{pmatrix}$

(c) $|A'|$ is the determinant of the inverse of A, while $|A|'$ is the inverse of the determinant of A.

(d) $|A'| = 2, |A|' = 3' = 2.$

(e) $|A| \cdot |A'| = 1$; hence, $|A'|$ is the inverse of $|A|$.

9. Partial answer:

$$\tilde{A}A = \begin{pmatrix} a_{22} & a_{12}^* \\ - & - \end{pmatrix} \begin{pmatrix} a_{11} & a_{12} \\ a_{21} & a_{22} \end{pmatrix} = \begin{pmatrix} a_{22}a_{11} + a_{12}^*a_{21} & a_{22}a_{12} + a_{12}^*a_{22} \\ - & - \end{pmatrix} = \begin{pmatrix} |A| & 5 \\ - & - \end{pmatrix}$$

11. Every matrix satisfies the conditions of one and only one of the categories of Theorem 7.12-2.

EXERCISE 7.13

1. (b) Singular. (c) $\begin{pmatrix} 4 & 2 \\ 5 & 1 \end{pmatrix}$ (d) (i) $\begin{pmatrix} 5 & 5 \\ 3 & 1 \end{pmatrix}$ (ii) $\begin{pmatrix} 4 & 4 \\ 2 & 2 \end{pmatrix}$ (e) Both singular.

3. $\begin{pmatrix} 5 & 2 \\ 4 & 4 \end{pmatrix}$, unique.

5. $\begin{pmatrix} 5 & 3 \\ 5 & 4 \end{pmatrix}$, unique.

7. Any matrix A with $a_{11} = a_{12} = 5$.

9. (a) $\begin{pmatrix} 1 & 3 \\ 2 & 1 \end{pmatrix}$, $\begin{pmatrix} 1 & 5 \\ 3 & 2 \end{pmatrix}$

EXERCISE 7.14

1. (a) $x = 1, y = 4$ (b) $x = 2, y = 3$ (c) $x = 4, y = 5$ (d) $x = 4, y = 1$

3. (a) and (b)

Table A7.14-3

x	1	2	3	4	5
(a) $y = x + 2$	3	4	5	1	2
(b) $y = 3x + 4$	2	5	3	1	4

(c) $x = 4$, $y = 1$.

5. There are an infinite number of values for x, hence a table similar to Table A7.14-3 cannot be constructed.

7. (a) $x = -\frac{1}{4}$, $y = \frac{3}{2}$ (b) $x = -3$, $y = 3$
(c) $x = 4$, $y = -5$ (d) $x = -1$, $y = 1$

EXERCISE 7.15

1. 1 tractor, 4 fire engines.

3. 1 jar of peach jelly, 3 jars of currant jelly.

EXERCISE 7.16

1. (b) Sasha has an ass

3. $\begin{pmatrix} a & a \\ h & n \end{pmatrix} M' = \begin{pmatrix} s & a \\ s & h \end{pmatrix}$, Hans has Anna's sash

EXERCISE 8.1

1. (a) $\sim s$ (b) $s \wedge \sim r$ (c) $p \rightarrow \sim q$ (d) $q \leftrightarrow r$

3. (a) Denial, *not*. Immediate and ultimate component: the dormouse sleeps.
(b) Conjunction, *and*. Immediate components: the dormouse sleeps, the gryphon does not dance. Ultimate components: the dormouse sleeps, the gryphon dances.

(c) Conditional, *if . . . , then*. Immediate components: the cat grins, the mock turtle does not sob. Ultimate components: the cat grins, the mock turtle sobs.

(d) Biconditional, *if and only if*. Immediate and ultimate components: the mock turtle sobs, the gryphon dances.

EXERCISE 8.2

1. (a) The sea is wet or the sands are dry, and the moon is bright.
(b) The sea is wet, or the sands are dry and the moon is bright.

3. No pattern.

5. No pattern.

7. (a) $\sim p \vee q$ (b) $\sim(p \wedge q)$ (c) $\sim p \leftrightarrow \sim q$ (d) $(p \wedge q) \vee \sim q$

9. (a) $\sim(p \vee q)$ (b) $\sim p \vee q$

11. (a) $p \leftrightarrow q$ (b) $(p \wedge q) \vee (\sim p \wedge \sim q)$

EXERCISE 8.3

1. Rephrase the theorem: If P, Q, and R are pairs and $P + Q = P + R$; then $Q = R$.

3. Rephrase the theorem: If P, Q, and R are pairs and it is not true that P is F or a divisor of F; then if $PQ = PR$, then $Q = R$.

EXERCISE 8.4

1. When the immediate component has truth value false, true.

3. When either or both immediate components are true.

5. When the immediate components have opposite truth values.

EXERCISE 8.5

1. (c) Note that p must be replaced by a true statement and q by a false one.

3. Table A8.5-3: (a) Line 4. (b) Line 3.

Table A8.5-3

$((p$	$\rightarrow$	$q)$	$\wedge$	$\sim q)$	$\rightarrow$	$\sim p$
T	T	T	F	F	T	F
F	T	T	F	F	T	T
T	F	F	F	T	T	F
F	T	F	T	T	T	T

5. (a) Table A8.5-5

$((p$	$\rightarrow$	$q)$	$\wedge$	$(p$	$\rightarrow$	$r))$	$\leftrightarrow$	$(p$	$\rightarrow$	$(q$	$\wedge$	$r))$
T	T	T	T		T	T	T		T		T	
F	T	T	T		T	T	T		T		T	
T	F	F	F		T	T	T		F		F	
F	T	F	T		T	T	T		T		F	
T	T	T	F		F	F	T		F		F	
F	T	T	T		T	F	T		T		F	
T	F	F	F		F	F	T		F		F	
F	T	F	T		T	F	T		T		F	

(b) Note that the instances of p, q, and r must have appropriate truth values.
(c) All instances are true.
(d) Yes.

7. Table A8.5-7

p	$\wedge$	$\sim p$
T	F	F
F	F	T

EXERCISE 8.6

1. The premise is true only in line 3 of Table A8.6-1, and in this line the conclusion is true.

Table A8.6-1

p	q	p	$\wedge$	$(p$	$\rightarrow$	$\sim q)$	$\sim q$
T	T		**F**		F	F	**F**
F	T		**F**		T	F	**F**
T	F		**T**		T	T	**T**
F	F		**F**		T	T	**T**

3. The conjunction of the premises is true only in line 4 of Table A8.6-3, and in this line the conclusion is true.

Table A8.6-3

p	q	$(p \rightarrow q)$	$\wedge$	$\sim q$	$\sim p$
T	T	T	**F**	F	**F**
F	T	T	**F**	F	**T**
T	F	F	**F**	T	**F**
F	F	T	**T**	T	**T**

5. No. **7.** No.

9. Yes. The truth table for Exercise 8.5.6 can be used.

EXERCISE 8.7

1. The two patterns have identical truth values in all lines of Table A8.7-1.

Table A8.7-1

p	q	$(p \vee \sim p)$	$\wedge\ q$
T	**T**	T	**T**
F	**T**	T	**T**
T	**F**	T	**F**
F	**F**	T	**F**

3. The two patterns have identical truth values in all lines of Table A8.7-3.

Table A8.7-3

p	q	r	$p \wedge$	$(q \vee r)$	$(p \wedge q)$	$\vee$	$(p \wedge r)$
T	T	T	**T**	T	T	**T**	T
F	T	T	**F**	T	F	**F**	F
T	F	T	**T**	T	F	**T**	T
F	F	T	**F**	T	F	**F**	F
T	T	F	**T**	T	T	**T**	F
F	T	F	**F**	T	F	**F**	F
T	F	F	**F**	F	F	**F**	F
F	F	F	**F**	F	F	**F**	F

5. No. **7.** Yes.

9. For each assignment of truth values to the ultimate components, each pattern has exactly one truth value.

11. When P is true, so is its equivalent pattern Q; and because Q is true, so is its equivalent pattern R. Hence, P and R are true together. Similarly, P and R are false together.

EXERCISE 8.8

1. The two patterns have opposite truth values in all lines of Table A8.8-1.

Table A8.8-1

p	q	p	$\wedge$	$\sim q$	p	$\rightarrow$	q
T	T		**F**	F		**T**	
F	T		**F**	F		**T**	
T	F		**T**	T		**F**	
F	F		**F**	T		**T**	

3. The two patterns have opposite truth values in all lines of Table A8.8-3.

Table A8.8-3

p	q	p	$\vee$	q	$\sim p$	$\wedge$	$\sim q$
T	T		**T**		F	**F**	F
F	T		**T**		T	**F**	F
T	F		**T**		F	**F**	T
F	F		**F**		T	**T**	T

5. No.

EXERCISE 8.9

1. Make truth tables as for the corresponding statement patterns:
(a) Commutativity of alternation.
(b) Associativity of conjunction.
(c) Associativity of alternation.
(d) Distributivity of conjunction with respect to alternation.
(e) Distributivity of alternation with respect to conjunction.
(f) Idempotence of conjunction.
(g) Idempotence of alternation.
(h) Commutativity of biconditional.
(i) Associativity of biconditional.

3. When one immediate component is false, then an alternation is true or false according as the other immediate component is true or false.

5. When one immediate component is true, then a biconditional is true or false according as the other immediate component is true or false.

EXERCISE 8.10

1. (a) 1 (b) None. (c) 2, 4 (d) 1, 2, 4, 5 (e) 1, 2, 3, 4 (f) 1, 2, 5

EXERCISE 8.11

1. F. 2 is the additive identity.

3. T. 4 has a multiplicative inverse or $4^* = 4'$.

5. T of 5. y is the additive identity.

7. T of 1, 2, 3, 4. If z is the additive identity, then z has a multiplicative inverse.

EXERCISE 8.12

1. Use Corollary 8.12-1(a) and make truth tables for the following statement patterns.

(a) $\sim(p \wedge q) \leftrightarrow (\sim p \vee \sim q)$
(b) $((p \vee q) \wedge \sim p) \rightarrow q$
(c) $((p \rightarrow q) \wedge (q \rightarrow r)) \rightarrow (p \rightarrow r)$

3. No. **5.** No. **7.** No.

EXERCISE 8.13

1. (a) $\exists x(Px)$ (b) $\forall x(Px)$ (c) $\exists x(\sim Px)$ (d) $\forall x(\sim Px)$

3. (a) When statement (a) is true, then there is at least one element that has an inverse. Each such element presents a counterexample to the statement that every element fails to have an inverse.

(b) Yes. The answer to part (a) asserts that when statement (a) is true, then statement (d) is false. Now show that when statement (a) is false, then statement (d) is true.

EXERCISE 8.14

1. (a) $\exists x(Px \wedge Qx)$ (b) $\sim\exists x(\sim Px \wedge Qx)$
(c) $\forall x(Px \rightarrow Qx)$ (d) $\exists x(\sim Px \wedge \sim Qx)$

EXERCISE 8.15

1. For Exercise 1: all T.
For Exercise 2: (a) F (b) T (c) T (d) T

3. (a) The antecedent is a conjunction; hence, both of its immediate components are true. The truth of $\exists x(Px)$ requires that the universe contain an element of which Px is true. Of this element, Qx must be true because by Rule 7 the conditional $Px \rightarrow Qx$ is always true. Hence, by Rule 6 $\exists x(Qx)$ is true.

(b) Both immediate components of the antecedent are true. The truth of $\forall x(Px)$ requires that open sentence Px be true of every element in the universe. It follows that Qx is true of every element because the conditional $Px \rightarrow Qx$ is always true. Hence, by Rule 7 $\forall x(Qx)$ is true.

EXERCISE 8.16

1. In the five-element system, all elements have additive inverses but not all have multiplicative inverses. For some elements the additive and multiplicative inverses are identical. No element has 5 for a multiplicative inverse.

3. $\exists x(\sim Px)$, $\forall x(Qx)$, $\forall x(\sim Rx)$, $\exists x(Sx)$

In the five-element system, some elements fail to have additive inverses but all have multiplicative inverses. For no elements are additive and multiplicative inverses identical. Some element has 5 for its multiplicative inverse.

EXERCISE 9.1

1. (a) $\{1, 2, 3, 4, 5, 6, 7\}$ (b) $\{1\}$ (c) $\varnothing$ (d) U

3. (a) See Exercise 2.5.6.
(b) The first immediate component is true of 1 and the second is true of 4.
(c) The two immediate components are never true together.
(d) The antecedent is false of 1 and 2, and the consequent is true of 5.

EXERCISE 9.3

1. $A = E, B = F, C = D$

3. $A = C = E = F, B = D$

EXERCISE 9.4

1. (a) D with M; F or G with any set; H with M; etc.
(b) D with E, H, J, K, or L; L with M, etc.

3. (a) $0 \in L$ but $0 \notin F$.
(b) A can contain no elements; hence, $A = \varnothing$.
(c) The sentence $\forall x(x \in A \rightarrow x \in \varnothing)$ must be true.

5. (a) G contains integers that are not members of E.
(b) A must contain all the elements in the universe; hence, $A = U$.
(c) The sentence $\forall x(x \in U \rightarrow x \in A)$ must be true.

EXERCISE 9.6

1.

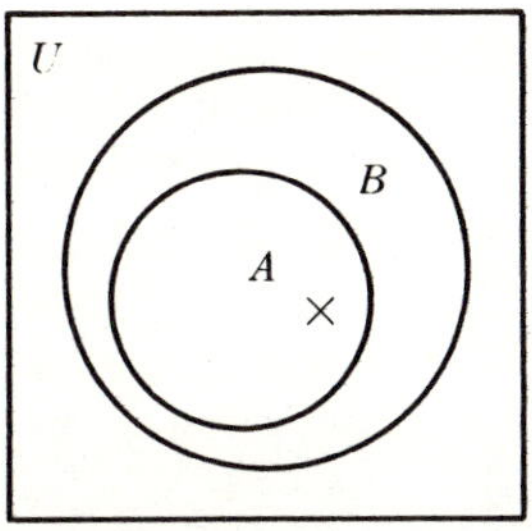

FIGURE A.5

3.

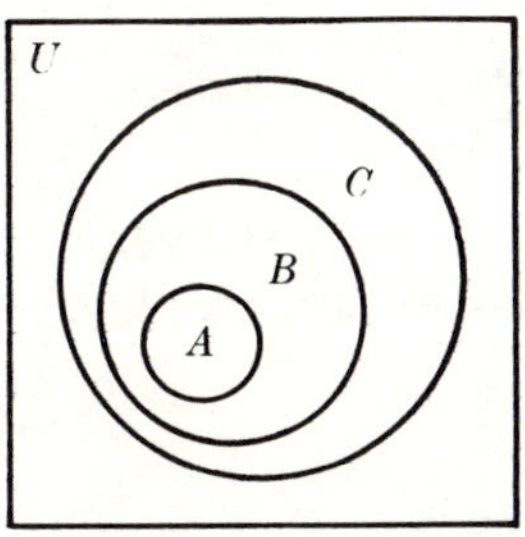

FIGURE A.6

EXERCISE 9.7

1. Each of the A and B circles contains a crossmark.

3. (a) No. (b) Yes. (c) No.

5. (a) D and H have element 1 in common. (b) Yes.
(c) E and H have element -1 in common. (d) No.

7. (a) When the set is not empty.
(b) When the set is empty.
(c) No. (d) No.

9. (a) Yes. (b) No.

EXERCISE 9.9

1. The two quantified sentences are true together and false together. When true, the first defines the relation A meets A and the second asserts that A is not empty.

When false, the first asserts that A does not meet A and the second asserts that A is empty.

EXERCISE 9.10

1. Valid. See Exercise 9.6.1.

3. Not valid. Construct a counterexample.

5. Valid. See Figure 9.5.

EXERCISE 9.11

1. $(A \cap B)' = \{1,2,4,6,7\}$.

3. $(A \cap B)' = A' \cup B'$, $(A \cup B)' = A' \cap B'$, $A + B = A' + B'$, $(A + B)' = A' + B = A + B'$.

EXERCISE 9.12

1. $A \cap B = \{3\}$, $A \cap C = \varnothing$, $B \cap C = \{5,7\}$

EXERCISE 9.13

1. For Theorem 9.13-2 the sets in (a) show that relations $A \subset B$ and $A \cap B = A$ are true together. The sets in (b) show that the two relations are false together.

3. (b) No.

5. (a) $\{2,4\}$ (b) $\{4\}$ (c) $\{4,8\}$ (d) $\{4\}$

7. (a) (b)

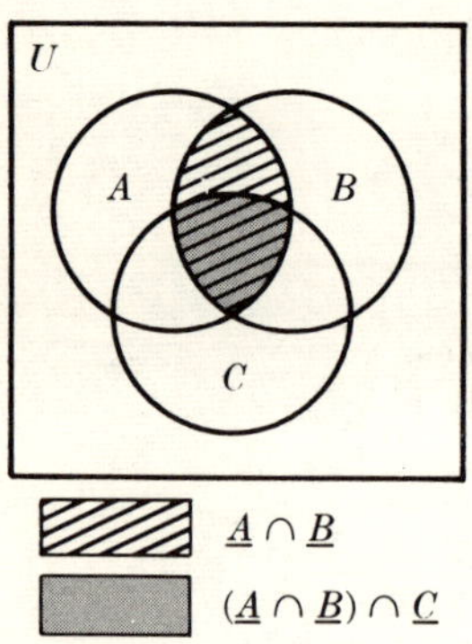

FIGURE A.7

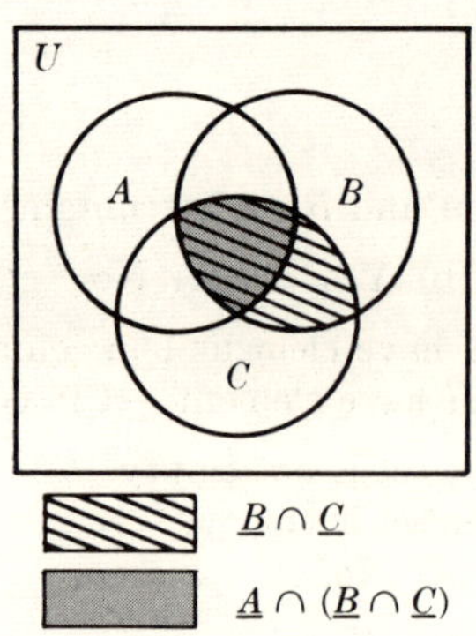

FIGURE A.8

(c) Associativity.

9. Use Corollary 9.13-2.

EXERCISE 9.14

1. When A is disjoint from C, then all subsets of A are disjoint from all subsets of C. Because $A \cap B \subset A$ and $C \cap D \subset C$, the relation between B and D is of no significance.

3. (b)

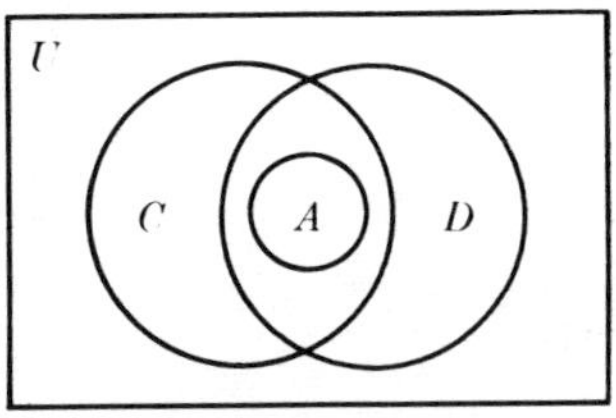

FIGURE A.9

EXERCISE 9.15

1. $E = \{1,2,3,4,5\}$, $F = \{2,4,6,8\}$, $G = \{4,8\}$, $H = \{3,6,9\}$

3. (a) The sets show that the two relations stated in the theorem are true together.
(b) The sets show that the two relations in the theorem are false together.

5. (a) (b)

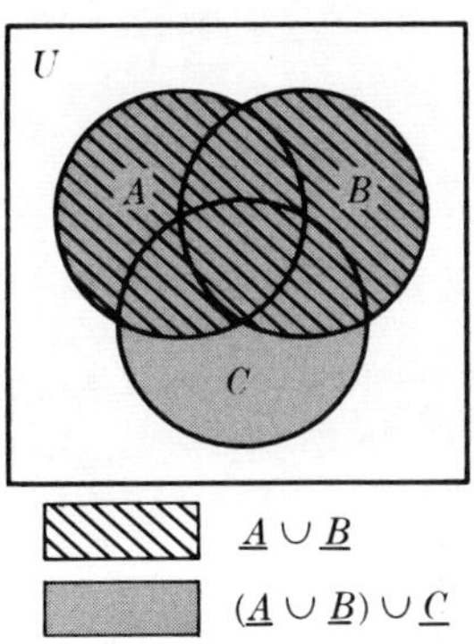

FIGURE A.10

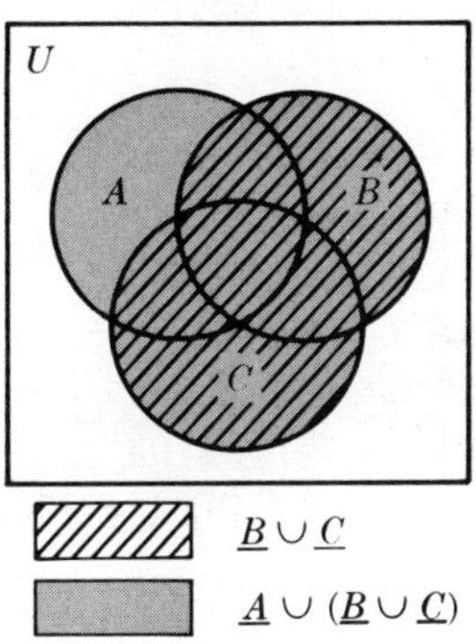

FIGURE A.11

7. Yes. $\varnothing$

9. (a) If $A \subset C$ and $B \subset D$, then $A \cup B \subset C \cup D$.
(c) Yes. $A \subset C \subset C \cup D$ and $B \subset D \subset C \cup D$.

11. (a) If A is disjoint from C and B is disjoint from D, then $A \cup B$ is disjoint from $C \cup D$.
(c) No.

EXERCISE 9.16

1. When A meets C, then any set containing A meets any set containing C. Because $A \cup B$ contains A and $C \cup D$ contains C, the relation between B and D is of no significance.

EXERCISE 9.17

1. Formula (1): $\{2,3,4,5,6\}$
Formula (2): $\{3,5,6\}$

EXERCISE 9.18

1. $E' = \{5\}$, $F' = \{1,4,5\}$, $G' = \{2,3\}$, $H' = \{2,3,4\}$.

3. (a)

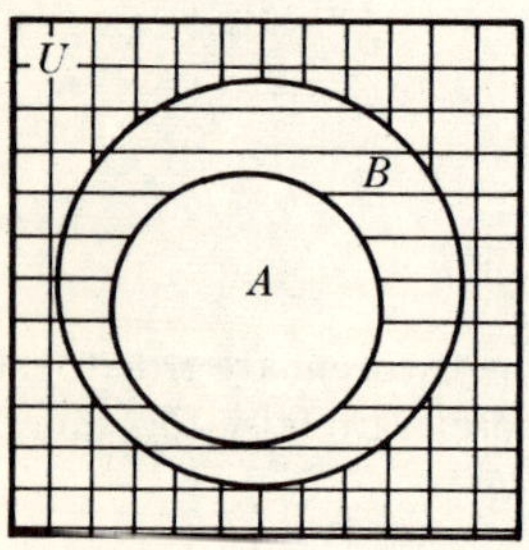

FIGURE A.12

(b) Yes.

5. (a) No. (b) No.

7. $E \cup G = F' \cup H' = (F \cap H)'$
$(E \cup G)' = E' \cap G' = F \cap H$
$E' \cup G' = (E \cap G)'$
$(F \cup H)' = F' \cap H'$

9. (a) Use Figure 9.9. (b) Use Figure A.13.

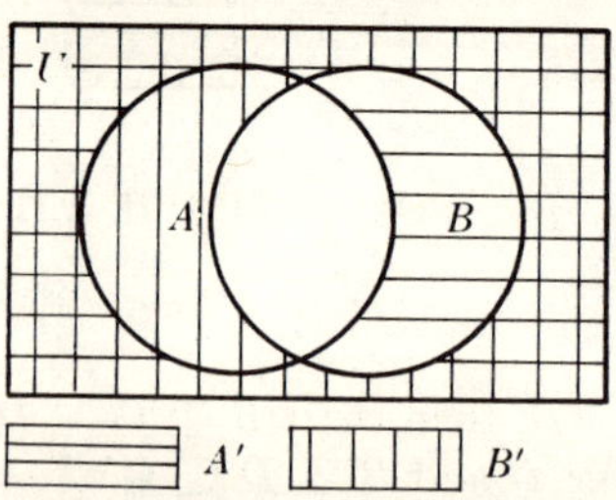

FIGURE A.13

(c) Yes.

11. (a)

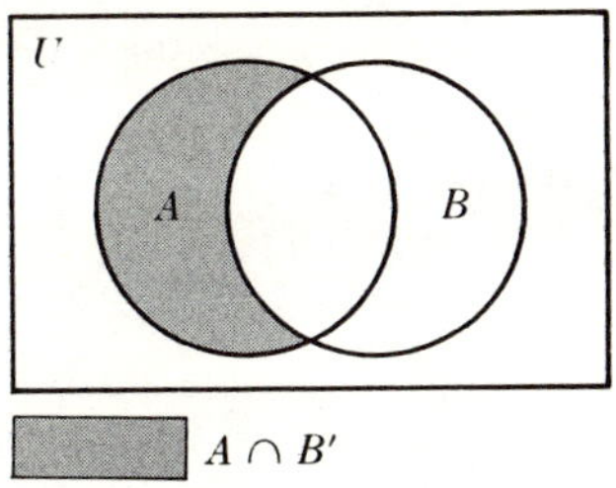

FIGURE A.14

(b) $A \cap B' = \varnothing$.

13. (a) The set $(A \cap B')'$ is represented in Figure A.14 by all of the rectangle except the shaded crescent. This is exactly the region that represents $A' \cup B$ in the diagram of Exercise 12(a).

(b) Yes.

EXERCISE 9.19

3. $B \subset A$

EXERCISE 9.20

1. Steps in justification of first law:

(1) $\sim(p \wedge q) \leftrightarrow (\sim p \vee \sim q)$
(2) $\forall x(\sim(x \in A \wedge x \in B) \leftrightarrow (x \notin A \vee x \notin B))$
(3) $\forall x(x \notin A \cap B \leftrightarrow (x \in A' \vee x \in B'))$
(4) $\forall x(x \in (A \cap B)' \leftrightarrow x \in A' \cup B')$
(5) $(A \cap B)' = A' \cup B'$

EXERCISE 9.21

1. (a) Use a diagram showing $A \cup B$ and $(A \cap B)'$.

(c) The shaded set contains all the elements that are members of A or of B but not of both A and B.

(d) Yes.

3. There are 16 items in A and 13 in B.

5. There are 22 items in B.

7. $(A' \cap B') \cup C$

9. By commutative and associative laws the left member becomes

$$(A \cup B)' \cap (A \cap B) \cap C \cap C' \qquad \text{where} \qquad C \cap C' = \varnothing$$

EXERCISE 9.22

1. (a) 1 (b) 1 (c) 0 (d) 1 (e) 0 (f) 0

5. The expressions have identical values for every assignment of values to sets R and S.

EXERCISE 9.23

1. The current flows when switches R and S are in opposite states.

EXERCISE 10.3

Step 1. SAS.
Step 2. Corresponding angles in congruent triangles.
Step 3. The whole is greater than any of its parts.
Step 4. Substitution of $\measuredangle ACB$ for its equal $\measuredangle MBP$.

EXERCISE 10.4

1. (a) SAS.
(b) $\measuredangle 3 = \measuredangle 4$ and $\measuredangle 7 = \measuredangle 8$.
(c) and (d) Use definition of right angle: if two adjacent angles are equal and supplementary, then they are right angles.

EXERCISE 10.5

1. Statement F: $\measuredangle CHG$ is less than $\measuredangle BGH$. Now $\measuredangle 1 + \measuredangle 4$ is less than $\measuredangle 2 + \measuredangle 4$.

3. The common perpendicular is a transversal.

5. (a) The sum of the summit angles is two right angles and the summit angles are equal.
(b) Use Figure 10.6.
Parallel: (i) MN is a common perpendicular.
(ii) AB is a common perpendicular.
Equal: $\measuredangle 3 = \measuredangle 5$ and $\measuredangle 1 = \measuredangle 7$ by Proposition 29; hence, $\triangle CAN \cong \triangle NMC$.
(c) Both quadrilaterals are rectangles.

EXERCISE 11.2

1. (a) Line AB is the common perpendicular to lines m and n. Use Theorem 11.2-3.
(c) $\measuredangle 8 < \measuredangle 9 < \measuredangle 10 < \measuredangle 11$.

3. Acute. **5.** Yes. **7.** $\measuredangle 1 < \measuredangle 2$.

9. (a) Use Theorems 11.2-2 and 11.2-3.
(b) Line ABD is a transversal of AC and BE.
(c) The exterior angle of a triangle is greater than the sum of the opposite and interior angles, and the sum of all the angles of a triangle is less than two right angles.

EXERCISE 11.3

1. Use Theorem 11.3-1.

3. (a) The lateral sides are greater than the median.
(b) The summit is greater than the base.

5. $\measuredangle BDE + \measuredangle 2 > \measuredangle A + \measuredangle C$.

7. As the size of the polygon increases, the angle sum decreases.

EXERCISE 11.4

1. Angles SRU and STW are angles of parallelism.

3. No.

5. First prove $\triangle CMG \cong \triangle DMH$ by ASA. Then prove $\triangle GMH$ isosceles.

7. The two quadrilaterals coincide and are the same rectangle.

9.

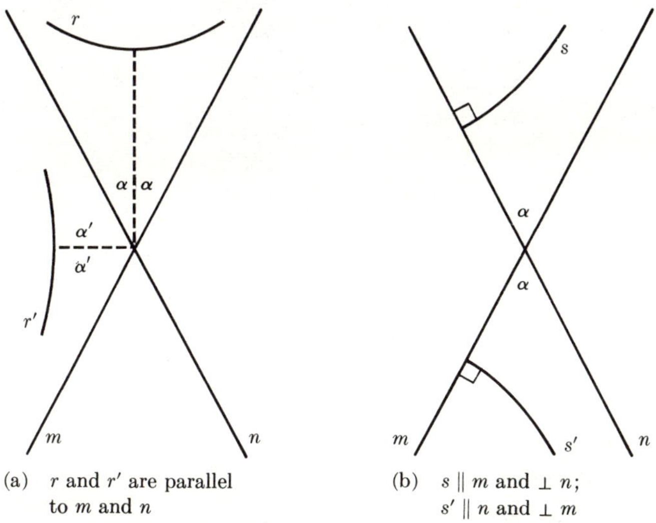

(a) r and r' are parallel to m and n

(b) $s \parallel m$ and $\perp n$; $s' \parallel n$ and $\perp m$

FIGURE A.15

EXERCISE 11.5

1. Yes. Draw construction line AD and use properties of isosceles triangles.

3. The angle sum of $MNDB$ is greater than the angle sum of $ABCD$.

5. (a) Yes. (b) No.

EXERCISE 12.2

1. Measure the length of the perpendicular dropped from the point to the line.

3. The triangles are isosceles and birectangular.

EXERCISE 12.3

1. Use Theorem 12.2-1.

3. No.

5. No. Construct a counterexample.

7. Let $\triangle ABC$ have right angles at A and B. Then sides AC and BC meet at the pole of AB.

9. (b) Summit and base meet at the two poles of the median MN.

11. For Saccheri quadrilateral $ABCD$:
(a) When $AC = BD = d$, then points C and D are both poles of AB on the same side of AB.
(b) When $AB = d$, then A is the pole of BD; hence, $AD = d$.

13. Yes.

15. Yes. Use Theorem 12.3-5.

EXERCISE 12.4

1. Never.

3. By Theorem 12.4-2 line AC extended passes through B where it makes a right angle with AB. Since BD is perpendicular to AB at B and since D lies on the same side of AB as C, line BD falls along line AC extended.

5. (a) Obtuse. (b) Greater than four right angles.

7. $\measuredangle 2 < \measuredangle 1$

9. $\measuredangle BDE + \measuredangle 2 < \measuredangle A + \measuredangle C$

11. (a) Each side adjacent to the fourth angle is smaller than its opposite side.
(b) The summit is shorter than the base.

EXERCISE 12.5

1. B is the pole of AC; hence, angles C, C', and C'' are right angles. $\measuredangle ABC'$ is acute, $\measuredangle ABC$ is right, $\measuredangle ABC''$ is obtuse.

3. In Figure A.15 let common perpendicular CDD' of lines ACB, ADB, and $AD'B$ have length less than $2d$. Use the methods of proof of Theorem 12.5-1 and of Exercise 2.

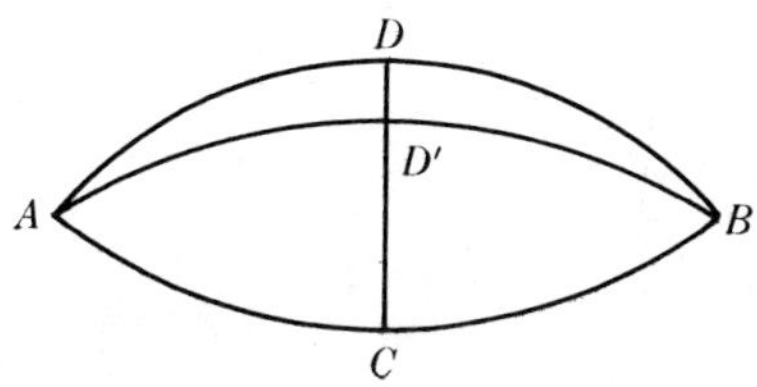

FIGURE A.16

5. The quadrilaterals are distinct.

EXERCISE 12.6

1. Yes.

3. The angle sum of $MNDB$ is less than the angle sum of $ABCD$.

5. (a) Yes. (b) No.

7. $CD = CE = d/2$, and base angles are acute.

EXERCISE 12.8

1. Euclid's Propositions 1–15 and 26. Theorems 10.4-1, 10.4-2, 10.4-3, and 10.4-4.

INDEX